Joint Workshop of KO2PI 2017 & 2nd International Conference on Mathematics, Science, Technology, Education, and their Applications (ICMSTEA 2016)

The Role of Mathematics, Sciences, Technology, and Education Towards ASEAN Economic Community and Global Challenges

Journal of Physics: Conference Series Volume 954

Makassar, Indonesia
3 – 4 October 2016 and 10 October 2017

ISBN: 978-1-5108-5929-6
ISSN: 1742-6588

Printed from e-media with permission by:

Curran Associates, Inc.
57 Morehouse Lane
Red Hook, NY 12571

Some format issues inherent in the e-media version may also appear in this print version.

Printed by Curran Associates, Inc. (2018)

For permission requests, please contact the Institute of Physics
at the address below.

Institute of Physics
Dirac House, Temple Back
Bristol BS1 6BE UK

Phone: 44 1 17 929 7481
Fax: 44 1 17 920 0979

techtracking@iop.org

Additional copies of this publication are available from:

Curran Associates, Inc.
57 Morehouse Lane
Red Hook, NY 12571 USA
Phone: 845-758-0400
Fax: 845-758-2633
Email: curran@proceedings.com
Web: www.proceedings.com

Table of contents

Volume 954

Joint Workshop of KO2PI & 2nd International Conference on Mathematics, Science, Technology, Education, and their Applications (2nd ICMSTEA)

3–4 October 2016, Makassar, Indonesia

Accepted papers received: 4 January 2018
Published online: 22 February 2018

Preface

Papers

IOP Conf. Series: Journal of Physics: Conf. Series **954** (2018) 011001 doi:10.1088/1742-6596/954/1/011001

Forewords from the Chairman of Committee

Assalamu'alaykum wa Rahmatullahi wa Barakatuh

First of all, I wish to extend a warm welcome to fellow delegates from the various countries and regions. I realize that you are fully dedicated to the sessions that will follow but I do hope you will also take time to enjoy sparkling Makassar with its tropical setting, friendly people, and wonderful cuisine.

This 2nd International Conference on Mathematics, Science, Technology, Education, and Their Application 2016 is organized by the Faculty of Mathematics and Natural Sciences, Universitas Negeri Makassar to bring all experts and researches in these field sharing their important thoughts and findings. The conference will be held in two days from 3rd – 4th of October 2016 with two keynote speakers, seven invited speakers, and more than 80 parallel speakers from different backgrounds.

Let me take this opportunity, on behalf of the committee members, to express my gratitude and sincere thanks to the keynote, invited and all parallel speakers for spending their valuable time with us in this conference. I do hope that your time in Makassar will be valuable and memorable.

Finally, I would like to thank all steering and organizing committee for their hard work and dedication to the success of this conference. I would like also to apologize to all of you should you find any inconvenience during this event.

Thank you very much,

Wassalamu'alaykum Warahmatullahi Wa barakatuh

Chairman of Committee,

Dr. Drs. A. Mushawwir Tayeb, M.Kes.

IOP Conf. Series: Journal of Physics: Conf. Series **954** (2018) 011001 doi:10.1088/1742-6596/954/1/011001

The 2nd ICMSTEA Speech
By The Dean of Mathematics and Natural Sciences Faculty
Universitas Negeri Makassar

Your excellency Rector of Universitas Negeri Makassar
Honorable Vice Rectors and Dean of All Faculties
Honorable Keynote Speakers
Distinguished all invited speakers from outstanding universities
Distinguished all speakers and guests
All participants,
Ladies & Gentlement,

Assalamu'alaykum Warahmatullahi Wabarakatuh. My greetings for all of you. May peace and God's blessing be upon us all. *Alhamdulillah*, all praises be to the Almighty God, Allah *subhanahu wata'ala.*

It is my pleasure to welcome you all to the opening of The 2nd International Conference on Mathematics, Science, Technology, Education & their Applications (2nd ICMSTEA). I am delighted to see that the Mathematics and Natural Science Faculty has again organized the second conference that capitalize on our strength and built on our commitment to promoting Mathematics, Science, Technology and Education.

I do hope that this conference would bring a great opportunity for all of us to strengthen our contribution to the advancement of our nation.

I would like to take this opportunity to thank the conference organizing committee for their diligent work. I would also like to thank participants, especially those of you coming from abroad, for joining us and sharing your valuable experiences. Should you find any inconveniences and shortcomings, please accept our sincere apologies.

Finally, let me wish you fruitful discussion and a very pleasant stay in Makassar.

Thank you,
Wassalamu'alaykum Warahmatullahi Wabarakatuh

Dean of Faculty of Mathematics and Natural Sciences
Universitas Negeri Makassar

Prof. Dr. Abdul Rahman, M.Pd.

IOP Conf. Series: Journal of Physics: Conf. Series **954** (2018) 011001 doi:10.1088/1742-6596/954/1/011001

Theme:
"The Role of Mathematics, Sciences, Technology, and Education towards ASEAN Economic Community and Global Challenges".

Speakers:

Professor Valerio Sbordoni, Ph.D.
Zoology dan Evolutionary Biology
University of Rome Tor Vergata, Italy

Associate Professor Mohd. Ekhwan Toriman, Ph.D.
Development and Environment Studies
Universitas Kebangsaan Malaysia

Professor Fahrul Zaman bin Huyop, B.Sc., M.Sc., Ph.D
Biochemistry and Biotechnology
Universitas Teknologi Malaysia

Prof. Alfred Edward Szmidt, Ph.D.
Molecular Population Genetics and Phylogeography
Kyushu University, Japan

Sitti Maesuri Patahuddin, Ph.D.
Mathematics Education
University of Canberra Australia,
President Director Institute for Educational Development (IFED)

Professor Driss Boutat
Mathematics
Loire Valley University, France

Prof. Birgit Loch
Mathematics
Swinburne University of Technology, Australia

Pelindung	: Rektor Universitas Negeri Makassar	
Penanggungjawab	: Prof. Dr. Abdul Rahman, M.Pd.	(Dekan FMIPA)
Pengarah	: Prof. Dr. rer. nat. H. Muharram, M.Si.	
	Dr. Muhammad Arsyad, M.T.	
	Dr. Kaharuddin Arafah, M.Si.	
	Prof. H. Muh. Arif Tiro, M.Pd, M.Sc, Ph.D.	(Ka. Prodi Statistika)
	Dr. Hisyam Ihsan, M.Si.	(Koord. ICP)
	Drs. Suwardi Annas, M.Si, Ph.D.	(Koord. Kersbang)
	Dr. H. Agus Martawijaya, M.Pd.	(Kajur Fisika)
	Dr. Djadir, M.Pd.	(Kajur Matematika)
	Dra. Hj. Sumiati Side, M.Si.	(Kajur Kimia)
	Drs. M. Nur Zakaria Leo, M.Si.	(Kajur Geografi)

Panitia Pelaksana

Ketua	: Dr. Drs. A. Mushawwir Tayeb, M.Kes	(Kajur Biologi)
Wakil Ketua	: Drs. Subaer, M.Phil, Ph.D.	
Sekretaris	: Fajar Arwadi, S.Pd, M.Sc.	
Wakil Sekretaris	: Hartati, S.Si, M.Si, Ph.D.	
Bendahara	: Dr. A. Mu'nisa, S.Si, M.Si.	
Wakil	: T.P. Harmah, S.E.	

Seksi Kesekretariatan

Koordinator	: Rachmawati Muhtar, S.Si, M.Si, Ph.D.
Wakil Koordinator	: Dr. Helmi, M.Si.
Anggota	: Drs. H. Syahruddin, M.Pd.
	Ahmad Zaki, S.Si, M.Si.
	Jusniar, S.Pd, M.Pd.
	St. Fatmah Hiola, S.P., M.Si.
	Nani Kurnia, S.Si, M.Si.
	Irwan, S.Si, M.Si.
	Herman, S.Pd, M.Pd.
	Kamaruddin, S.Sos, M.Si.
	Sitti Rahma Yunus, S.Pd, M.Pd.
	Husain, S.Si, M.Pd.
	Ahmad Fudhail Majid, S.Pd, M.Si.

Seksi Acara

Koordinator	: Dr. A. Asmawati Azis, M.Si.
Wakil Koordinator	: Drs. Abd. Haris Bakri, M.Si.
Anggota	: Pariabti Palloan, S.Si, MT.
	Dr. Ir. Hj. Rosdiana Ngitung, M.P.
	Dr. Hasri, M.Si.
	Rosmini Maru, S.Pd, M.Si, Ph.D.
	Sitti Busyrah Muchsin, S.Pd, M.Pd.
	Syahrullah Asyari, S.Pd, M.Pd.
	Hj. Muliana, SE, M.Pd.
	Drs. Haksani, M.Pd.

Seksi Perlengkapan

Koordinator	: Dr. Awi Dassa, M.S.
Wakil Koordinator	: Dr. Muhiddin P, M.Pd.
Anggota	: Dr. Alimuddin Ali., M.Si.
	Sukarna, S.Pd, M.Si.
	Drs. Ibrahim, M.Si.
	Faisal, S.Pd, M.Pd.
	Sawiah, SE.
	Syamsuddin Mas'ud
	Hardin, S.Si, S.Pd, M.Pd.

Seksi Publikasi dan Dokumentasi

Koordinator	: Dr. Adnan BG, M.S.
Wakil Koordinator	: Dr. Ahmad Yani, M.Si.
Anggota	: Nasrul Ihsan, S.Si., M.Si.
	Ansari Saleh Ahmar, S.Si, M.Sc.
	A. Rahmat Saleh, S.Pd, M.Pd.
	Andi Irma Suryani, S.Pd, M.Si.
	Jafaruddin, S.Pd, M.Pd.
	Andi Irhamsyah, S.Si, M.Si.

Seksi Konsumsi

Koordinator	: Dr. Ir. Hilda Karim, M.Si.
Wakil Koordinator	: Dr. Bunga Dara Amin, M.Ed.
Anggota	: Dr. Ernawati Sjahruddin Kaseng, S.Pi, M.Si.
	Dra. Syamsiah, M.Si.
	Dra. Nurhayati, M.Si.
	Nurlina, SH.

Tim Reviewer

Koordinator	: Prof. Oslan Jumadi, S.Si, M.Phil, Ph.D.
Wakil Koordinator	: Muh. Abdy, S.Si, M.Si, Ph.D.
Anggota	: Prof. Dr. Jasruddin M, M.Si.
	Prof. Muris, M.Si, Ph.D.
	Prof. Dr. Ir. Yusmina Hala, M.Si.
	Prof. Dr. Usman Mulbar, M.Pd.
	Prof. Dr. Muh. Danial, M.Pd.

Tim Penyunting

Koordinator	: Dr. Syafruddin Side, M.Si.
Wakil Koordinator	: Dr. Ir. Wiharto Caronge, M.Si.
Anggota	: Uca, S.Si, M.P., Ph.D.
	Iwan Dini, S.Si, M.Si.
	Hartono, S.Si, S.Pd, M.Biotech.
	Bustang, S.Pd, M.Sc.
	Prasetyo, S.Pd.
	Muh. Husnul Khuluq, S..Pd., M.Sc.
	Said Fachry Assagaf, S.Pd, M.Sc.
	Sahlan Sidjara, S.Si, M.Sc.

IOP Conf. Series: Journal of Physics: Conf. Series **954** (2018) 011002 doi:10.1088/1742-6596/954/1/011002

Peer review statement

All papers published in this volume of *Journal of Physics: Conference Series* have been peer reviewed through processes administered by the proceedings Editors. Reviews were conducted by expert referees to the professional and scientific standards expected of a proceedings journal published by IOP Publishing.

Physical Fitness For Futsal Referee Of Football Association In Thailand

Acting Sub L.t.Thaweesub Koeipakvaen

Naresuan University, Thailand

thaweesub@yahoo.co.th

Abstract. The purpose of the research to study physical fitness for futsal Referee of Football Association in Thailand and Compare of the Physical Fitness for first level, second and third futsal Referee of Football Association in Thailand. The population of first level, second and third level futsal Referee of Football Association in Thailand 107 person. The sample were futsal Referee of Football Association in Thailand 97 person. First level futsal Referee 22 person. Second level futsal Referee 11 person. Third level futsal Referee 64 person. The instrument used was futsal physical fitness test of Football Association in Thailand. Endurance Test (1,000 meter), Speed Test (4x10 meter) 2 time, and Agility Test (80 meter) 2 time. The statistic for data analysis were one way Anova, Percentage, Mean, Standard Deviation and F-test. The results were as the follow: (1) the result comparing F-test first level futsal referee with level second and first level futsal referee with third level. the statically significant different at the 0.05 level, and the result comparing Endurance as the follow the Physical Fitness for first level, second and third futsal Referee. the statically significant different at the 0.05 level.

Keyword*:* Physical Fitness, Futsal Referee, Football Association in Thailand.

1. Introduction

Physical fitness is one of important components of human life with which man can do activities efficiently. Being healthy is what everyone expects in life. Physical fitness comprises two aspects of health-oriented and body movement-or sport-oriented fitness. The latter refers to fitness related to the ability to control body movement and is called motor fitness. Physical fitness test is a way to measure and evaluate the performance of various body systems such as body structure, musculoskeletal system, circulatory system, respiratory system, and nerve and muscle systems. Physical fitness test is very important and necessary to find out the physical fitness condition of an individual. It can also be used to create proper physical fitness programs or the right method of building physical fitness. In addition, the result of physical fitness test can be used to know one's weakness and thus improve his/her physical fitness.

There are many types of physical fitness test and each of them has different aspects of coverage. Each type will result in certain level of different aspects of physical fitness. It will help test takers to know their weakness and decide which activities and exercises fit their physical condition. Physical fitness test is crucial for futsal referees of the Football Association of Thailand under the Royal Patronage. With the physical fitness test, referees will be able to know their physical fitness level and determine which aspects of physical fitness they need to develop. In addition, the result of physical fitness test is used as a standard required to determine which futsal referee level they will belong to.

Therefore, the researcher realized the importance of promoting physical fitness test of futsal referees of the Football Association of Thailand under the Royal Patronage. The current study was conducted to develop the physical fitness of the futsal referees of the Football Association of Thailand under the Royal Patronage to reach their optimum development.

The current research aimed to : examine the physical fitness of futsal referees of the Football Association of Thailand under the Royal Patronage, and Compare the physical fitness of futsal referees of level 1, Level 2, and Level 3 of the Football Association of Thailand under the Royal Patronage

2. Method

2.1. Population and Sample

The population in this research included a total of 107 futsal referees from Level 1, Level 2, and Level 3, of the Football Association of Thailand under Royal Patronage in 2015. The samples were 97 futsal referees of the Football Association of Thailand under the Patronage. They were classified into three levels of futsal referees distributed as follows. 22 futsal referees of level 1, 11 futsal referees of level 2 and 64 futsal referees of level 3.

2.2. Variables of the Study

The independent variable of the study was level of futsal referees of the Football Association of Thailand under the Royal Patronage. Level 1, Level 2 and Level 3. The dependent variables of the study on futsal referees of the Football Association of Thailand under the Royal Patronage include: Age, Gender

2.3. Scope of the Study

The physical fitness test of futsal referees of the Football Association under the Royal Patronage consists of the following. Endurance test: running 1000 meters Speed test: shuttle runs of 4x10 meters Agility test: running 80 meters Speed test: shuttle runs of 4x10 meters and Agility test: running 80 meters. Referees who cannot pass the physical fitness test item 3.1.1 (running 1000 meters) are not eligible to take the other test items. Referees who cannot pass the physical fitness test items 3.1.2, 3.1.3, 3.1.4, and 3.1.5 in the first trial will be given another chance to take the test items again. Failure in the second chance means failure in the test.

2.4. Research Instrument

The instrument used in this research is the physical fitness test of futsal referees of the Football Association of Thailand under the Royal Patronage. It consists of 5 items as follows:

- Endurance Test of running 1000 meters in not more than 4 minutes and 10 seconds with a break for not more than 15 minutes.
- Speed Test of shuttle runs of 4x10 meters in not more than 11 seconds with a break for not more than 5 minutes.
- Agility test of running 80 meters (a 30-meter forward run, a 10-meter left lateral run, a 10-meter right lateral run, a 10-meter backward run, and a 20-meter forward run) in not more than 20.50 seconds with a break for not more than 5 minutes.
- Speed Test of shuttle runs of 4x10 meters in not more than 11 seconds with a break for not more than 5 minutes.
- Agility test of running 80 meters (a 30-meter forward run, a 10-meter left lateral run, a 10-meter right lateral run, a 10-meter backward run, and a 20-meter forward run) in not more than 20.50 seconds with a break for not more than 5 minutes.

The research instrument used in this study was developed following the steps below. The instrument was consulted to experts to enhance the content validity. It involves 5 experts that include:

- Dr. Panya Hanlumyuong, Director of Physical Education and Recreation Department, former international football referee
- Assoc. Prof. Dr. Panya Sangkawadee, Department of Physical Education and Exercise Science, Faculty of Education, Naresuan University
- Assoc. Prof. Nitidech Cherdput, official of futsal competition, Football Association of Thailand under the Royal Patronage
- Surasak Thongsuwimon, official of futsal competition, Football Association of Thailand under the Royal Patronage
- Adulyachart Kanthama, official of futsal competition, Football Association of Thailand under the Royal Patronage

Table 1. Test Criteria

Sequ-ence	Test items			**standard**	**Score**
	Endurance running 1000 meters	Agility (80 meters)	Speed Running 4X10 meters		
1	Lower - 3.40 minutes	Lower - 18.00 seconds	Lower - 09.50 seconds	Excellent	5
2	3.40-3.49 minutes	18.00 -18.99 seconds	09.50-09.99 seconds	Good	4
3	3.50-3.99 minutes	19.00-19.99 seconds	10.00-10.49 seconds	Fair	3
4	4.00-4.10 minutes	20.00-20.99 seconds	10.50-10.59 seconds	Poor	2
5	4.10 – higher	21.00 – higher	11.00 – higher	Very poor	1

1. The maximum duration of passing the endurance physical fitness test is not more than 4.10 minutes.
2. The maximum duration of passing the agility test is not more than 21.50 seconds.
3. The maximum duration of passing the speed test is not more than 11 seconds.

2.5. Data Collection

The data in this study were collected by the researcher following the steps as follows:

- Studying the method, equipment, facilities, and the place of the test.
- Asking for permission to the head of referees of the Football Association of Thailand under the Royal Patronage.
- Finding research assistants to collect the data, tools and equipment, and discussing and demonstrating how the data collection should be performed.
- Preparing the venue of the test and sheets needed to note down the data.
- Coordinating all parties involved in the data collection before it was performed.
- Explaining and demonstrating the procedures of the test to bring understanding to the parties involved.
- Noting the data of every test taker on the sheets and inputting the data into the computer for data analysis.

2.6. Data Analysis

The data were collected from every item of the physical fitness test and were taken into a statistical analysis that includes:

- Finding the mean and standard deviation values of the physical fitness test of each futsal referee

- Comparing the scores of the physical fitness test of the futsal referees with the criteria of the International Football Federation
- Presenting the results of the analysis in tables with some analytical descriptions
- Comparing the results of the pysical fitness test of referees of level 3 and those of level 2 using the one-way analysis of variance (ANOVA).

3. Result and Discussion

3.1. Result

Part 1 Overall physical fitness of futsal referees of the Football Association of Thailand under the Royal Patronage of level 1, Level 2, and Level 3

Table 2 The average duration ($\overline{X}$) and standard deviation of the endurance physical fitness test of futsal referees of the Football Association of Thailand under the Royal Patronage of level 1, Level 2, and Level 3

Level	Number of samples	Percentage	($\overline{X}$)	(S.D.)	Level
Futsal referees of level 1	22	22.68	3.75	0.28	Moderate
Futsal referees of level 2	11	11.34	3.60	0.15	Moderate
Futsal referees of level 3	64	65.98	3.55	0.16	Moderate
Total	97	100.00	3.60	0.21	Moderate

Table 3 above shows the average duration and standard deviation of the 97 samples' performance in the endurance physical fitness test. The average duration of endurance physical fitness of the referees is 3.6 seconds, which is categorized as moderate. Referees of Level 3 had the lowest average duration at 3.55 seconds, which also lies in the moderate level, followed by those of Level 2. The highest duration belonged to referees of Level 1 at 3.6 seconds.

Table 3 The average duration ($\overline{X}$) and standard deviation of the speed physical fitness test of futsal referees of the Football Association of Thailand under the Royal Patronage of level 1, Level 2, and Level 3

Level	Number of Samples	Percentage	($\overline{X}$)	(S.D.)	level
Futsal referee, level 1	22	22.68	9.60	1.42	High
Futsal referee, level 2	11	11.34	9.85	0.23	High
Futsal referee, level 3	64	65.98	9.84	0.37	High
Total	97	100.00	9.78	0.74	High

The result shows that the overall average duration of all the samples in the first speed test is 9.78 seconds, which is considerably high. Referees of Level 1 showed the lowest average duration in the speed test of physical fitness at 9.60 seconds. On the other hand, referees of Level 3 ranked the second in their average duration at 9.84 seconds, followed by referees of level 2 in the topic position at 9.85 seconds.

Table 4 The average duration ($\overline{X}$) and standard deviation of the first agility physical fitness test of futsal referees of the Football Association of Thailand under the Royal Patronage of level 1, Level 2, and Level 3

Level	Number of Samples	Percentage	($\overline{X}$)	(S.D.)	level
Futsal referee, level 1	22	22	19.04	1.02	Medium
Futsal referee, level 2	11	10.00	19.21	0.65	Medium
Futsal referee, level 3	64	64.0	18.99	0.84	High
Total	97	100.00	19.03	0.86	Medium

The result shows that the overall average duration of the whole samples in the first agility test is 19.03 seconds and is considered is medium. Referees of Level 3 showed the lowest average duration at 18.99 seconds, followed by those of Level 1 and Level 2 at 19.04 and 19.21 respectively. While the performance of referees of Level 3 is considered high, the other two levels are medium.

Table 5 The average duration ($\overline{X}$) and standard deviation of the second speed physical fitness test of futsal referees of the Football Association of Thailand under the Royal Patronage of level 1, Level 2, and Level 3

Level	Amount	Percentage	($\overline{X}$)	(S.D.)	level
Futsal referee, level 1	22	22.68	10.32	2.02	Medium
Futsal referee, level 2	11	11.34	9.89	0.21	High
Futsal referee, level 3	64	65.98	9.88	0.34	High
Total	97	100.00	9.98	1.00	High

The result shows that the overall average duration of the test takers in the second speed test is high, at 9.98 seconds. Referees of Level 3 generated the lowest average duration at 9.88 while referees of Level 2 generated the second average duration at 9.89. Lastly, referees of Level 1 made the highest average duration at 10.32 seconds. While referees of level 1's achievement is considered medium, that of the other two levels are high.

Table 6 The average duration ($\overline{X}$) and standard deviation of the second agility physical fitness test of futsal referees of the Football Association of Thailand under the Royal Patronage of level 1, Level 2, and Level 3

Level	Amount	Percentage	($\overline{X}$)	(S.D.)	level
Futsal referees, level 1	22	22.68	18.94	0.68	High
Futsal referees, level 2	11	11.34	18.88	0.50	High
Futsal referees, level 3	64	65.98	18.80	1.26	High
Total	97	100.00	18.84	1.08	High

The result shows that the overall average duration of the samples in the second agility test is high, at 18.84 seconds. Referees of Level 3 made the lowest average duration at 18.0

IOP Conf. Series: Journal of Physics: Conf. Series **954** (2018) 012001 doi:10.1088/1742-6596/954/1/012001

seconds, followed by referees of level 2 and level 1 at 18.88 and 18.94 respectively. All of these scores are considered high.

3.2. The Comparison between the Physical Fitness of Referees of Level 1, Level 2, and Level 3 of the Football Association of Thailand under the Royal Patronage

Table 7 The Analysis of Variance of the Endurance Physical Fitness of Futsal Referees of Level 1, Level 2, and Level 3 of the Football Association of Thailand under the Royal Patronage

Test	Sources of variance	df	SS	MS	F	p-value
Durability	Between group	2	0.69	0.34	9.23	0.00*
	In group	94	3.50	0.40	-	-
	total	96	4.18	-	-	-

The result shows that the endurance physical fitness of futsal referees of level 1, level 2, and level 3 demonstrates a statistically significant difference.

Table 8 Comparative Analysis of the Endurance Physical Fitness Test of Futsal Referees of Level 1, Level 2, and Level 3 of the Football Association of Thailand under the Royal Patronage.

Level	1	2	3
1	-	0.16*	0.20*
2	-	-	0.05
3	-	-	-

The result shows that the physical fitness test of endurance of futsal referees of level 1, level 2, and level 3 results in a statistically significant difference.

Table 9 The Analysis of Variance of the Speed Physical Fitness of Futsal Referees of level 1, level 2, and level 3 of the Football Association of Thailand under the Royal Patronage

Test	Sources of variance	df	SS	MS	F	p-value
Speed	Between group	2	0.97	0.49	0.89	0.42
	In group	94	51.46	2.55	-	-
	Total	96	52.43	-	-	-

The result shows that the variance of the physical fitness of speed of futsal referees of level 1, level 2, and level 3 is similar.

Table 10 The Analysis of Variance of the Agility Physical Fitness of Futsal Referees of Level 1, Level 2, and Level 3 of the Football Association of Thailand under the Royal Patronage

Test	Sources of variance	Df	SS	MS	F	p-value
Agility	Between group	2	0.43	0.21	0.29	0.75
	In group	94	70.08	0.75	-	-
	total	96	70.51	-	-	-

The result shows that there is no difference in the agility physical fitness between futsal referees of level 1, level 2, and level 3.

Table 11 The Analysis of Variance of the Second Speed Physical Fitness of Futsal Referees of Level 1, Level 2, and Level 3 of the Football Association of Thailand under the Royal Patronage

Test	Sources of variance	Df	SS	MS	F	p-value
Speed	Between group	2	3.22	1.61	0.63	0.20
	In group	94	92.97	0.99	-	-
	Total	96	96.19	-	-	-

The result shows that there is no difference in the speed physical fitness of futsal referees of level 1, level 2, and level 3.

Table 12 The Analysis of Variance of the Second Agility Physical Fitness of Futsal Referees of Level 1, Level 2, and Level 3 of the Football Association of Thailand under the Royal Patronage

Test	Sources of variance	Df	SS	MS	F	p-value
Agility	Between group	2	0.31	0.15	0.13	0.88
	In group	94	111.46	1.19	-	-
	Total	96	111.76	-	-	-

The result shows there is no difference in the physical fitness of agility of futsal referees of level 1, level 2, and level 3.

3.3. Discussion

The overall results of the endurance test of physical fitness of the 97 futsal referees of level 1, level 2, and level 3 of the Football Association of Thailand under the Royal Patronage demonstrate an average duration of 3.60 seconds to complete the test. Futsal referees of level 3 shows the lowest average duration at 3.55 seconds. It may be benefited the factor of age of the referees. Getchell states that physical fitness is the ability to operate with maximum efficiency of heart, blood vessels, and muscles which results in the perfect health and the ability to perform routine tasks enthusiastically [1]. Greenberg, Dintiman and Oakes state that physical fitness is the ability to fulfill the demands of life and support an energy to respond to unexpected events [2]. It consists of five basic elements: cardiovascular endurance, muscular strength, endurance muscle elasticity and shape. Corbin states that the physical fitness requires well-maintained organs to avoid sickness or disease [3]. It includes good teeth, good eyes, good ear, and ability to work hard with full performance. Futsal referees of level 2 show 3.60 seconds of average duration, followed by those of level 1 at 3.75 seconds. The first speed test of physical fitness of futsal referees of Football Association of Thailand under Royal Patronage shows that the average duration is 9.78 seconds. Futsal referees of level 1 show the lowest average duration at 9.49 seconds. According to Wuest and Butcher (cited in Wiwatchai Worabaworn 2002, P. 102), physical fitness is important for children and adults because it is related to health and physiology of an individual, along with the development of the intelligence. One's performance depends on the training program, the amount of training, and a frequency of practice. Futsal referees of level 3 and 2 show the average of time at 9.84 and 9.85 respectively. The first agility test of physical fitness of futsal referees of level 1, level 2, and level 3 of the Football Association of Thailand under the Royal Patronage in overall shows that the average duration to complete the test is 19.03 seconds. Futsal referees of level 3 show the lowest average duration at 18.99 seconds. According to Getchell (Getchell, 1976, pp. 8-10 cited in Benjamard Kerdmalai), the physical fitness is the ability to operate with maximum efficiency of heart, blood vessels, and muscles which results in the perfect health and the ability to perform routine tasks enthusiastically. Futsal referees of level 1 and level 2 show average durations at 19.04 and 19.21 respectively. The second speed test of physical fitness test of futsal referees of level 1, level 2, and level 3 of the Football Association of Thailand under the Royal Patronage results an average duration at 9.98 seconds. Futsal referees of level 3 show average duration at 9.88 seconds. According to Worasak Pienshop, speed is the ability to contract any part of muscles in a short time, for example, short range running [4]. This development of speed activity may be used as the activity to improve strength and speed. Futsal referees in level 2 shows average duration at 9.89 seconds, followed by those of level 1 at 10.32 seconds. The second agility test of physical fitness test of futsal referees of level 1, level 2, and level 3 of the Football Association of Thailand under the Royal Patronage results in average duration at 18.84. Futsal referees of level 3 generated the lowest average duration at 18.80 seconds. According to Worasak Pienshop, agility is the ability to change the position or direction effectively [5]. The ability to shrink muscles results in the effective movement, for example, the ability to squat, do shuttle run, or zigzag running quickly. The activity which causes muscles to cooperate or change the direction of the body helps in the development of the agility. Futsal referees of level 2 demonstrated average duration at 18.88 seconds, followed by those of level 1 at 18.94 seconds. The comparative result of the physical fitness between futsal referees of level 1, level 2, and level 3 of the Football Association of Thailand under the Royal Patronage. The variance of the endurance physical fitness of futsal referees of level 1, level 2, and level 3 shows a statistically significant different at 0.05. There is no difference in the first speed test of physical fitness of futsal referees between level 1, level 2, and level 3. There is no difference in the first agility test of physical fitness of futsal referees between level 1, level 2, and level 3. There is no difference in the second speed test of physical fitness of futsal referees between level 1, level 2, and level 3. There is no difference in the second agility test of physical fitness of futsal referees between level 1, level 2, and level 3. The comparison of the results of the endurance physical fitness test shows a statistically significant difference between referees of level 1, level 2, and level 3, at 0.05.

IOP Conf. Series: Journal of Physics: Conf. Series **954** (2018) 012001 doi:10.1088/1742-6596/954/1/012001

4. Conclusion

The general information of physical fitness of futsal referees of level 1, level 2, and level 3 of the Football Association of Thailand under Royal Patronage

- Futsal referees of level 3 showed the lowest average duration in endurance physical fitness. Referees of level 1 made the highest average duration while those of level 2 is in between them.
- Futsal referees of level 1 showed the lowest average duration in the first speed test of physical fitness. Those of level 3 ranked the second and those of level 2 made the highest average duration.
- The futsal referees of level 3 had the lowest average duration in the first agility test of physical fitness. Those of level 1 ranked the second and those of level 2 made the highest average duration.
- The second agility test of physical fitness resulted in that futsal referees of level 3 had the lower average duration, followed by those of level 2 and level 1.

The result of the endurance physical fitness test of futsal referees of level 1, level 2, and level 3 shows a statistically significant difference.

- The results of the first speed physical fitness test of futsal referees of level 1, level 2, and level 3 are similar.
- There is no difference in the first agility physical fitness test between futsal referees of level 1, level 2, and level 3.
- There is no difference in the second speed physical fitness test between futsal referees of level 1, level 2, and level 3.
- There is no difference in the second agility physical fitness test between futsal referees of level 1, level 2, and level 3.
- The comparative analysis of the results of the durability physical fitness test shows a statistically significant difference between futsal referees of level 1, level 2, and level 3, at 0.05.

References

[1] Getchell. *(1979)*. Physical Fitness: A way of Life. New York: John Wiley and Sons, Inc.
[2] Greenberg, Dintiman; Oakes. *(1998)*. Physical Fitness and Wellness. 2nd ed. Boston:Allyn and Bacon.
[3] Corbin. *(1970)*. Concept in Physical Education. W.M.C. Brown Company Publishers.
[4] Worasak Pienshop. (1991). The Code Articles of Prof.Worasak Pienshop. Bangkok: Commemorative of retirement September 21st 1991.
[5] Worasak Pienshop. (2003). Teaching Methods of Physical Education. Bangkok: Thaiwattanapanich.A reference

IOP Conf. Series: Journal of Physics: Conf. Series **954** (2018) 012002 doi:10.1088/1742-6596/954/1/012002

Pedagogical Content Knowledge: Teacher's Knowledge of Students in Learning Mathematics on Limit of Function Subject

Ma'rufi[1], I Ketut Budayasa[2], & Dwi Juniati[2]

[1]Training and Education, Universitas Cokroaminoto Palopo, Indonesia
[2]Training Faculty of Mathematics and Science Universitas Negeri Surabaya

marufi.ilyas@gmail.com

Abstract. This research aims at describing the profile of high school teacher's Pedagogical Content Knowledge in learning mathematics from the perspective of teaching experience. Pedagogical Content Knowledge (PCK) covers teacher's knowledge of subject matter, knowledge of pedagogy, and knowledge of students. The subject of this research was two high school mathematics teachers who have different teaching experience. The data were obtained through interview and observation then analyzed qualitatively. The focus of this research is the novice teacher's PCK deals with knowledge of students. Knowledge of Student is defined as teacher's knowledge about the students' conception and misconception on limit of function material and teacher's ability to cope with students' difficulty, mistake, and misconception. The result of this research shows that novice teacher's ability in analyzing the cause of students' difficulty, mistake, and misconception was limited. Novice teacher tended to overcome the students' difficulty, mistake, and misconception by re-explaining the procedure of question completion which is not understood by the students.

1. Introduction

Teacher is a key component in a learning as well as in learning mathematics. A teacher should master the mathematic content taught and know how to transfer the lesson so the students can understand it easily. Knowledge of content and knowledge of pedagogic are not enough to obtain effective teaching practice without knowledge of students, curriculum, education objectives, and teaching materials. A teacher should know and comprehend a particular topic which is easy or difficult for the students and a conception or misconception which the students may have related to a particular topic. Shulman explained that *Pedagogical Content Knowledge* (PCK) is a kind of specific knowledge as the basic knowledge for teacher which includes the connection of various knowledge and skill of representation, analogy, examples, demonstration of a material in order to be understood by the students [1].

According to Lee, teacher's way to transfer mathematic knowledge to students with understandable technique is the core of PCK [2]. In addition, it is explained that teacher's PCK is an important element to be an effective mathematics teacher. Teacher's mastery on the lesson or mathematic content is important for teacher's success in teaching. A teacher cannot be expected to explain a mathematic concept, if he does not have a complete comprehension of that concept [3]. On the other side, mastering the lesson that will be taught is not enough to gain the objective intended on students [4]. Thus, the teacher who is expected can teach effectively is a teacher who master the lesson that will be taught and able to explain that material so it can be understood by the students.

Ball and Bass identified teacher's knowledge deal with the students' difficulties and leaning strategies which suitable to overcome those problems in learning mathematics as a part of PCK [5]. Meanwhile, Ma [6] gave a definition that PCK in mathematics learning is a teacher's ability using his knowledge of mathematics to explore the mathematic topics and present the content of the topics in the methods that can build a successful mathematics learning. Furthermore, Ma explained that a teacher should be able to demonstrate the width, depth, and interconnection of content knowledge and utilize all of them to express mathematic solution. Kilic divided PCK into four components namely: (1) *Knowledge of Subject Matter*, (2) *Knowledge of Pedagogy*, (3) *Knowledge of Students*, and (4) *Knowledge of Curriculum* [7].

Based on the definitions above, it can be concluded that knowledge of content or knowledge of main material owned by a mathematics teacher should be transformed by utilizing various resources such as text book with the concept presentation that is easy to be understood by the students. Besides, a teacher in transforming knowledge of content should use different representations, help students to make connection between different representations to solve mathematic problems, identify students' wrong thinking, be able to give respond toward students' questions.

PCK is a specific knowledge which integrates knowledge of mathematic content, knowledge of pedagogic, and knowledge of students. This PCK is an important knowledge in teaching because it can help teacher to anticipate students' learning misconception and mistake and be ready to give alternative model or explanation to overcome those misconception and mistake. An, Kulm, and Wu identified four aspects of PCK [8]. They are (1) building students' idea in mathematics, (2) overcoming students' misconception, (3) involving students in learning mathematics, and (4) supporting students' thought of mathematics.

Novak & Gowin stated that a misconception is the interpretation of concepts in a statement which is not acceptable [9]. Then, Smith in Sessa and Roschelle (in [10]) stated that a misconception can be easily fixed through an isolation. Moore stated *"one method of remediation is by "explaining"* [10]. While Lucariello stated that misconception can really hamper the learning for some reasons [11]. First, the students generally do not realize that the knowledge they have is wrong. Besides, a misconception can be very strong in the students' thought. Also, new experience that is integrated through a wrong conception can be disturbing in understanding a new information.

Next, Moore explained that the causes of misconception are: (1) the schema that is owned is wrong and fail to be changed, (2) the schema is wrong in accommodating a new information, (3) the new concept or schema corrupts the previous schema and causes confusion [10]. Another causes of misconception are the development stage of cognition is not appropriate with the concept learned, the student's reasoning is limited and wrong, the student's ability in grasping and understanding the concept learned, and the student's interest to learn the concept.

Zevenbergen, Dole, and Wright explained that a good teaching involves teacher's knowledge of students' thought about mathematic concept and knowledge of how to direct them to construct a more complex, complete, and strong knowledge using organized activity, habit, learning environment [12]. Students' misconception happened in mathematics learning is because of the less of students' understanding on mathematic concept, included on the material of quadrate function. The teacher should have knowledge of the students' mistake and misconception on the material taught so he can be more focus in teaching process by adopting appropriate model, method, strategy, and approach.

Mathematics teacher's PCK on algebra function limit material in this research is a kind of knowledge which integrates knowledge of content, knowledge of pedagogy, and knowledge of students in learning mathematics on algebra function limit material. The components of PCK in this research are (1) *Knowledge of Subject Matter,* (2) *Knowledge of Pedagogy,* (3) *Knowledge of Students.* This article describes the novice teachers' knowledge of students in overcoming the students' misconception on algebra function limit material.

2. Method

This research is a descriptive research using qualitative approach. The subject of this research was two senior high school mathematics teachers who teach at grade XI Social in the same school. The two teachers have the qualification of Mathematics Education Bachelor. The criteria in determining the subject are: (1) teaching at grade XI Social in the same school, (2) having minimum 5 years teaching experience, (3) willing to give relevant data, included the learning observation in each class. Subject 01 with 15 years teaching experience (experienced teacher) is symbolized as S01, subject 02 with 7 years teaching experience (novice teacher) is symbolized as S02, since upraised as civil servant.

The data were collected through the observation of the learning and interview the subjects. The algebra function limit was held in five meetings according to the subjects' Lesson Plan and the sixth meeting was the test. The researchers did the observation on five meetings of the subjects. The students' answer sheets were examined, then it was found that the problems on the test which mostly answered wrong by the students were number 3 and number 5. The students' answers on number 3 and 5 have different mistakes. The interview with the subjects was based on the questionnaire adapted from An, Kulm, and Wu [8]. The questionnaire contained the result of students' work in answering the problems given. The questions given to the teachers refer to the result of students' work in order to explore the teachers' PCK deals with the teachers' knowledge in overcoming the students' misconception. The questionnaire consists of two issues based on the result of the students' test to investigate the teachers' PCK of the knowledge about students on the topic of algebra function limit. The supporting instrument used was in the form of questionnaire.

Problem 1: The students at grade XI Social were given problem $\lim_{x\to3}\frac{9-x^2}{4-\sqrt{x^2+7}}$. The result of student Dp and Sr were as follow:

a. The answer of Dp: $\lim_{x\to3}\frac{9-x^2}{4-\sqrt{x^2+7}} = \frac{9-3^2}{4-\sqrt{3^2+7}} = \frac{9-9}{4-\sqrt{16}} = \frac{9-9}{4-4} = \frac{0}{0}$

$$\lim_{x\to3}\frac{9-x^2}{4-\sqrt{x^2+7}} = \lim_{x\to3}\frac{9-x^2}{4-\sqrt{x^2+7}} \times \frac{4+\sqrt{x^2+7}}{4+\sqrt{x^2+7}}$$
$$= \lim_{x\to3}\frac{(9-x^2)(4+\sqrt{x^2+7})}{16-x^2+7}$$
$$= \lim_{x\to3}\frac{9-3^2(4+\sqrt{3^2+7})}{16-3^2+7}$$
$$= \lim_{x\to3}\frac{9-9\,(4+\sqrt{16})}{16-9+7}$$
$$= \lim_{x\to3}\frac{9-9\,(4+4)}{16-9+7} = \frac{0\,(8)}{7+7} = \frac{0}{14} = 0$$

b. The answer of Sr: $\lim_{x\to3}\frac{9-x^2}{4-\sqrt{x^2+7}} = \frac{9-3^2}{4-\sqrt{3^2+7}} = \frac{9-9}{4-\sqrt{16}} = \frac{0}{0}$

$$\lim_{x\to3}\frac{9-x^2}{4-\sqrt{x^2+7}} = \lim_{x\to3}\frac{9-x^2}{4-\sqrt{x^2+7}} \times \frac{4+\sqrt{x^2+7}}{4+\sqrt{x^2+7}}$$
$$= \frac{9-3^2\times4+\sqrt{3^2+7}}{4-\sqrt{3^2+7}\times4+\sqrt{3^2+7}}$$
$$= \frac{9-9\times4+\sqrt{16}}{4-\sqrt{16}\times4+\sqrt{16}}$$
$$= \frac{9-9\times4+4}{4-4\times4+4}$$
$$= \frac{4}{4} = 1$$

Problem 2: The students worked on the problem $\lim_{x\to\infty}\frac{(x^2+1)^2}{4x^4-3x+1}$. The result of the student Al and Ni were as follow:

a. The answer of Al

$$\lim_{x\to\infty}\frac{(x^2+1)^2}{4x^4-3x+1} = \lim_{x\to\infty} = \frac{(x^2+1)^2}{4x^4-3x+1}$$

$$= \frac{x^4 + 2}{4x^4 - 3 + 1}$$
$$= x^4 + 2 + 4x^4 - 3 + 1$$
$$= 4x^8 + 0$$
$$= 4x^8$$

b. The answer of Ni

$$\lim_{x\to\infty} \frac{(x^2+1)^2}{4 \times 4 - 3x + 1} = \lim_{x\to\infty} \frac{\left(\frac{x^2}{x^4} + \frac{1}{x^4}\right)\left(\frac{x^2}{x^4} + \frac{1}{x^4}\right)}{\frac{4x^4}{x^4} - \frac{3x}{x^4} + \frac{1}{x^4}}$$
$$= \lim_{x\to\infty} 4x^4 - 3x + 1$$
$$= \lim_{x\to\infty} 4\infty^4 - 3\infty + 1$$
$$= \infty + 1 = \infty$$

Based on the result of the students' work, to trace the teachers' knowledge of the students' mistake and misconception, the following questions were given to the subjects:

1. What is probably thought by each student?
2. What might be the cause of misconception on each student?
3. How is the alternative way to overcome the misconception of the four students?
4. What kind of question/task will be given to each student to overcome the misconception?
5. How to explain those problems to make the students easier to understand the problems?

3. Result and Discussion

The PCK of the novice teacher deals with the knowledge to overcome the students' mistake and misconception on algebra function limit material is described as follow:

3.1. The novice teacher knowledge of the students' mistake and misconception on algebra function limit

The novice teacher identified that the students' mistake and misconception based on the students' work describe what might be thought by the students. Generally, the students' answers for problem 1 are the same because initialized with substitution. In the beginning of limit material, to determine the value of function limit, the first step done is substitution. Then the two students got the result of indefinite form (0/0). They continued by choosing multiplication of complex number because the form of the function limit is fraction and its denominator contains irrational form. Thus, the steps to determine the value of function limit is already correct. But, when finishing the problem, the students did the mistake namely forget to give parenthesis in doing two syllables multiplication operation.

Student *Dp* was wrong when multiplied $(9 - x^2)$ with $(4 + \sqrt{x^2 + 7})$ that forget to put parenthesis so the result got was $9 - x^2(4 + \sqrt{x^2 + 7})$ as well as the result of the multiplication of the denominator in which *Dp* forget to put parenthesis after the subtraction operation so the result got was $(4 - x^2 + 7)$ which should be $4 - (x^2 + 7)$. The procedural mistake also happened that student *Dp* did wrong in doing calculation because forget to give parenthesis. That case also showed that student *Dp* has not understood mathematics basic rule about the use of parenthesis and its function. Student *Sr* also did the same mistake with student *Dp* namely calculation case. The difference was that student *Sr* directly did substitution after multiplying the complex number. Basically, student *Sr* and *Dp* did the same mistake because the purpose of the multiplication of complex number is make it easy to eliminate the same factor later on, but in the two students' alternative answers there was no factor eliminated. The mistake of the two students was also a concept mistake because the function of complex number multiplication has not been understood that is when doing complex number multiplication the same factor will be later obtained, but the two students' answers did not get to that stage.

Two students faced concept and procedure mistake on the second problem. *Al* and *Ni* did not understand that the concept of function limit for x approaches is illimitable. When substituting directly, the result got is $\frac{\infty}{\infty}$, so the algebra manipulation should be done before direct substitution of x value approaches illimitable. For the alternative answer of the first student, *Al* did not show substitution way, did not divide it with the highest degree and also did not multiply it with complex factor, what is done was only combining the denominator and the numerator. The first step done by *Al* was trying to scatter or finish the quadrate but it was still wrong. *Al* was wrong in scattering $(x^2+1)^2$ because *Al* directly drew the degree so the result became (x^4+2). *Al* might think that to simplify $(x^2+1)^2$ is enough by multiplying 2 with the constanta number in parenthesis. Then *Al* continued the problem solving by taking the denominator to the numerator. Meanwhile, *Ni* has actually known that the direction of the solving is dividing by the highest degreed variable, but $(x^2+1)(x^2+1)$ which should be multiplied first was not done by *Ni*. When finding $(x^2+1)^2$, it should be scattered first before dividing it by the highest degree that becomes (x^2+1) multiplied by (x^2+1).

The novice teacher identified the students' mistake and misconception based on the result of their work on the test of algebra function limit by understanding what might be thought by them about algebra function limit. It is in line with Kilic's statement that a teacher needs to identify the students' misconception and difficulty by proposing questions or using appropriate task [7].

3.2. The novice teacher knowledge about the cause of the students' mistake and misconception on algebra function limit material

The novice teacher described that the cause of the mistake and misconception is that the students commonly have not understood the steps to determine the value of algebra function limit, the tendency of the students to memorize the problem and its solving steps. So when the students solve the problem without seeing the example, they will forget how to solve the limit problem. As an example, the students sometimes remember that the problem like $\lim_{x\to 3}\frac{9-x^2}{4-\sqrt{x^2+7}}$, will be multiplied with complex form but they do not understand the purpose of using complex number multiplication. As a result, they will directly substitute the limit value after multiplying it with the complex form, while indeed it should be factorized first because if it is substituted first the result will remain as indefinite value or 0/0. The misconception happened because the students have not understood the purpose of complex form multiplication so for a problem like $\lim_{x\to 3}\frac{9-x^2}{4-\sqrt{x^2+7}}$ it is solved by directly substituting without thinking to direct it to eliminate the similar factors of the denominator and the numerator.

The next is problem 2 that was the students' answers on the problem of calculating the value of function limit of $\lim_{x\to\infty}\frac{(x^2+1)^2}{4\times 4-3x+1}$. The novice teacher described that *Al*'s misconception in determining the value of illimitable function was because *Al* did not understand the steps to solve such problem so the writing of the solving was not clear from the beginning. While *Ni* might forget that for a problem like $\lim_{x\to\infty}\frac{(x^2+1)^2}{4\times 4-3x+1}$, $(x^2+1)^2$ should be scatter first becomes $(x^2+1)(x^2+1) = x^4 + 2x^2 + 1$, then it is divided by the highest degree to obtain the value of function limit $\lim_{x\to\infty}\frac{(x^2+1)^2}{4\times 4-3x+1}$.

The novice teacher knew the cause of the students' mistake and misconception based on the result of their work on the test of algebra function limit material. According to Kilic, a teacher needs to determine the source of a student's difficulty and mistake to be corrected accurately [7]. The research result of Tanisli & Kose showed that a pre-service elementary school mathematics teacher understood the students' thinking process related to the knowledge of variable, similarity, and equation but found it difficult to explain the cause of the students' thought [13].

3.3. The novice teacher knowledge about the alternative way to overcome the students' mistake and misconception on algebra function limit material

The novice teacher described that to overcome the students' mistake and misconception in completing the problem $\lim_{x\to 3}\frac{9-x^2}{4-\sqrt{x^2+7}}$, the steps of solving the problem should be re-explained. The first thing that should be explained is the purpose of the multiplication with complex form namely $\lim_{x\to 3}\frac{9-x^2}{4-\sqrt{x^2+7}}\cdot\frac{4+\sqrt{x^2+7}}{4+\sqrt{x^2+7}} = \lim_{x\to 3}\frac{(9-x^2)(4+\sqrt{x^2+7})}{16-(x^2+7)} = \lim_{x\to 3}\frac{(9-x^2)(4+\sqrt{x^2+7})}{9-x^2}$, after that algebra manipulation or factorization is done to obtain the same denominator and numerator that is $\lim_{x\to 3}\frac{(9-x^2)(4+\sqrt{x^2+7})}{9-x^2} = \lim_{x\to 3}(4+\sqrt{x^2+7})$ so it can be simplify then directly substituted $= 4 + \sqrt{3^2+7} = 8$.

The novice teacher overcame the students' mistake by re-explaining that when we substitute the result got is indefinite value or 0/0. Then, it is explained to the students that the denominator and the numerator contain a factor which results in zero when it is substituted. When we eliminate the factor which results in zero, the limit value is gained because the value resulted is no longer in indefinite value like $\lim_{x\to 3}\frac{9-x^2}{4-\sqrt{x^2+7}}$ which becomes a factor resulting zero that is (x-3) because the limit approaches 3, in order that the function form does not result in 0/0, the value (x-3) that will be eliminated by form multiplication or factorization of one function form to get the similar form (x-3) on the denominator and the numerator, so it can be eliminated or similar to one.

Next, to cope with the students' mistake and misconception on the problem $\lim_{x\to\infty}\frac{(x^2+1)^2}{4x4-3x+1}$, the novice teacher re-explained the procedure of problem solving namely the first step is scattering $(x^2+1)^2$ to gain $\lim_{x\to\infty}\frac{x^4+2x^2+1}{4x^4-3x+1}$, then each syllable is divided by the highest degreed variable, $\lim_{x\to\infty}\frac{\frac{x^4}{x^4}+\frac{2x^2}{x^4}+\frac{1}{x^4}}{\frac{4x^4}{x^4}-\frac{3x}{x^4}+\frac{1}{x^4}} = \lim_{x\to\infty}\frac{1+\frac{2}{x^2}+\frac{1}{x^4}}{4-\frac{3}{x^3}+\frac{1}{x^4}} = \frac{1+0+0}{4-0+0} = \frac{1}{4}$. The students' mistake and misconception can also be overcame by giving them another suitable problem, by reviewing the suitable solving stages so they can correct their misconception.

The novice teacher overcame the students' mistake and misconception by re-explaining the concept or procedure of the correct problem solving and giving another example similar with the problem. This is fit with Moore's statement that one remediation method is by explaining [10]. While Richland explained that using analogy example is one way for the teacher to fix the students' misconception [14]. Analogical reasoning has important role and is an effective strategy in mathematics learning [15]-[19].

One aspect of PCK in this research is the novice teacher knowledge about the students which covers their mistake and misconception, the cause of the mistake and misconception, and the alternative way to cope with the students' mistake and misconception on algebra function limit material. The research result of Lee showed that teaching experience is also important factor in mathematics teacher PCK [2]. PCK contains the element of teaching material, how to teach it, and knowledge of students. Baker & Chick emphasized that the knowledge of lesson material, knowledge of pedagogic, and knowledge of students are important parts of PCK [20]. The knowledge of students is the core of PCK and considered as one of the important component of PCK [1][8].

4. Conclusion

PCK is a kind of specific knowledge which integrates the mastery of mathematics content, how to teach it, and knowledge of students. Knowledge of students consists of teacher's knowledge about the students' conception and misconception on algebra function limit material. The teaching experience is one of the important factors of teacher's PCK. The novice teacher knew the students' mistakes and misconception based on the result of the students' work on the test of algebra function limit material by understanding the students' thought. The novice teacher knew the cause of the students' mistake and misconception that the students did not understand the concept and solving procedure of algebra function limit material clearly. The novice teacher overcame the students' mistake and misconception by re-explaining the concept or solving procedure correctly and giving similar example.

References

[1] Shulman, L. S. 1986. Those who understand: Knowledge growth in teaching. *Educational Researcher,* **15**, 2, pp. 4 – 14.

[2] Lee, Joohi, 2010. Exploring Kindergarten Teachers' Pedagogical Content Knowledge of Mathematics: *Department of Curriculum and Instruction-EC4, College of Education and Health Profesional, University of Texas at Arlington, Science Hall.*

[3] Yeo, K.K.J. 2008. Teaching Area and Perimeter Mathematics-Pedagogical Content Knowledge in Action. In M. R, Brown & K. Maker (Eds). Procceding of the 31th Annual Conference of the Mathematics Education Research Group Australia. Merga, pp. 621-627

[4] Isiskal, M. & Cakirogu, E. 2011. The nature of Prospective Mathematics Teachers' Pedagogical Content Knowledge: The Case of Multiplication of Fraction. Journal Math Teacher Education. **14**, 3, pp.213-230.

[5] Ball, D. L., & Bass, H. 2000. Interweaving content and pedagogy in teaching and learning to teach: Knowing and using mathematics. In J. Boaler (Ed.), *Multiple perspectives on mathematics teaching and learning* (pp. 83–104). Westport, UK: Ablex

[6] Ma, Liping, 1999. Knowing and Teaching Elementary Mathematics:*Teachers' Understanding of Fundamental Mathematics in China and the United States*. New York: Routledge.

[7] Kilic, Hulya. 2011. Preservice Secondary Mathematics Teachers' Knowledge of Students. Turkish Online *Journal of Qualitative Inquiry*, **2**, 2.

[8] An, S., Kulm, G., & Wu, Z. 2004. The Pedagogical Content Knowledge of Midle School,Mathematics Teacher in China and The USA. *Journal of Mathematics Teachers Education*, **7**, pp. 145-172

[9] Novak & Gowin. 1984. *Learning How to Learn*. New York: Cambridge University Press

[10] Moore, Gordon March, 2006. *Mathematical Misconceptions in Schooll Children*. Identification Impact and Remediation with a Computer Based Assesment System Master's Dissertation.

[11] Lucariello, Juan. 2004. How Do I Get. My Students Their Alternative Conception (Misconception) for Learning. *City University of New York.*

[12] Zevenbergen, Dole,Wright. 2004. Teaching Mathematics in primary Schools. Sydney: Allen & Unwin.

[13] Tanisli, D & Kose, N.Y. 2013. Pre-Service mathematics Teachers' Knowledge of Students about the Algebratic Concepts. *Australian Journal of Teacher Education*, **38**, 2.

[14] Richland, Lindsey E., Holyoak, Keith J., and Stigler, James W. 2004. Analogy Use in Eighth-Grade Mathematics Classrooms. *Cognition and Instruction*, **22**, 1, pp. 37–60

[15] Arsyad, N., Rahman, A., & Ahmar, A. S. 2017. Developing a self-learning model based on open-ended questions to increase the students' creativity in calculus. *Global Journal of Engineering Education*, **9**, 2, pp. 143-147.

[16] Ahmar, A. S., & Rahman, A. 2017. Development of teaching material using an Android. *Global Journal of Engineering Education*, **19**, 1.

[17] Mulbar, U., Rahman, A., & Ahmar, A. S. 2017. Analysis of the ability in mathematical problem-solving based on SOLO taxonomy and cognitive style. *World Transactions on Engineering and Technology Education*, **15**, 1, pp. 74-77.

[18] Rahman, A., & Ahmar, A. S. 2016. Exploration of Mathematics Problem Solving Process Based on The Thinking Level of Students in Junior High Schoo. *International Journal of Environmental & Science Education*, **11**, 14.

[19] Rahman, A., Ahmar, A. S., & Rusli. 2016. The Influence of Cooperative Learning Models on Learning Outcomes Based on Students' Learning Styles. *World Transactions on Engineering*

and Technology Education, **14**, 3, pp. 74-77.
[20] Baker, M & Chick, H. 2006. *Pedagogical Content Knowledge for Teching Premary mathematics*: A case study of two teachers.

IOP Conf. Series: Journal of Physics: Conf. Series **954** (2018) 012003 doi:10.1088/1742-6596/954/1/012003

A New Image Encryption Technique Combining Hill Cipher Method, Morse Code and Least Significant Bit Algorithm

Dicky Nofriansyah[1*], Sarjon Defit[2], Gunadi W Nurcahyo[2], G Ganefri[3], R Ridwan[3], Ansari Saleh Ahmar[4] and Robbi Rahim[5]

[1]Doctoral Student, Universitas Negeri Padang, Padang, Indonesia
[1]Department of Information System, STMIK Triguna Dharma, Medan, Indonesia
[2]Faculty of Computer Science, Universitas Negeri Padang, Padang, Indonesia
[3]Faculty of Engineering, Universitas Negeri Padang, Padang, Indonesia
[4]Department of Statistics, Universitas Negeri Makassar, Makassar, Indonesia
[5]School of Computer and Communication Engineering, Universiti Malaysia Perlis, Malaysia

*dickynofriansyah@ymail.com

Abstract. Cybercrime is one of the most serious threats. Efforts are made to reduce the number of cybercrime is to find new techniques in securing data such as Cryptography, Steganography and Watermarking combination. Cryptography and Steganography is a growing data security science. A combination of Cryptography and Steganography is one effort to improve data integrity. New techniques are used by combining several algorithms, one of which is the incorporation of hill cipher method and Morse code. Morse code is one of the communication codes used in the Scouting field. This code consists of dots and lines. This is a new modern and classic concept to maintain data integrity. The result of the combination of these three methods is expected to generate new algorithms to improve the security of the data, especially images.

1. Introduction

The rapid development of technology today makes the data storage process easier. However, many people now have doubts about the security when the computer stores the data. It can't be separated from the occurrence of various wiretapping and monitoring actions by unauthorized or irresponsible parties so that secrecy is not maintained in securing data. Many ways to secure data one of them by using Cryptography and Steganography method[1] [2] [3].

Cryptography is the science to the confidentiality of the message. While Steganography is the art of hiding messages in digital media in such a way that other people do not realize there is a message in the media. In Cryptography and Steganography then the data that is considered secret will be disguised in such a way that if the data can be obtained, then it will not be understood by unauthorized parties. One method that can be used in cryptographic techniques is Least Significant Bit (LSB) and Hill Cipher[1]. Hill Cipher is one of the key symmetric cryptography algorithms. Hill Cipher algorithm uses the m x m sized matrix as the key for encryption and decryption. While the LSB method is a simple steganography method and easy to implement [1, 2].

IOP Conf. Series: Journal of Physics: Conf. Series **954** (2018) 012003 doi:10.1088/1742-6596/954/1/012003

Morse code is a system of representation of letters, numbers, and punctuation by using code signals. Morse code was created by Samuel F.B. Morse and Alfred Vail in 1835. Morse code is also used and studied in scouting or scouting. In the scouting world, Morse code is delivered using a scout flashlight or whistle. Morse code is performed using short whistle-length whips to represent the point and blow the whistle with long duration to represent the line. In the order of bits in a byte (1 byte = 8 bits), there is the least significant bit (MSB) and least significant bit (LSB). Example byte 11010010, the bit number 1 (first, underlined) is the MSB bit, and the bit number 0 (last, highlighted) is the LSB bit. The suitable bit to replace is the LSB bit because the change only changes the byte value one or higher or one lower than the previous value

2. Methodology

Data is a record of facts. Data is a plural form of datum, derived from the Latin meaning something given. In everyday use, the data means a statement received as is.

Figure 1. Cryptography Concept

Cryptography from Greek. According to language, the term cryptographic word is divided into two, crypto and graphia. Crypto means secret and graphia means writing. According to its terminology, cryptography is the science and art of keeping messages safe when messages are sent from one place to another several mechanisms develop in modern cryptography[13][14][15] that is:

a. Hash function. The hash function is a function that mapping a message of any length to a special text called message digest with a fixed length. The hash function is used as the test value (check value) on the data integrity mechanism.
b. Encryption with a symmetric key. The encryption with the symmetric key is the encryption that the encryption key and the decryption key are equal. The key to symmetric encryption is assumed to be confidential only to those who encrypt and decrypt who knows its value.
c. Encoding with asymmetric key. Encoding with asymmetric keys or often called public key encryption is encrypted with different encryption and decryption keys. An encryption key also called a public key, is open. Meanwhile, the decryption key which is also known as the private key is closed or secret.

Steganography is a branch of science that studies about how to hide a "secret" information in other information. Steganography has a history similar to cryptography, both of which were widely used during the war era. The steganography difference with encryption lies in how the process of concealment of data and the result of the process. Encryption operates randomizing the original data to produce encrypted data that is entirely random and distinct from the original[13], whereas steganography hides data in other data to be hosted before and after the hiding process is almost the same [4].

IOP Conf. Series: Journal of Physics: Conf. Series **954** (2018) 012003 doi:10.1088/1742-6596/954/1/012003

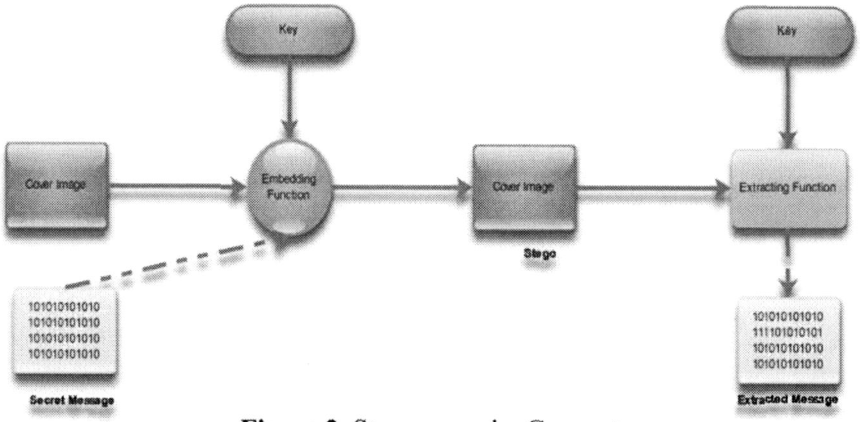

Figure 2. Steganography Concept

3. Results and Discussion

The use of cryptographic techniques with hill chipper algorithm[12], Morse code and steganography with LSB method is expected to protect and secure secret messages when the message is delivered through internet media from unauthorized parties; this is to avoid the occurrence of crimes such as taking or tapping of the secret message delivered it. This LSB method is quite vulnerable to the presence of steganography-analysis that suspects the existence of confidential information in it. If the steganography-analysis detects the presence, then the message will soon be known. Another case if the message is encrypted first using hill chipper algorithm before the insertion process. If steganography-analysis succeeds in expressing the message in the stego-image, the message is still not easily known because it is still in an encrypted state.

An improvement security by combining cryptographic techniques with hill chipper algorithms and steganography with the LSB method simultaneously will result in a better secret message security level. The secret message security process that is done is to insert the message into a container (cover) that is in the form of digital images using steganography technique, where the message before inserted already encrypted first using cryptographic techniques.

Phase 1: Hill Cipher and Converted Morse code

Before the secret message is inserted into the digital image media, the message will be encrypted using hill cipher algorithm. This encryption process requires a key that is the same length as the message text, and then the key will repeatedly be treated along the number of message text. This encryption process will produce ciphertext ready to be inserted into digital image media[5].

Following the problem limitation, the image file used is a 24-bit bitmap format. The secret message that can be accommodated in the digital image media depends on the capacities of the imagery. Example in the 24-bit digital image file measuring 400 x 267 pixels, there are 40050 pixels, each pixel (dot) in the image consists of three primary colors, namely red, green and blue (RGB), where each base color is presented with a size of 8 bits (1 byte). Total capacity of the container media is 40050 x 3 = 120150 bytes. Since each byte can insert only 1 bit of its LSB, the size of the data inserted in the image is 40050: 8 = 5006, 25 bytes (1 byte = 1 character). The size of this data should be reduced by the length of the filename because the concealment of the secret message hides not only the contents of the message but also the filename[6].

The message insertion process is done, first the message is encrypted with Hill cipher algorithm. For the use of hill chipper algorithm for this data encryption, matrix is used with order 2 * 2, with boundary value of each element between 0-255. The matrix multiplication result is then converted into Morse code in point and row model. Since Morse code recognizes the letters A to Z and 0 to 9, then the data character is additionally converted into binary data form [7] [8]. From the value, will be taken the binary value of each character that will then be inserted into the image by using the method LSB (Least Significant Bit). As an illustration, is suppose from the following matrix:

a. Element (1, 1) the binary value of 7 is 0111.
b. Element (1, 2) the binary value of 18 is 10010.
c. Element (2, 1) the binary value of 23 is 10111.
d. Element (2, 2) the binary value 11 is 1011.

The key matrix to be inserted is:
Element (1, 1) = 00000111.
Element (1, 2) = 00010010.
Element (2, 1) = 00010111.
Element (2, 2) = 00001011.
If an image has the raster data as below:

00110011	01010101	11111111
R	G	B

Figure 3. Examples of Raster Data

After inserted key element matrix (1, 1) by replacing 4-bit R value and value B from the right with binary value element (1,1) will become new pixel data.

00110000	01010101	11110111
R	G	B

Figure 4. New Data Pixel

Phase 2: Least Significant Bit Procedure
The technique of inserting data in this image using the LSB (Least Significant Bit) method, where the data will be inserted into the lower bits (4 bits below) [3] [9] [10]. But, on the odd bits of the RGB value of a pixel (the smallest data unit that makes up the image). 2-bits into the value of R and 2 more bits into the B value, so one character will need 2-pixel containers. Suppose the steps to insert the character "I", assuming already done data randomization by Hill Cipher method and Converting to Morse Code [11] ; it will do the stages as follows:

Table I. Converting Morse code to Ascii Code

Plaintext	Morse Code	Ascii Code
A	.-	46,45
K	-.-	45,46,45
U	..-	46,46,45

After that will be converted into binary values so the resulting calculation as follows:
46 = 00101101, 45 = 00101100
45 = 00101100, 46 = 00101101, 45 = 00101100

46 = 00101101, 46 = 00101101, 45 = 00101100
Then take the value of each pixel of the image to be inserted. As in the picture below:

Character Length ←	Px1	Px2	Px3	Px4	Px5	
	Px6	Px7	Px8	Px9	Px10	
	Px11	Px12	Px13	Px14	Px15	→ Data Text
	Px16	Px17	Px18	Px19	Px20	
Key ←	Px21	Px22	Px23	Px24	Px25	

Figure 5. Illustration of a Pixel Image

From the above pixels px1 (pixel 1) - px5 (pixel 5) is used to hold the length of the character, which serves as the capture limit of data that has been inserted by other characters. Then px6 (pixel 6) - px20 (pixel 20) according to the data holding requirement is used to hold the character data, and then px21 (Pixel 21) onwards is used to store the elements of the encryption matrix. Suppose the data of each pixel is as follows:

a. Px6 -> R = 145, G = 22, B = 128
b. Px7 -> R = 150, G = 22, B = 129
c. Px8 -> R = 155, G = 21, B = 130
d. Px9 -> R = 220, G = 19, B = 131
e. Px10 -> R = 215, G = 30, B = 125
f. Px11 -> R = 214, G = 16, B = 150

Then from those values will also be converted into binary values, so the following values appear:

a. Px6 -> R = 10010001, G = 00010110, B = 10000000
b. px7 -> R = 10010110, G = 00010110, B = 10000001
c. Px8 -> R = 10011011, G = 00010101, B = 10000010
d. Px9 -> R = 11011100, G = 00010011, B = 10000011
e. px10 -> R = 11010111, G = 00011110, B = 01111101
f. px11 -> R = 11010110, G = 00010000, B = 10010110

Inserts the binary value of the character to the binary value of the above pixels.

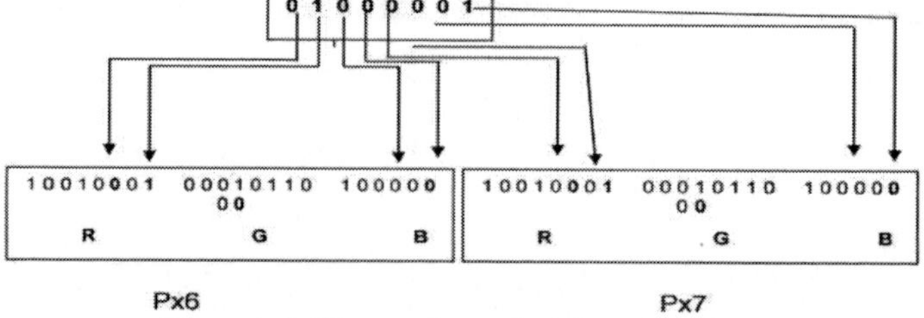

Figure 6. Illustration of a Message Insertion

In this research, at least pixel for an image is 22-pixels obtained from text data, or plain text is at least one character. In the process of the program to facilitate the calculation of each odd data, the author adds one character so that the amount of data becomes even. Data owned now into two characters where the length of the character of the data is 2. The amount of data that will be added to the duration of the binary value of 2 characters is 8 times 2 is 16. So, also added with the number of matrices 2 x 2 is 4. The total minimum media is: Number of text data + Number of binary data + Number of matrices is 2 + 16 + 4 is 22-pixels.

4. Conclusion

From the results of design steganography security applications, then obtained a conclusion is By using the LSB and Hill Cipher method and Morse Code was able to secure data better in comparison using only 1 algorithm. Looking at the characteristics of these three methods, they can be combined into a new concept or model in securing the picture message data.

REFERENCES

[1] Z. E. Dawahdeh, *et al.*, 2017. A new image encryption technique combining Elliptic Curve Cryptosystem with Hill Cipher, *J. of King Saud University - Comp. Inf. Sci.*.

[2] Q. Liu, *et al.*, 2008. Image complexity and feature mining for steganalysis of least significant bit matching steganography, *Inf. Sci.*, **178**, pp. 21-36.

[3] S. Wang, *et al.*, 2015. Least significant qubit (LSQb) information hiding algorithm for quantum image, *Measurement,* 73, pp. 352-359.

[4] H. Sajedi, 2016. Steganalysis based on steganography pattern discovery, *J. of Inf. Security Appl.,* **30**, pp. 3-14, 2016.

[5] Y. Sazaki and R. S. Putra, 2016. Implementation of Affine Transform Method and Advanced Hill Cipher for securing digital images, in *2016 10th International Conference on Telecommunication Systems Services and Applications (TSSA)*, pp. 1-5.

[6] A. A. M. Khalaf, *et al.*, 2016. A triple hill cipher algorithm proposed to increase the security of encrypted binary data and its implementation using FPGA, in *2016 18th International Conference on Advanced Communication Technology (ICACT)*, pp. 1-1.

[7] B. Purnama and A. H. H. Rohayani, 2015. A New Modified Caesar Cipher Cryptography Method with LegibleCiphertext From a Message to Be Encrypted, *Procedia Comp. Sci.,* **59**, pp. 195-204.

[8] D. C. Mishra, *et al.*, 2015. Security of RGB image data by affine hill cipher over SLn(qF) and Mn(qF) domains with Arnold transform, *Optik - Int. J. for Light Electron Optics,* **126**, pp. 3812-3822.

[9] Z.-H. Wang, *et al.*, 2012. Optimizing least-significant-bit substitution using cat swarm optimization strategy, *Inf. Sci.,* **192**, pp. 98-108.

[10] W.-L. Xu, *et al.*, 2016. An improved least-significant-bit substitution method using the modulo three strategy, *Displays,* **42**, pp. 36-42.

[11] C.-H. Yang, 2003. An interactive morse code emulation management system, *Comp. Math. Appl.,* **46**, pp. 479-492.

[12] A. Putera, U. Siahaan, and R. Rahim, 2016. Dynamic Key Matrix of Hill Cipher Using Genetic Algorithm, *Int. J. Secur. Its Appl.*, **10**, 8, pp. 173–180.

[13] H. Nurdiyanto, R. Rahim, and N. Wulan, 2017. Symmetric Stream Cipher using Triple Transposition Key Method and Base64 Algorithm for Security Improvement, *J. Phys. Conf. Ser.*, **930**, 1, pp. 12005.

[14] R. Rahim, M. Dahria, M. Syahril, and B. Anwar, 2017. Combination of the Blowfish and Lempel-Ziv-Welch algorithms for text compression, **15**, 3, pp. 292–297.

[15] R. Rahim, 2017. Man-in-the-middle-attack prevention using interlock protocol method, *ARPN J. Eng. Appl. Sci.*, **12**, 22, pp. 6483–6487.

IOP Conf. Series: Journal of Physics: Conf. Series **954** (2018) 012004 doi:10.1088/1742-6596/954/1/012004

Comparative Study of Learning Using E-Learning and Printed Materials on Independent Learning and Creativity

Niken Wahyu Utami & Abdul Aziz Saefudin
Univeristas PGRI Yogyakarta
niken@upy.ac.id*

Abstract. This study aims to determine: 1) differences in students taking independent learning by using e-learning and the students who attend the learning by using the print instructional materials ; 2) differences in the creativity of students who follow learning with e-learning and the students who attend the learning by using the print instructional materials ; 3) differences in learning independence and creativity of students attend learning with e-learning and the students who attend lessons using printed teaching materials in the subject of Mathematics Instructional Media Development. This study was a quasi-experimental research design using only posttest control design. The study population was all students who take courses in Learning Mathematics Media Development, Academic Year 2014/2015 100 students and used a random sample (random sampling) is 60 students. To test the hypothesis used multivariate analysis of variance or multivariable analysis of variance (MANOVA) of the track.The results of this study indicate that 1) There is a difference in student learning independence following study using the e-learning and the students who attend lessons using printed teaching materials in the lecture PMPM ($F = 4.177$, $p = 0.046 < 0.05$) ; 2) There is no difference in the creativity of the students who complete the learning by using e -learning and students to follow the learning using printed teaching materials in the lecture PMPM ($F = 0.470$, $p = 0.496 > 0.05$) ; No difference learning independence and creativity of students attend learning by using e-learning and the students who attend the learning using printed teaching materials in the lecture PMPM ($F = 2.452$, $p = 0.095 > 0.05$). Based on these studies suggested that the learning using e -learning can be used to develop student creativity, while learning to use e -learning and teaching materials can be printed to use to develop students' independence.

1. Introduction

Mathematics Education Study Program is one of the courses in the Faculty of Teacher Training and Education at PGRI University Yogyakarta. One of the compulsory subjects in the Mathematics Education Study Program is the Mathematics Learning Mathematics course (PMPM). The course of PMPM is organized with the aim of producing graduates who are competent in the field of mathematics education, especially in producing mathematics learning media. The subject of PMPM learns about everything about mathematics learning media, the final project is a mathematics learning media in the form of software (using Macromedia Flash 8 Professional) or hardware that can be used in learning in school.Making learning media of mathematics demands high creativity so that obtained by learning media with good quality. Nevertheless, based on observations made on student work outcomes this year, especially in the form of software, the creativity of students is not good, that is seen from the still not variations of the types of animation, object / form, or simulation of learning used in explaining certain material. Ahmar and Rahman doing research about development of teaching material using an Android [1], the result of this research that teaching material using Android is effective to learn. They doing research in cognitive style [2], learning style [3], and the relationship with problem solving [4] in the learning process. So far, the learning course of Mathematics Learning Media Development (PMPM) is still dominated by lecturers [5].

Learning a lot using a lecture method that is assisted by the media slide material in explaining the learning materials to students. After that, usually the lecturer provide direction and guidance to students to practice in the computer lab. With such learning, enabling student self-reliance in understanding and practicing the material obtained tend to be less because very dependent on the lecturer. In turn, this also allows students' creativity in producing good mathematics learning media is also not running optimally. Preliminary analysis of student creativity is also done through observation of the results of work students who follow the lectures PMPM Academic Year 2013/2014. The result of observation on student work showed that the result of the work in the form of software, the making of object images in the learning media is also less smooth. It is possible the lack of student flying hours in creating / drawing objects using tools in Macromedia Flash 8 Professional. In addition, this is also possible because students do not practice independently in drawing objects or making animations used in the software [6]. Therefore, the independence of student learning is needed in the manufacture of software learning media mathematics. Innovative learning relevant to current conditions is student centered learning. That is, it is learning that emphasizes to students that knowledge should be built (constructed) by students themselves. Students can be directly involved to gain experience learning from lecturing process. Moreover, learning to students includes learning to adults (andradogy learning) to enable students to learn independently either with or without the help of others (lecturers).

Independence of learning is important for the creativity of students in producing the work to develop optimally [7]. Learning using e-learning or web-based learning and learning using printed materials provide students the opportunity to hone their creativity and learning independence in solving problems related to the learned concepts . Web-based learning or learning using e-learning is one of innovative learning that has characteristics such as the existence of more communication (interactivity), independence, accessibility, and enrichment of various materials. This characteristic allows students to learn independently and optimize their creativity in learning. While learning using printed materials that contain material and guidance in understanding and practicing the lecture material also provides opportunities for students to learn independently. The requested tasks and materials provided in the printed materials provide an opportunity for students to explore their creativity in producing the work

IOP Conf. Series: Journal of Physics: Conf. Series **954** (2018) 012004 doi:10.1088/1742-6596/954/1/012004

/ product. With these characteristics, it is possible that learning using e-learning and learning using printed materials provides an opportunity to increase students' learning independence and creativity. Based on the various descriptions, it is necessary to study the application of learning using e-learning or web-based learning and learning using printed materials in lectures PMPM [8].

Through the application of these two lessons are expected to increase the independence of learning and creativity of students in developing and producing the media of learning mathematics.

2. Research Methods

The research procedure in this research includes three stages, namely preparation stage, implementation, data analysis, and reporting. Stages of Preparation, The purpose of this stage is to prepare the research tool after analyzing the learning problems. Activities at this stage include analysis of learning problems, concept analysis, instrument preparation, and device preparation. The description is as follows. Analysis of learning problems, The problems that are often encountered by students in lecturing the Development of Mathematics Learning Media (PMPM) is the independence of learning and creativity of students who have not been optimal[9]. The materials taught in PMPM lecturing activities, namely functions, and types of learning media, selecting and determining learning media, designing instructional media, creating and using learning media [10]. From the material taught, conducted research about lectures on the material of making computer-based learning media using Flash 8. For this learning, divided into several materials, namely Introduction to Flash, Drawing objects with Flash, Animation Each Frame, Motion Tween Animation, Animation Shape Tween, Symbol Button, [11] Full Media View Using Various Animations and Buttons (Symbol Button). The research instruments compiled in this study are self-directed questionnaire and creativity test.

Learning tools are arranged in research activities include the syllabus PMPM Course, PMPM Course module, and PMPM Course webpage (learning.upy.ac.id). Implementation phase of this research is experimental test and implementation of quasi experimental research [12][13]. The implementation includes: a) independent questionnaire test of learning independence and creativity tests, b) the application of learning, ie the application of learning using e-learning and print materials in the classroom to improve students' learning independence and creativity in the experimental class and control class; c) provides a questionnaire of learning independence and student creativity tests in both classes [14]. The experiment class is in class 4A3 and control class in 4A1 class. The instrument test is done in 4A2 class.

3. Result and Discussion

Prior to Monova test, prerequisite test is performed first. The prerequisite test conducted in this research includes the data distribution normality test, homogeneity test of variance, and multicollinearity test. The data normality test showed that the Kolmogorov-Smirnov test statistic value for the learning independence variable has a significance level of 0.122 (greater than 0.05) and student creativity has a significance level of 0.183 (greater than 0.05). That is, the data is otherwise normally distributed.

The homogeneity test of variance shows that the statistical value of Levene statistic test for the data of learning independence variable has a significance level of 0.003 (less than

0.05) so that the data is not normally distributed, but for the creativity variable data has a significance level of 0.397 (more than 0.05) otherwise normally distributed. The multicollinearity test results show that the VIF value coefficient is close to 1 for all independent variables. Similarly, the tolerance value is close to 1 for all independent variables. This means that between the same bound variables do not occur multicollinearity. After fulfilling the specified requirements, then a multivariable of variance analysis (MANOVA) is performed. MANOVA test results are as follows [3].

There is a difference of student learning independence that follow the learning by using e-learning and students who follow the learning by using printed material in PMPM lecture (F = 4,177, p = 0,046 <0,05). There is no difference of creativity of students who follow the learning by using e-learning and students who follow the learning by using printed materials in PMPM lecture (F = 0,470, p = 0,496> 0,05). There is no difference of learning independence and creativity of students follow the learning by using e-learning and students who follow the learning by using printed materials in PMPM lecture (F = 2.452, p = 0,095> 0,05).

Based on the results of this study, the group of students who were given learning by using printed materials, after being given the treatment had an average score of learning independence of 70.87 with a standard deviation of 10.73 highly qualified. The group of students who used e-learning learning after being given the treatment had an average learning independence score of 66.1 with a standard deviation of 6.92 that was highly qualified. Qualification of the average score of independence of student learning both groups in the same qualifications so as to not have a significant difference. Even so, in general the average score of learning independence of student group learning given by using printed materials is relatively better compared to the group of students who use learning with e-learning. The group of students who were given learning using printed materials, after being given the treatment had an average score of creativity of 68.4 with a standard deviation of 23.28 that qualified creatively. The group of students who used e-learning learning after being given the treatment had an average creativity score of 72.33 with a standard deviation of 21.10 that qualified creatively.

Qualification scores average creativity of students of both groups in the same qualification so as to have no significant differences. In general, however, the average score of students' learning creativity using e-learning learning is relatively better compared to that of students who use learning with printed materials. Based on the results of descriptive analysis and MANOVA one-way analysis, it can be taken a justification that learning using printed and learning materials using e-learning there is no difference in influencing students' learning independence and creativity in PMPM lectures. The reason is that students who follow the learning by using print and e-learning materials are both doing media development activities independently. Lecturers only provide guidance when students have difficulty in the learning process. As a result, student learning independence is well established. Meanwhile, students' creativity in learning between students using printed materials and students using e-learning differs significantly even in creative qualifications. This is possible because students who learn to use e-learning will find it easier to develop ideas by way of direct browsing on the internet. As a result, the ideas obtained are not monumental, but tend to be varied and the students try to develop the idea either by imitation (imitation) or adaptation of those ideas. This result is in line with Rusman's opinion (2012: 293). He stated that the advantages of learning done with e-learning is the web as a medium equipped with a hyperlink that allows

IOP Conf. Series: Journal of Physics: Conf. Series **954** (2018) 012004 doi:10.1088/1742-6596/954/1/012004

to access information at random which impact on our speed to obtain information on the web. Students who learn through e-learning certainly have these advantages compared if he learns through printed materials. Thus, students are easier to obtain creative ideas in the process of developing innovative learning media.

4. Conclusion

The results of independent questionnaire and creativity test showed that the independence questionnaire and creativity test can be used in the research because it is valid and reliable. There is a difference of student learning independence that follow the learning by using e-learning and students who follow the learning by using printed material in PMPM lecture (F = 4,177, p = 0,046 <0,05). There is no difference of creativity of students who follow the learning by using e-learning and students who follow the learning by using printed materials in PMPM lecture (F = 0,470, p = 0,496> 0,05). There is no difference of learning independence and creativity of students follow the learning by using e-learning and students who follow the learning by using printed materials in PMPM lecture (F = 2.452, p = 0,095> 0,05).

References

[1] Ahmar, A. S., & Rahman, A. 2017. Development of teaching material using an Android. *Glob. J. Eng. Educ.*, **19**, 1.
[2] Mulbar, U., Rahman, A., & Ahmar, A. S. 2017. Analysis of the ability in mathematical problem-solving based on SOLO taxonomy and cognitive style. *World Trans. Eng. Technol. Educ.*, **15**, 1, pp. 74-77.
[3] Rahman, A., Ahmar, A. S., & Rusli. 2016. The Influence of Cooperative Learning Models on Learning Outcomes Based on Students' Learning Styles. *World Trans. Eng. Technol. Educ.*, **14**, 3, pp. 74-77.
[4] Rahman, A., & Ahmar, A. S. 2016. Exploration of Mathematics Problem Solving Process Based on The Thinking Level of Students in Junior High Schoo. *Int. J. Environ. Sci. Educ.*, **11**, 14.
[5] Alessi, S.M. & Trollip, S.R. 2001. *Multimedia for learning: Methods and development* (3rd ed.). Massachusetts: Allyn & Bacon.
[6] Depdiknas. 2008. *Panduan penyusunan bahan ajar.* Jakarta: Direktorat Pembinaan Sekolah Menengah Atas
[7] Goodman and Smart. 1999. *Emotional Intelligence.* New York: Bantam Books.
[8] Grieshober, W. E. 2004. *Continuing a Dictionary of Creativity Terms & Definition.* New York: International Center for Studies in Creativity State University of New York College at Buffalo. http://www.buffalostate.edu/orgs/cbir/ReadingRoom/theses/Grieswep.pdf, 28 January 2013
[9] Hargis, J. nd. *The Self-Regulated Learner Advantage: Learning Science on Internet.* http://www.jhargis.com/
[10] Hackbarth, S. 1996. The Educational technology handbook. New Jersey: Educational Technology Publication Inc.
[11] Karnes & Bean. 2009. Methods *and* Materials *for* Teaching *the* Gifted 2009. USA: Prufrock Press Inc
[12] Muhammad Ali dan Muhammad Asrori. 2009. *Psikologi Remaja Perkembangan*

IOP Conf. Series: Journal of Physics: Conf. Series **954** (2018) 012004 doi:10.1088/1742-6596/954/1/012004

Peserta Didik. Jakarta: Bumi Aksara.

[13] Ronger, R. 1990. *The 19 Habits of Highky Successful People: Powerful Strategies for Personal Triumphs.* Malaysia: Wynwood Press.

[14] Rusman. 2012. *Belajar dan Pembelajaran Berbasis Komputer: Mengembangkan Profesionalisme Guru Abad 21.* Bandung: Alfabeta.

IOP Conf. Series: Journal of Physics: Conf. Series **954** (2018) 012005 doi:10.1088/1742-6596/954/1/012005

Design of a Blended Learning Environment Based on Merrill's Principles

Janner Simarmata*[1], Asari Djohar[2], Janulis Purba[2], Enjang A Juanda[2]

[1]State University of Medan, Indonesia
[2]Faculty of Vocational and Technology Education, Indonesia University of Education, Indonesia

*jannersimarmata@unimed.ac.id

Abstract. Designing blended learning courses requires a systematic approach, in instructional design decisions and implementations, instructional principles help educators not only to specify the elements of the course, but also to provide a solid base from which to build the technology. The blended learning course was designed based on Merrill's First Principles of Instruction with five phases. This paper helps inform educators about how to develop appropriate learning styles and preferences according to students' learning needs.

1. Introduction

Internet technology advances and computer information technology and networked learning made it possible to design and utilize new generation learning environments that are realistic, authentic, and engaging [1]. Today, blended learning technology is one of the most promising applications in education, since blended learning can be considered a sort of electronic learning or its extension; The main difference is the necessity of face-to-face communication of students with each other and with the teacher [2].

Olivier & Trigwell explains that blended learning was originally used for learning that seeks to combine face-to-face learning with online learning [3]. This mixed learning is derived from two words, blend and learning. The word blend means combining things and learning demonstrates the assimilation of new knowledge. With blended learning [4], teachers can take advantage of online resources in the learning process and this helps students become more active and effective learners, [5][6] including effort and time-saving as well as cost effectiveness. Moreover, blended learning can be associated with problem solving [7][8], learning style, and learning achievement. Blended learning can be used to support problem solving [9][10] based learning which results will give description about learning style [11][12] and learning achievement from students [13]. Online resources learning have studied by Ahmar and Rahman, from this study, they have conclusion that online resources can improve the knowledge of the students [14].

2. Methodology

There are several reasons why using blended instruction is improved pedagogy, easy access to knowledge, more interaction among learners, personal presence, cost effectiveness, and ease of revision of learning content [15]. While Singh and Reed [16] proposed there were six combinations of blended instruction is: (a) offline and online learning, (b) self-paced, live, and collaborative learning, (c) structured and unstructured learning, (d) custom content with off-the-shelf content, (e) work and learning, and (f) ingredients blending synchronous physical formats, synchronous online formats, and self-paced, asynchronous formats.

To design effective instruction it must follow some of the principles as presented by Merrill [16], see figure 1 below.

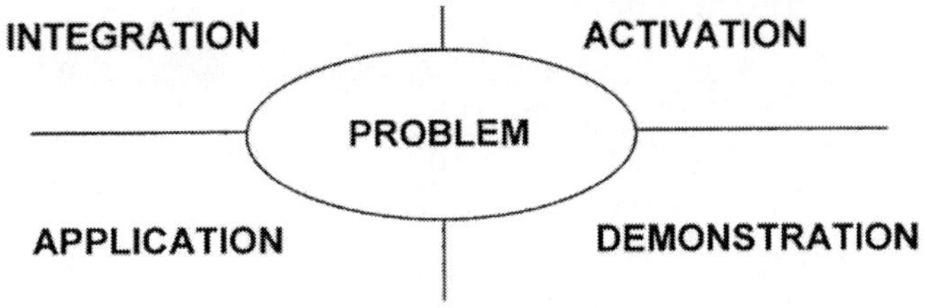

Figure 1. Phases for Effective Instruction

Here is a brief explanation of these five steps:

1. Learning is promoted when learners are engaged in solving real-world problems **[PROBLEM]**.
2. Learning is promoted when existing knowledge is activated as a foundation for new knowledge **[ACTIVATION]**.
3. Learning is promoted when new knowledge is demonstrated to the learner **[DEMONSTRATION]**.
4. Learning is promoted when new knowledge is applied by the learner **[APPLICATION]**.
5. Learning is promoted when new knowledge is integrated into the learner's world **[INTEGRATION]**.

Figure 2 is the blended learning design steps that include: problems, activation, demonstration, application, and integration with using Merrill's Principles of Instruction [17].

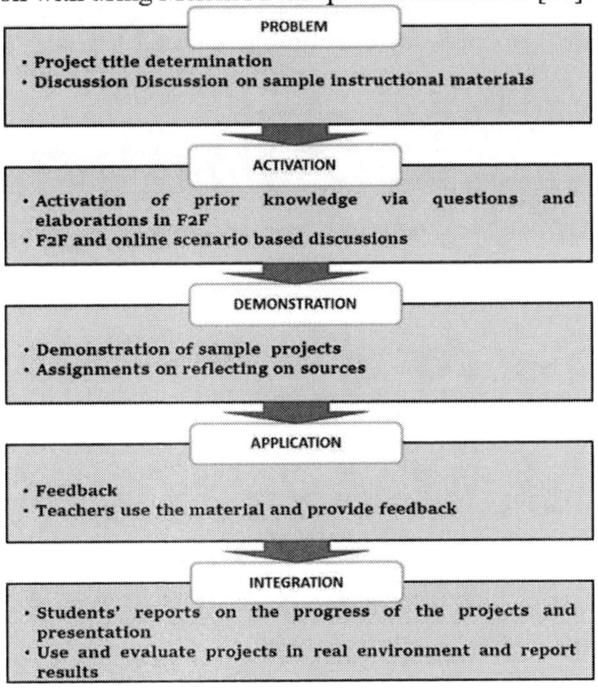

Figure 2. The design of blended learning using Merrill's Principles of Instruction is adapted from Gedik, N *et all* [18]

IOP Conf. Series: Journal of Physics: Conf. Series **954** (2018) 012005 doi:10.1088/1742-6596/954/1/012005

3. Result and Discussion

Blended learning designs developed through use of Merrill's Principles of Instruction are problems, activation, demonstration, application, and integration. Figure 3 is the result of blended learning design, in blended learning teachers can upload lesson material, create project task information. Students can also comment on teacher-uploaded subject matter similar to a facebook app. Besides, student with student, student with teacher can discuss each other through the discussion form (chat).

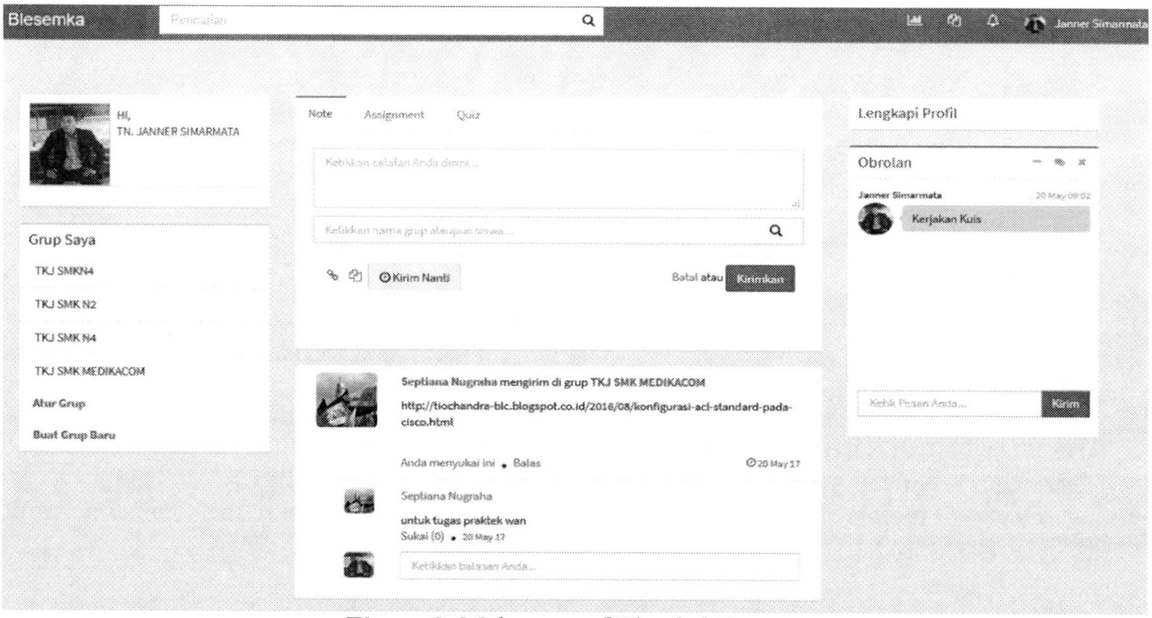

Figure 3. Main page of Blended Course

Tasks assigned by a teacher can be uploaded to the blended learning system, to be evaluated by teacher and to give the final assessment of each meeting, the figure 4 below is the result of teacher evaluation and progress of each student.

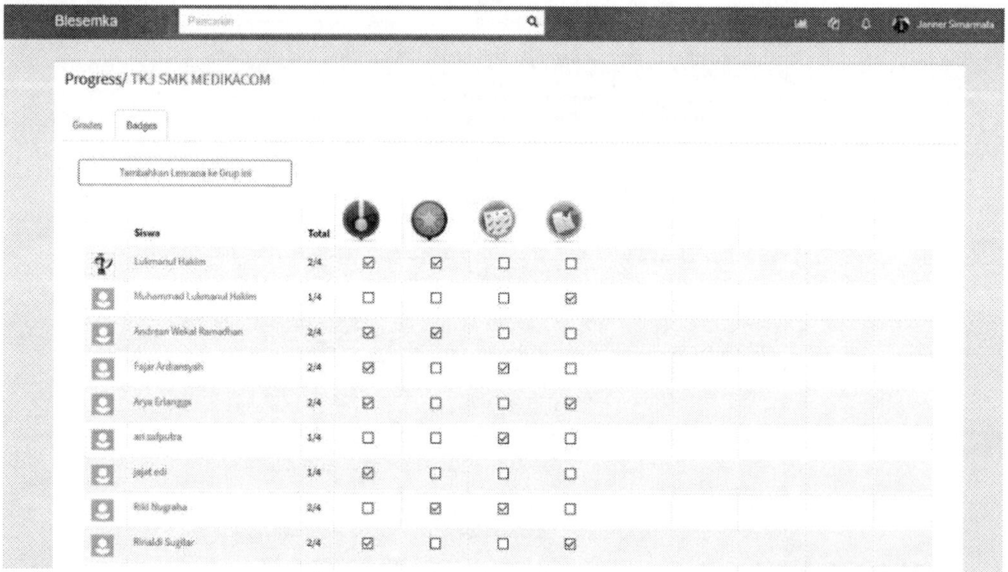

Figure 4. Track student progress

4. Conclusion
The focus of this study is mainly on how to design a supportive blended course as an attempt to transit from traditional (F2F) delivery methods to more advanced methods involving technology. It can be argued that designing a Blended learning is not as simple as combining an online environment with a F2F course. Beyond that basic integration, the selection of a best environment for implementation is also required.

References
[1] Caner, M, 2010, A blended learning model for teaching practice course, *Turkish Online J. Distance Educ.*, **11,** 3.
[2] Matukhin, D, and Zhitkova, E, 2015, Implementing Blended Learning Technology in Higher Professional Education, *Procedia-Social Behav. Sci.*, **206**, pp. 183–188.
[3] Oliver, M, and Trigwell, K, 2005, Can 'blended learning'be redeemed?, *E-learning Digit. Media*, **2**, 1, pp. 17–26.
[4] Al Zumor A. W. Q., Al Refaai I. K., Eddin E. A. B., and Al-Rahman F. H. A., 2013, EFL Students' perceptions of a blended learning environment: Advantages, limitations and suggestions for improvement, *English Lang. Teach.*, **6**, 10, p. 95.
[5] Aldalalah, O. A. and Gasaymeh, A.-M. M., 2014, Perceptions of Blended Learning Competencies and Obstacles among Educational Technology Students in Light of Different Anxiety Levels and Locus of Control., *Contemp. Educ. Technol.*, **5**, 3, pp. 218–238.
[6] Vaughan, N., 2007, Perspectives on blended learning in higher education, *Int. J. ELearning*, **6**, 1, p. 81.
[7] Oliver, R., 2015, Using a Blended Learning Approach to Support Problem-Based Learning with First Year Students in Large Undergraduate Classes, in *Proceedings of the 2005 conference on Towards Sustainable and Scalable Educational Innovations Informed by the Learning Sciences: Sharing Good Practices of Research, Experimentation and Innovation*, pp. 848–851.
[8] Yeen-Ju, H. T., Mai, N., and Selvaretnam, B., 2015, Enhancing problem-solving skills in an authentic blended learning environment: a Malaysian context, *Int. J. Inf. Educ. Technol.*, **5**, 11, p. 841.
[9] Rahman, A. and Ahmar, A.S., 2016, Exploration of Mathematics Problem Solving Process Based on The Thinking Level of Students in Junior High School, *Int. J. Environ. Sci. Educ.*, **11**, 14, pp. 7278–7285.
[10] Mulbar, U., Rahman, A., and Ahmar, A. S., 2017, Analysis of the ability in mathematical problem-solving based on SOLO taxonomy and cognitive style, *World Trans. Eng. Technol. Educ.*, **15**, 1, pp. 68–73.
[11] Rahman, A., Ahmar, A. S., and Rusli, 2016, The influence of cooperative learning models on learning outcomes based on students' learning styles, *World Trans. Eng. Technol. Educ.*, **14**, 3, 2016.
[12] Rahman, A. and Ahmar, A. S., 2017, Relationship of learning styles and learning achievement of mathematics based on genders, *World Trans. Eng. Technol. Educ.*, **15**, 1.
[13] Arsyad, N., Rahman, A., and Ahmar, A. S., 2017, Developing a self-learning model based on open-ended questions to increase the students' creativity in calculus, *Glob. J. Eng. Educ.*, **19**, 2, pp. 143–147.
[14] Ahmar, A. S. and Rahman, A., 2017, Development of teaching material using an Android, *Glob. J. Eng. Educ.*, **19**, 1.
[15] Osguthorpe, R. T. and Graham, C. R., 2003. Blended learning environments: Definitions and directions., *Q. Rev. Distance Educ.*, **4**, 3, pp. 227–233.
[16] Singh, H. and Reed, C., 2001, A white paper: Achieving success with blended learning, *Cent. Softw.*, **1**, pp. 1–11.
[17] Merrill, M. D., 2002, First principles of instruction, *Educ. Technol. Res. Dev.*, **50**, 3, pp. 43–59.

IOP Conf. Series: Journal of Physics: Conf. Series **954** (2018) 012005 doi:10.1088/1742-6596/954/1/012005

[18] Gedik, N., Kiraz, E., and Ozden, M. Y., 2013, Design of a blended learning environment: Considerations and implementation issues, *Australas. J. Educ. Technol.*, **29**, 1.

Block Architecture Problem with Depth First Search Solution and Its Application

Robbi Rahim*[1,6], Dahlan Abdullah[2,6], Janner Simarmata[3,6], Andri Pranolo[4,6], Ansari Saleh Ahmar[5,6], Rahmat Hidayat[6,7], Darmawan Napitupulu[6,8], Heri Nurdiyanto[6,9], Bayu Febriadi[10] and Z Zamzami[10]

[1]School of Computer Engineering and Communication, Universiti Malaysia Perlis, Malaysia
[2]Department of Informatics, Universitas Malikussaleh, Aceh, Indonesia
[3]Universitas Negeri Medan, Medan, Indonesia
[4]Department of Informatics, Universitas Ahmad Dahlan, Indonesia
[5]Department of Statistics, Universitas Negeri Makassar, Indonesia
[6]Komunitas Kolaborasi Publikasi Indonesia (KO2PI), Indonesia
[7]Department of Information Technology, Politeknik Negeri Padang, Indonesia
[8]Research Center for Quality System and Testing Technology, Indonesian Institute of Sciences, Indonesia
[9]Department of Informatics, STMIK Dharma Wacana, Indonesia
[10]Department of Information System, Universitas Lancang Kuning, Indonesia

*usurobbi85@zoho.com

Abstract. Searching is a common process performed by many computer users, Raita algorithm is one algorithm that can be used to match and find information in accordance with the patterns entered. Raita algorithm applied to the file search application using java programming language and the results obtained from the testing process of the file search quickly and with accurate results and support many data types.

1. Introduction

Block Architecture Problem is one of the classic problems in the field of Artificial Intelligence studies, this problem is illustrated in boxes arranged into 2 or 3 piles of beams as initial conditions. The goal of this problem is to get a pile of boxes as desired. An allowed operation in the process of completion is to decrease (x), which means that box x is derived from a certain stack and place it (x, y), meaning that the x box is placed above the y box, provided that the x and y boxes must is at the top of a pile.

Block Architecture Problem can be solved with some algorithms such as A* (A Star), Breadth First Search and Depth First Search[1]–[3]. The Depth First Search[1], [3], [4] algorithm is chosen for solving this problem because the search process of the solution will continue to achieve the desired goal[1], [5], and the Depth First Search algorithm is more efficient[6] for search space with multiple branches because it does not need to evaluate all the nodes at a certain level[6] and also the algorithm

Depth First Search requires relatively small memory because only the nodes on the active path are stored[2], [7].

Depth First Search algorithm used in the completion of Block Architecture Problem will be applied to the application created using Visual Basic.Net programming language to simulate the solution of the problem, the experiment is perform on 8 blocks with the number of piles of block as much as 3 pieces of piles, namely: stack-1, 2 and 3, the conclusion of Block Architecture Problem with applications could be one of solution form state and space problem.

2. Methodology

Searching[8]–[11] is one of the most important processes in searching for a solution, the block architecture problem in this research is solved by applying the Depth First Search algorithm and then implemented the application created by using Visual Basic.Net, the operation and action of the problem and the solution can be seen in the following example :

1. Initial condition, there are 3 pieces named C, A, B, C is the lowest pile and A is the top pile, see figure 1 below:

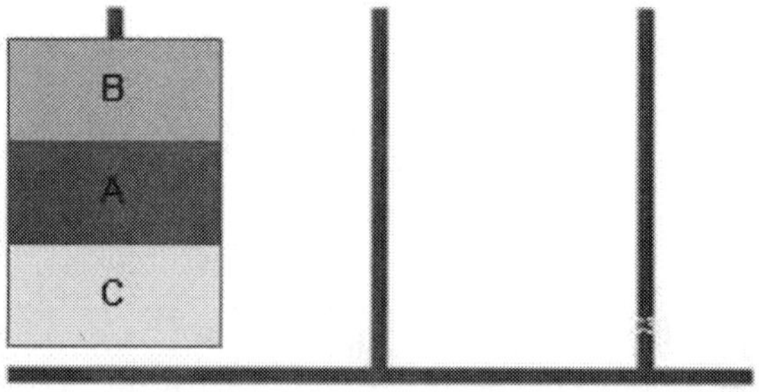

Figure 1. Initial Condition

2. Move block B from pole-1 to pole-3. The rule is that the block must be the topmost block of the 1st pole and occupy the topmost position on the 3rd pole.

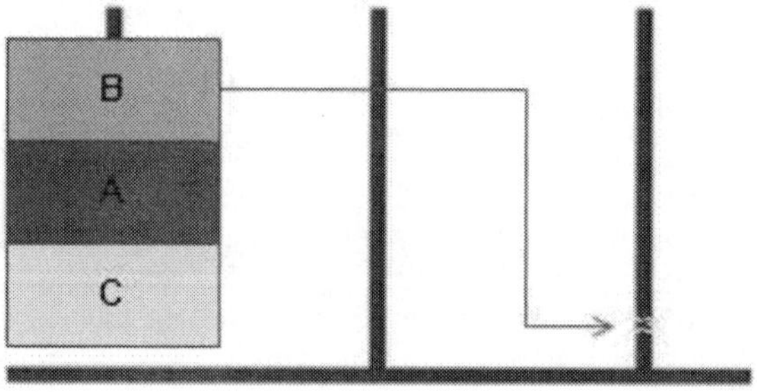

Figure 2. Moving block from pole-1 to pole-3

3. Move block A from pole-1 to pole-2. The rule is that the block must be the topmost block of the 1st pole and the A beam occupies the topmost position on the 2nd pole

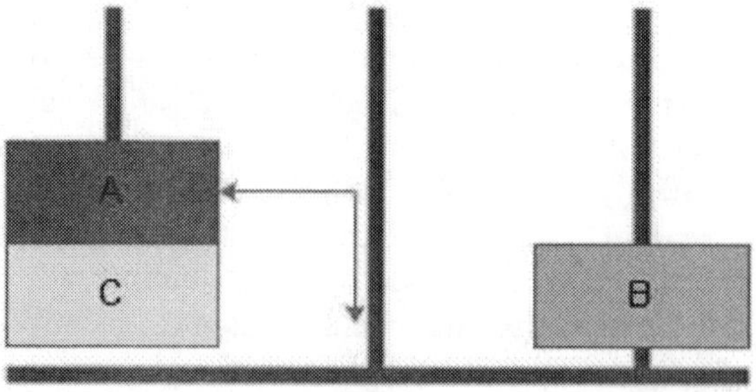

Figure 3. Moving block from pole-1 to pole-2

IOP Conf. Series: Journal of Physics: Conf. Series **954** (2018) 012006 doi:10.1088/1742-6596/954/1/012006

4. Move block B from pole-3 to pole-2. The rule is that the block B must be the topmost block from pole-3 and the B block occupies the topmost position on the 2nd pole.

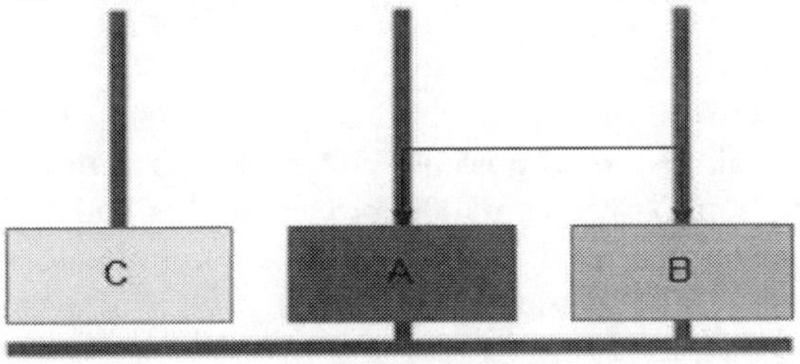

Figure 4. Moving block from pole-3 to pole-2

The above process is a manual process of how to move the block letter from pole-1 to pole-3 with existing rules, manual testing in figure 1 to figure 4 with 3 blocks and it can be more than 3 blocks, the result block after moving as follows:

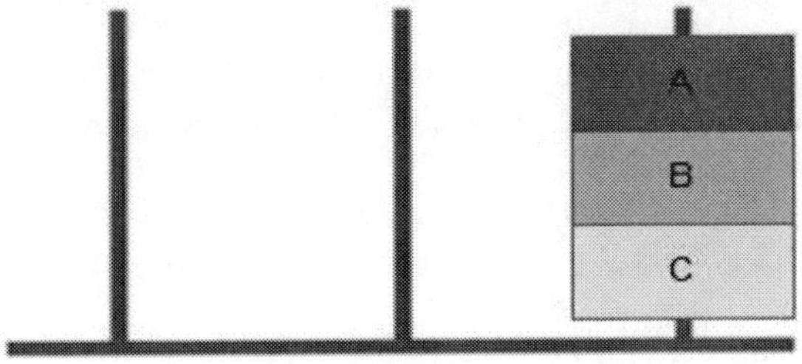

Figure 5. Goal State

Figure 5 is a manual process of moving block letter from pole-1 to pole-3, from initial state to goal state requires many steps with different variations to completion process, and the depth first search algorithm could use to find any solution.

3. Result and Discussion

Depth First Search algorithm used in the accomplishment of Block Architecture Problem and applied to application could be used well to solve the problem, figure 6 is an application proposed with example of initial state and goal state

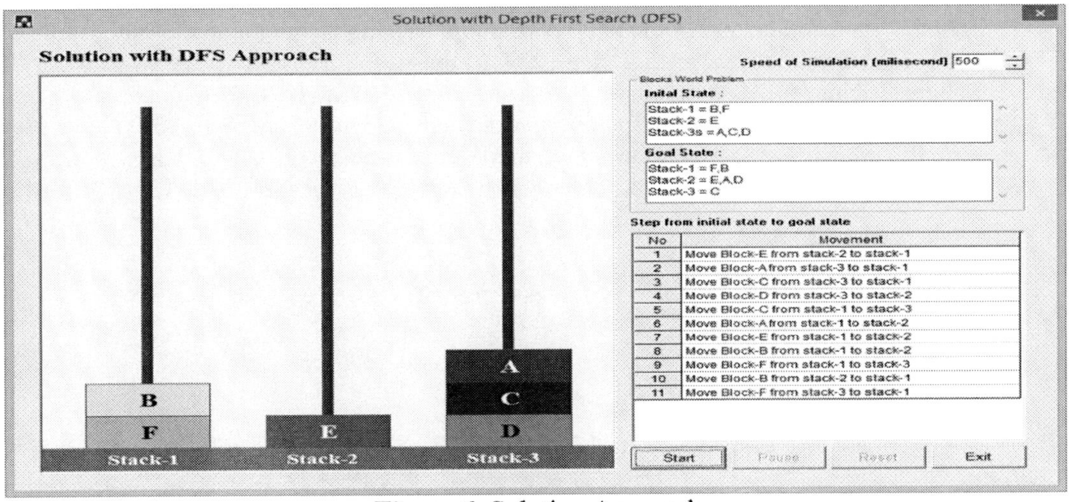

Figure 6. Solution Approach

IOP Conf. Series: Journal of Physics: Conf. Series **954** (2018) 012006 doi:10.1088/1742-6596/954/1/012006

Figure 6 shows each steps for transfer blocks from each pole, the conclusion process will be shown in simulation with 1 second interval for each block when switching, the depth first search algorithm used can be applied well to the simulation created.

4. Conclusion

Block Architecture Problem can be solved well using the Depth First Search algorithm because the search process is perform almost in all possible nodes, the simulation can be used up to 9 blocks with possible solutions up to approximately 95000 process states and takes a very long time to get the fastest solution, future development allows to apply or compare algorithms such as BFS, Heuristics, Dynamic Programming to get the best results for completion Block Architecture Problem

References

[1] R. Marinescu, 2010. Best-first vs. depth-first AND/OR search for multi-objective constraint optimization, in *Proceedings - International Conference on Tools with Artificial Intelligence, ICTAI*, **1**, pp. 439–446.

[2] A. S. M. Lumenta, 2014. Perbandingan Metode Pencarian Depth-First Search, Breadth-First Search dan Best-First Search pada Permainan 8-Puzzle, *e-Journal Tek. Elektro Komput.*.

[3] C. Mencía, M. R. Sierra, and R. Varela, 2013. Depth-first heuristic search for the job shop scheduling problem, *Ann. Oper. Res.*, **206**, 1, pp. 265–296.

[4] C. Rhee, Y. D. Liang, S. K. Dhall, and S. Lakshmivarahan, 1994. Efficient algorithms for finding depth-first and breadth-first search trees in permutation graphs, *Inf. Process. Lett.*, **49**, 1, pp. 45–50.

[5] U. A. Acar, A. Charguéraud, and M. Rainey, 2015. A Work-efficient Algorithm for Parallel Unordered Depth-first Search, *Proc. Int. Conf. High Perform. Comput. Networking, Storage Anal.*, pp. 67:1--67:12.

[6] P. A. Dow and R. E. Korf, 2009. Duplicate avoidance in depth-first search with applications to treewidth, in *IJCAI International Joint Conference on Artificial Intelligence*, pp. 480–485.

[7] E. Zunic, A. Djedovic, and B. Zunic, 2016. Software solution for optimal planning of sales persons work based on Depth-First Search and Breadth-First Search algorithms, in *2016 39th International Convention on Information and Communication Technology, Electronics and Microelectronics, MIPRO 2016 - Proceedings*, pp. 1248–1253.

[8] R. Rahim, I. Zulkarnain, and H. Jaya, 2017. A review: search visualization with Knuth Morris Pratt algorithm," in *IOP Conference Series: Materials Science and Engineering*, **237**, 1, pp. 12026.

[9] R. Rahim, I. Zulkarnain, and H. Jaya, 2017. Double hashing technique in closed hashing search process," *IOP Conf. Ser. Mater. Sci. Eng.*, **237**, 1, pp. 12027.

[10] R. Rahim, S. Nurarif, M. Ramadhan, S. Aisyah, and W. Purba, 2017. Comparison Searching Process of Linear, Binary and Interpolation Algorithm, *J. Phys. Conf. Ser.*, **930**, 1, pp. 12007.

[11] Rahim, R., Ahmar, A.S., Ardyanti, A.P., and Nofriansyah, D., 2017. Visual Approach of Searching Process using Boyer-Moore Algorithm, *J. Phys. Conf. Ser.*, **930**, 1, pp. 012001.

Exploring Practical Responses of M3LC for Learning Literacy

Nasrullah[1] & Baharman[2]

[1,2]Universitas Negeri Makassar, Makassar, Indonesia

nasrullah@unm.ac.id

Abstract. This study aims to explore the responses of participants toward Mathematics-Language Literacy Learning Courseware (M3LC) for learning literacy. There are five practical aspects concerned by involving 30 participants in the focus group discussion. In the beginning, participants were given some response sheet and introduced to M3LC by watching learning video of M3LC. At the end, they were asked to concern about response sheet and give comments related what they saw during the introduction session. The results show that the responses of users' agree and strongly agree are still higher than those of users' disagree or strongly disagree, with below 30% of responses are in the fair category. It means that the participants tend to give a positive opinion that M3LC is a useful courseware since it is qualified to satisfy 5 practical aspects, including knowledge use, knowledge construction, evaluation practice, social programming, and valuing to support literacy learning. In future, the implementation of using this courseware can be enhanced to further recognition of literacy level so that students can be well-prepared before starting learning activities in the classroom.

Keywords: Practical Response, M3LC, Learning Literacy

1. Introduction

Literacy skills are fundamental competencies for learners in the education system since it is needed to support the development of human resources. Students in the school need advanced levels of literacy to conduct their works, run their profession, act as excellence citizens, and perform their private lives [1]. In line with this opinion, providing a strong childhood literacy education in the early years leads to better results later [2]. OECD initiated Education at a Glance that led to the fulfillment of user needs, ranging from governments seeking to review academic and academic policies that require data for further analysis for the general public's will [3]. The implementation of Program for International Student Assessment (PISA) by OECD in Indonesia brings some useful information related to kinds of literacies, two of them such as mathematics and reading. At the same kind of assessment given to students in Indonesia, there is TIMSS that is assessing students' literacy in mathematics and science. According to international benchmark of TIMSS 2011, learning achievement of mathematics for Indonesian graders VIII that concern to measure content and cognitive dimension is in low level [4].

The main problems that students of Indonesia have been in terms of understanding of mathematical concepts, interpreting in the related concept, making a connection between mathematical concepts, and reading data [4]. The case like this is related with the explanation that behind of the construct of literacy there are emerging conceptual and procedural knowledge that support written language, including mathematics writing and language of talk [5]. Next, this becomes compulsory for the educational institution to be considered how to deliver learning literacy for learners in the school, in particular mathematics and language literacy.

It is not easy to cultivate mathematical and language literacy because in learning it is full of challenges that are not only one, even complex due to the many factors involved. Early learners need to understand why people read, write, and work out the settlement of a problem in order to be motivated to excel in their own literacy development. Through active involvement in the literacy process, children learn how to use their knowledge and skills flexibly and in combination with the entire domain of development [6].

Now, there are many sophisticated interactive types of equipment that are useful for supporting learning activities. Harnessing the existence of technology and its development give a positive contribution to method or approach of learning that can be applied to learners. Nugroho stated that the use of game sequences and geometry series in the classroom can increase students' interest in mathematics by 20.57% and improve mathematics learning achievement of 10.86%[7]. Similarly, Al-astal & Zaydah have been developing e-books in order to support mathematical thinking skills and the acquisition of mathematical concepts for graders 5th in Gaza [8] and also Ahmar & Rahman have been developing teaching material using Android [9] and they study how to improve the learning process with cognitive style [10], learning style [11], and problem solving [12].

One of beneficial and powerful technology using computer and internet is eXe-learning. It is an interactive application that teachers can be engaged to create innovative learning programs. This program is connected in a comprehensive manner such as hyperlink setting. The display can be shown on the page can be in the form of HTML, so the form is web-based on the user's computer. The variety of features contained in the eXe-learning program so that the combination of theory, images, video, and games is possible to be presented. Similarly, for interactive or non-interactive activities can be constructed so as to form an environment that supposes students to learn. Moreover, the literature to be developed directed can trigger students' knowledge, especially for math and language subjects.

Based on eXe-Learning system, Mathematics-Language Literacy Learning Courseware (M3LC) is constructed as an interactive learning equipment that is provided for students who want to enhance their literacy skills, in particular mathematics and language literacy. To what extend this product can address the main purpose of construction the courseware. By developing the annotations Shadish Cook and Leviton (in [13]), this article would like to identify practical response in terms of knowledge use, social programming, knowledge construction, valuing, and practical evaluation.

2. Methods

The aim of this study was to explore the responses of participants toward Mathematics-Language Literacy Learning Courseware (M3LC) in the learning literacy. In order to obtain data of practical response, there were 30 participants getting involved in the focus group discussion activities. The participants who were getting engaged in the discussion are student-teachers, teachers, and instructors of mathematics in high school. They are considered to have expertise in the field of mathematics and education. In the beginning, participants were given some response sheets and introduced to M3LC. During the introduction of the courseware, learning video with the application of M3LC was played in front of participants. At the end, they were asked to concern about response sheet and gave comments related what they saw during the introduction session.

At least 5 meetings of FGD were conducted to discuss practical of M3LC. These meetings were organized into five topics discussion. We started the meetings based on the topics as follows social programming, the use of knowledge, benefits, knowledge construction, and practical evaluation. Then, their responses were collected using Response Sheet (RS). For details, this RS contained the indicators of each topic. See the table below.

IOP Conf. Series: Journal of Physics: Conf. Series **954** (2018) 012007 doi:10.1088/1742-6596/954/1/012007

Table 1. Five Topics Used to Evaluate M3LC Based on the Response of Participants

Aspects of Response	Indicators
Social Programming (SP)	Direct the user to understand the learning topic well
	Friendly use as a learning medium
	Support learning activities "anywhere"
	Support learning activities "anytime"
	Support electronic-based learning activities
	Support as a learning tool
	Support as an interactive learning tool
Knowledge Use (KU)	Direct the use of learners' knowledge
	Help in improving the understanding of learning materials
	Organize tasks systematically
	Record the assignment given by the teacher
	Bridge the continuity of learning activities from each meeting
	Bring beneficial learning activities
	Trigger students to complete their assigned tasks
	Lead learners to build knowledge in completing tasks given
Valuing (V)	Promote learners' self-confidence in building knowledge through the tasks assigned
	Facilitate learners to learn individually
	Support to assess learners' performance in learning activities
	Support the transparency of student performance appraisal in learning activities
	Support learners identifying potential environments
	Build an academic atmosphere in learning activities
	Address the ability to interpret the problems given to learners
Knowledge Construction (KC)	Support the communication system in the learning activities
	Direct the establishment of a fellow networking system of learners
	Generate a network system between learners and educators
	Trigger learners to think logically
Evaluation Practice (EV)	Facilitate learners to think comprehensively
	Guide learners to think critically
	Teach learners to think about what is done

IOP Conf. Series: Journal of Physics: Conf. Series **954** (2018) 012007 doi:10.1088/1742-6596/954/1/012007

The response data obtained from the participants was quantitative data. In the response sheet, the display of indicators was written by providing five choices. They were strongly agree (score 5), agree (score 4), fair (score 3), disagree (score 2), and strongly disagree (score 1). The graphical form was selected to present the responses data [14] and percentage (%) to describe the tendency of responses which are divided into five groups of responses. These groups of responses based on the topics of discussion comprise Knowledge Use (KU), Social Programming (SP), Knowledge Construction (KC), Valuing (V), and Practical Evaluation (PE).

3. Results and Discussion

The preview of M3LC designed in this study is shown in picture below.

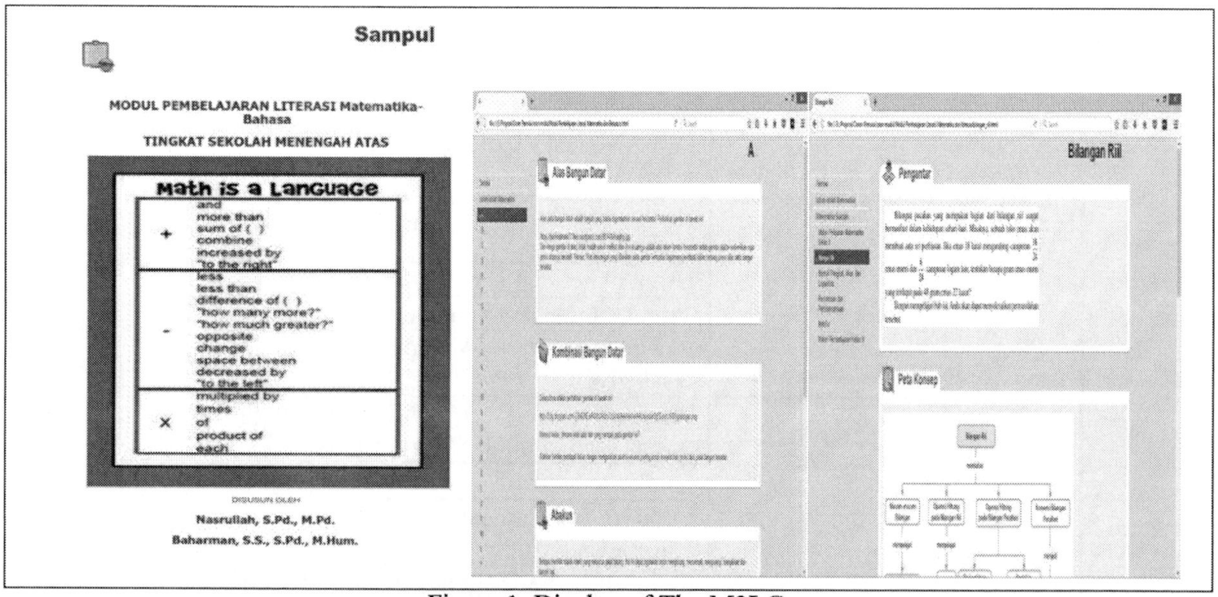

Figure 1. Display of The M3LC

M3LC is the product eXe-Learning based courseware constructed as an interactive learning equipment that is provided for students who want to enhance their literacy skills, in particular, mathematics and language literacy. Focus Group Discussion was conducted to explore the tendency of participants' responses toward the courseware in terms of knowledge use (KU), knowledge construction (KC), evaluation practice (EP), social programming (SP), and valuing (V). The results of data analysis of users' responses in regard to the advantages of the M3LC are summarized in figure 2.

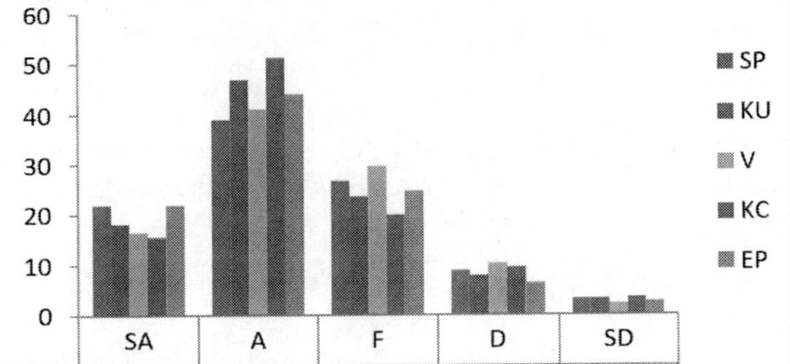

	SA	A	F	D	SD
SP	22,07	38,97	26,76	8,92	3,29
KU	18,26	46,89	23,65	7,88	3,32
V	16,51	41,04	29,72	10,38	2,36
KC	15,65	51,3	20	9,57	3,48
EP	22,02	44,04	24,77	6,42	2,75

Figure 2. Representation of Practitioners' Responses about M3LC

Figure 2 shows the representation of the participants' responses related to the practical evaluation of M3LC. The responses of FGD's participants in category disagree or strongly disagree are not more than 10.40%. It brings the information that more than 50% of their responses support the practical aspects of M3LC to be applied in the learning activities, although below 30% of responses are in fair category.

Based on this statistical information, it shows that M3LC brings standard of valuable courseware in that students can take some advantages related to learning literacy as the main purpose it was constructed. It is in line with the statement that internet based programs and community networking proficiencies have augmented in acceptance, enlightening their potential [15]. Then, the use of WeChat address to promise pedagogical prospective for helping language learning in various ways, in particular, in assisting semi-synchronous collaboration through media-based rich environment [16].

4. Conclusions

The potential of Mathematics-Language Literacy Learning Module (M3LC) can be explained in 5 practical aspects, including knowledge use, knowledge construction, evaluation practice, social programming, and valuing. These advantages value of M3LC is supported by responses from the practitioners in the classroom. Their responses show that the tendency of agree and strongly agree arguments are still higher than those of arguments in disagree or strongly disagree category, with below 30% of responses are in fair category. In the other words, the respondents tend to give positive opinion that this courseware is helpful and useful to support literacy learning. In future, the implementation of using this courseware can be enhanced to further recognition of literacy level so that students can be well-prepared before starting learning activities in the classroom.

Acknowledgement

The author expresses his gratitude to *Kemenristekdikti* for research grant of *Dosen Pemula* Which supports the implementation of this research. We also thank teachers and students in the Senior High

IOP Conf. Series: Journal of Physics: Conf. Series **954** (2018) 012007 doi:10.1088/1742-6596/954/1/012007

School in Makassar who have been getting involved in all of the research activities until the last of meeting.

References

[1] W. David, W. Thomas, and A. James, 1999, Adolescent literacy A postioll, **6**.
[2] C. Deeson, 2001, Early literacy, *Br. J. Educ. Technol.*, **32**, 4, p. 499.
[3] O. Indicators, Education at a Glance 2016. 2016.
[4] Setiadi, H., 2012, Kemampuan Matematika Siswa SMP Indonesia, Jakarta.
[5] V. Purcell-gates, 2001, Emergent Literacy Is Emerging Knowledge of Written, Not Oral, Language,. 92, pp. 7–22.
[6] C. S. Brown, 2014, Language and Literacy Development in the Early Years: Foundational Skills that Support Emergent Readers, *Lang. Lit. Dev. Early Years*, **24**, pp. 35–48.
[7] S. Nugroho, 2014, Pemanfaatan Mobile Learning Game Barisan dan Deret Geometri Untuk Meningkatkan Minat dan Hasil Belajar Matematika Sma Kesatrian 1, **I**.
[8] I. H. Al-astal, and A. A. A. Zaydah, 2015, The Effectiveness of an E-Book on Developing Mathematical Thinking Skills and Acquisition of Mathematical Concepts among 5th Graders in Gaza, **116**, 21, pp. 23–30.
[9] Ahmar, A. S., & Rahman, A. Development of teaching material using an Android. *Global Journal of Engineering Education*, vol. 19, no. 1, 2017.
[10] Mulbar, U., Rahman, A., & Ahmar, A. S. 2017. Analysis of the ability in mathematical problem-solving based on SOLO taxonomy and cognitive style. *World Trans. Eng. Technol. Educ.*, **15**, 1, pp. 74-77.
[11] Rahman, A., Ahmar, A. S., & Rusli. 2016. The Influence of Cooperative Learning Models on Learning Outcomes Based on Students' Learning Styles. *World Trans. Eng. Technol. Educ.*, **14**, 3, pp. 74-77.
[12] Rahman, A., & Ahmar, A. S. 2016. Exploration of Mathematics Problem Solving Process Based on The Thinking Level of Students in Junior High Schoo. *Int. J. Environ. Sci. Educ.*, **11**, 14.
[13] W. F. Heinecke, L. Blasi, N. Milman, and L. Washington, 1999, New Directions in the Evaluation of the Effectiveness of Educational Technology, *Conf. Educ. Technol.*, **1980**, pp. 1–7.
[14] E. Minter and M. Michaud, 2003, G3658-13 Ed Minter Mary Michaud 2003.
[15] R. Crawford, 2013, Ascilite Evolving technologies require educational policy change : Music education for the 21st century, **29**, 5, pp. 717–734.
[16] Y. Wang, J. Han, and N. Chen, 2016, Exploring the affordances of WeChat for facilitating teaching, social and cognitive presence in semi-synchronous language exchange, **32**, 4, pp. 18–37, 2016.

Keylogger Application to Monitoring Users Activity with Exact String Matching Algorithm

Robbi Rahim[1*], Heri Nurdiyanto[2], Ansari Saleh A.[3,4], Dahlan Abdullah[5], Dedy Hartama[6], and Darmawan Napitupulu[7]

[1]School of Computer Engineering and Communication, Universiti Malaysia Perlis, Malaysia
[2]Department of Informatics Engineering, STMIK Dharma Wacana, Metro Lampung, Indonesia
[3]Department of Statistics, Universitas Negeri Makassar, Makassar, Indonesia
[4]AHMAR Institute, Makassar, Indonesia
[5]Department of Informatics, Universitas Malikussaleh, Aceh, Indonesia
[6]Department of Information System, STIKOM Tunas Bangsa, Pematang Siantar, Indonesia
[7]Research Center for Quality System and Testing Technology, Indonesian Institute of Sciences, Indonesia

*usurobbi85@zoho.com

Abstract. The development of technology is very fast, especially in the field of Internet technology that at any time experiencing significant changes, The development also supported by the ability of human resources, Keylogger is a tool that most developed because this application is very rarely recognized a malicious program by antivirus, keylogger will record all activities related to keystrokes, the recording process is accomplished by using string matching method. The application of string matching method in the process of recording the keyboard is to help the admin in knowing what the user accessed on the computer.

1. Introduction

Security [1] [2] [3] is essential in an operating system, and there are many software vendors that provide antivirus software and other protection software so that the computer is free from threats or unwanted actions such as phishing, malware, keylogger and others.

Keylogger [4] is a tool that most progressed because this application is very rarely recognized as a malicious program by antivirus [5], the function of keylogger will record all activities related to the emphasis on the keyboard keys or activity on the operating system [4] [5]. Email addresses and passwords even up to a Facebook account with passwords can be stolen with keylogger apps without the realization of computer users.

String matching [6] [7] [8] is a method of character matching from input process like a keyboard, and matching process is done by checking every incoming character. The use of string matching algorithms on keylogger applications used to record user activity will be more comfortable and faster.

IOP Conf. Series: Journal of Physics: Conf. Series **954** (2018) 012008 doi:10.1088/1742-6596/954/1/012008

2. Methodology

Keylogger is a device either hardware or software used to monitor keyboard keystrokes [9] [10]. A keylogger will usually store the monitoring results of the keyboard keystrokes into a log/record/record file. Some specific keylogger can even send the recording to specific e-mail periodically [5] [11]. Keylogger can be used for the benefit of useful or even can be used for the benefit of evil. Right interests include monitoring employee productivity [12] for law enforcement and the search for evidence of the crime. Wrong interests include data theft and passwords [4] [12], see figure 1 below how keylogger work

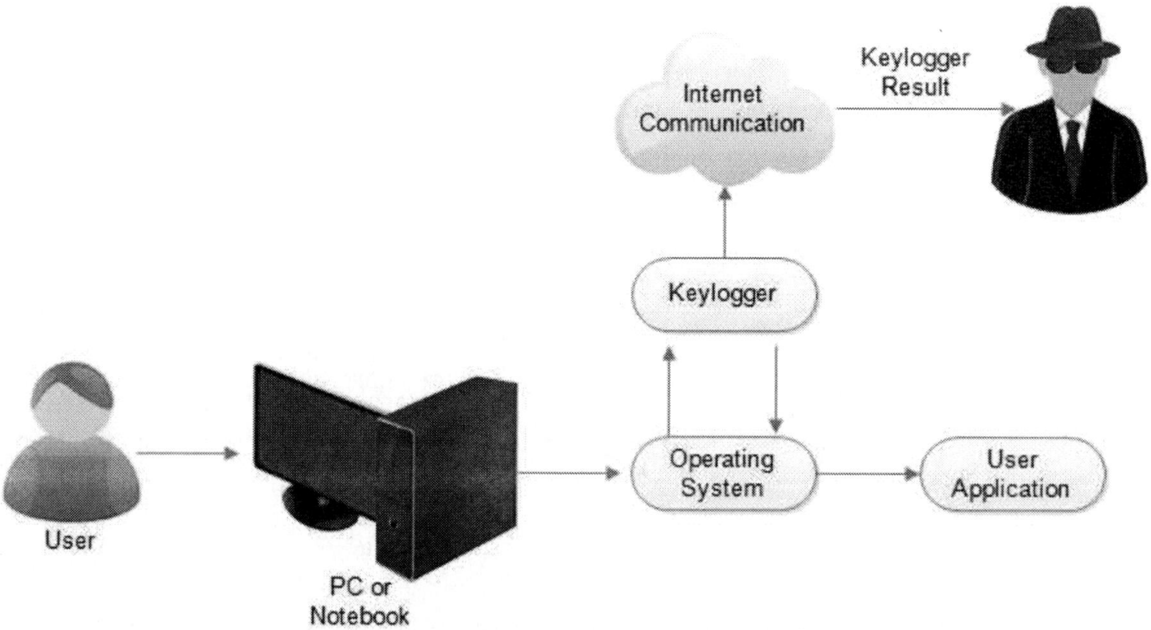

Figure 1. Keylogger Process in User Activity

The exact string matching algorithm is used to match keyboard input variables with input received from the keyboard, and exact string matching is a method that can be done to match input string variable with string comparison variable, for testing of exact string matching algorithm on keylogger application can be assumed in the following example:

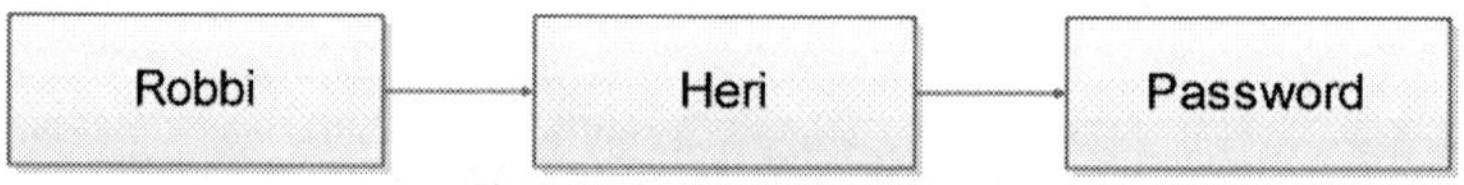

Figure 2. Sample Word

The word will be identified as a word entered by the user, the word is "Password" as shown in Figure 2, the word is then changed to ASCII form, after changed the result is "80 97 115 115 119 111 114 100", the ASCII results characters will be compared to the ASCII code contained in the keylogger system, if found then the system will record the ASCII characters and convert them into letters.

The keylogger application in this article is designed using the Visual C # programming language and runs on the Windows 8.1 operating system, and the experiment is used in Microsoft Word applications and some of the activities on the browser.

IOP Conf. Series: Journal of Physics: Conf. Series **954** (2018) 012008 doi:10.1088/1742-6596/954/1/012008

3. Result and Discussion

Keylogger experiment by applying exact string matching algorithm accomplished by using application designed using Visual C # with an interface as follows:

Figure 3. Main Keylogger Interface

Figure 3 shows the results of application interfaces designed to record activity on the user's computer, at the time this article is created the recording process is also being done to prove the application is running well and the application of exact string matching algorithm can be made well, to facilitate the examination of keylogger results all the recording will be stored in an encrypted file and cannot be accessed unless the party that embed the keylogger application, see Figure 4 below

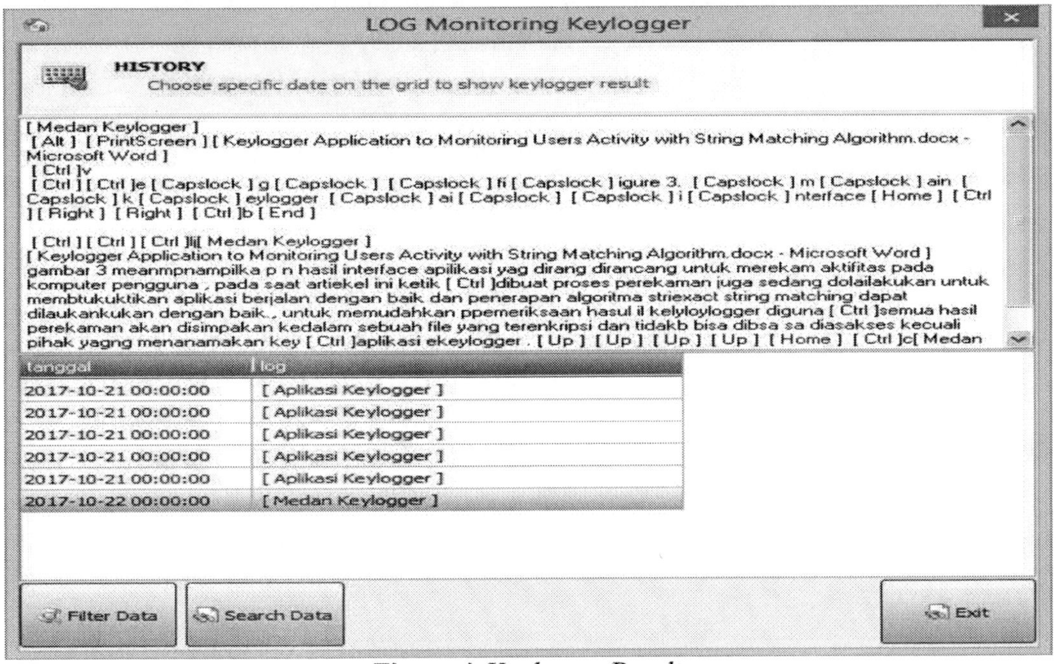

Figure 4. Keylogger Result

Recording of user activity when using software that exists in windows or online activity using a browser can be recording a well, and even function on a keyboard like a shortcut can also be recorded, and all of the keylogger results can be accessed directly by keylogger owner.

4. Conclusion

Keylogger applications designed by implementing the Exact String Matching algorithm can record all user activities related to the keyboard, and the results are stored automatically in a dedicated database that can only be accessed by the keylogger owner, the next development of the keylogger application can record the activity on the virtual keyboard or remote activity on the user's computer.

References

[1] R. Rahim and A. Ikhwan, "Study of Three-Pass Protocol on Data Security," *International Journal of Science and Research (IJSR),* vol. 5, no. 11, pp. 102-104, 2016.

[2] R. Rahim, "128 Bit Hash of Variable Length in Short Message Service Security," *International Journal of Security and Its Applications,* vol. 11, no. 1, pp. 45-58, 2017.

[3] A. P. U. Siahaan and R. Rahim, "Dynamic Key Matrix of Hill Cipher Using Genetic Algorithm," *International Journal of Security and its Applications,* vol. 10, no. 8, pp. 173-180, 2016.

[4] R. Venkatesh and R. K. Sekhar, "User Activity Monitoring Using Keylogger," *Asia Journal of Information Technology,* vol. 15, no. 23, pp. 4758-4762, 2015.

[5] D. Y. Pawade, A. Lahigude and D. Reja, "Review Report On Security Breaches Using Keylogger And Clickjacking," *International Journal of Advance Foundation and Research in Computer (IJAFRC),* vol. 2, no. Special Issue (NCRTIT 2015), pp. 55-59, 2015.

[6] K. K. Soni, R. Vyas, and A. Sinhal, "Importance of String Matching in Real World Problems," *International Journal Of Engineering And Computer Science,* vol. 3, no. 6, pp. 6371-6375, 2014.

[7] N. R. Dalal and P. Jadhav, "A Composite Algorithm for String Matching," *International Journal of Modern Trends in Engineering and Research (IJMTER),* vol. 2, no. 7, pp. 68-73, 2015.

[8] G. Pandey and G. L. Prajapati, "Applying Bi-Directional Search Strategy in Selected String Matching Algorithms," *International Journal of Computer Applications,* vol. 143, no. 11, pp. 40-43, 2016.

[9] N. Adhikary, R. Shrivastava, A. Kumar, S. K. Verma, M. Bag and V. Singh, "Battering Keyloggers and Screen Recording Software by Fabricating Passwords," *I. J. Computer Network and Information Security,* vol. 2012, no. 5, pp. 13-21, 2012.

[10] M. Dadkhah, M. D. Jazi, C. Ana-Maria and E. Barati, "An Introduction to Undetectable Keyloggers with Experimental Testing," *International Journal of Computer Communications and Networks,* vol. 4, no. 3, pp. 1-5, 2014.

[11] N. Pathak, A. Pawar, and B. Patil, "A Survey on Keylogger: A malicious Attack," *International Journal of Advanced Research in Computer Engineering & Technology (IJARCET),* vol. 4, no. 4, pp. 1465-1469, 2015.

[12] P. Tuli and P. Sahu, "System Monitoring and Security Using Keylogger," *International Journal of Computer Science and Mobile Computing,* vol. 2, no. 3, pp. 106-111, 2013.

Implementation Cryptography Data Encryption Standard (DES) and Triple Data Encryption Standard (3DES) Method in Communication System Based Near Field Communication (NFC)

Ratnadewi[1], Roy Pramono Adhie[1], Yonatan Hutama[1], A. Saleh Ahmar[2,3], M I Setiawan[4]

[1]Electrical Engineering Department, Universitas Kristen Maranatha , Bandung, Indonesia
[2]Department of Statistics, Universitas Negeri Makassar, Makassar, Indonesia
[3]AHMAR Institute, Makassar, Indonesia
[4]Narotama University, Surabaya, Indonesia

ratnadewi.bandung@gmail.com

Abstract. Cryptography is a method used to create secure communication by manipulating sent messages during the communication occurred so only intended party that can know the content of that messages. Some of the most commonly used cryptography methods to protect sent messages, especially in the form of text, are DES and 3DES cryptography method. This research will explain the DES and 3DES cryptography method and its use for stored data security in smart cards that working in the NFC-based communication system. Several things that will be explained in this research is the ways of working of DES and 3DES cryptography method in doing the protection process of a data and software engineering through the creation of application using C++ programming language to realize and test the performance of DES and 3DES cryptography method in encrypted data writing process to smart cards and decrypted data reading process from smart cards. The execution time of the entering and the reading process data using a smart card DES cryptography method is faster than using 3DES cryptography.

Keywords: Cryptography, Data Encryption Standard (DES), Triple Data Encryption Standard (3DES), NFC

1. Introduction

In a process of communication, there are exchanging the messages. Delivery of a message that is sensitive and must be protected in order to avoid the theft of information by parties who are not responsible. One way to protect the delivery of a message is to use cryptographic methods. One method that is most commonly used cryptography in securing the delivery of messages, especially messages in the form of text data, is a cryptographic method Data Encryption Standard (DES). DES cryptographic methods developed by scientists in the field of computer network security in the US computer company, the International Business Machines Corporation (IBM) since 1974. [1] DES cryptographic methods developed by the cryptographic algorithm Lucifer made by Horst Feistel and were first published in 1971.

Cryptographic methods Triple Data Encryption Standard (3DES), can also be called Triple Data Encryption Algorithm (TDEA), is a further development of DES cryptographic methods. Cryptographic 3DES method was first published in 1998 and was made to replace the DES cryptographic method as standard cryptographic text data in the world. Difference between the two is 3DES using triple DES-step process used for encryption and decryption process by using a three key combination for each process.

Near Field Communication (NFC) protocol and interface for data exchange process in a simple wireless communication system devices connected in near distance. NFC is one of the development of communications technology Radio Frequency Identification (RFID) and belong to the category of High frequency (HF) and has a working frequency of 13.56MHz. Examples of devices that was support NFC based communication process is a contactless smart card and smart card reader. Smart card reader is used as a device supporting the process of data writing and data reading contactless smart cards.

2. Theoretical Basis

2.1. Data Encryption Standard (DES)

DES cryptography method is a text data cryptographic method with the type of symmetric - key block cipher and consists of two processes, namely the process of encryption and decryption process. [2] Each of that process has an eight-step process. Decryption process is done by reversing the steps - steps encryption process. Key used in DES cryptography method has 56 bits of effective length.

The process of encryption using DES cryptographic method consists of the eight-step process, [3] [4], that are:

1. Convert plaintext and the key that will be processed into binary bits. Plaintext and the key that has been converted and then broken down into data blocks form with each of the block has a 64 bits (eight bytes) length. If the message is in the form of the alphabet or symbols, it must be first converted into decimal and hexadecimal form following the ASCII character table, and then are converted into binary bit.
2. Randomize the bits in plaintext data block based on Initial Permutation (IP) table, so that the bit sequence randomized compared to bits sequence of early plaintext block. The bits sequence after the second step followed the results from the IP table, with the first bit derived from the 58th sequence bit of early plaintext blocks, and then the second bit derived from the 50th sequence bits until the 64th sequence bit derived from the seventh. sequence bits.
3. The scrambling of key bits based on permuted Choice 1 (PC - 1) table. Results from the PC - 1 has a 56 bit length because the last bits of each byte of the key (8, 16, 24, 32, 40, 48, 56 and 64 bits) that acts as the parity bit is not used again in the next step process. Once completed the results of PC - 1 then divided into C_0 and D_0, with C_0 is 28 leftmost bits and D_0 is 28 rightmost bits from PC – 1 results bits sequence.
4. Shift bits to the left (left shift) at C_i and D_i as much as one or two times, with the value of i based on encryption process round that consists of 16 rounds. The result of the shift bits from every round of the C_i and D_i are then combined into C_iD_i with a length of 56 bits. After that the C_iD_i key bits are randomized based on PC - 2 (permuted Choice 2) table until produce the Ki variable.
5. Running the data expansion process of Ri-1 with a length of 32 bits (starting from the R0 of second process step results) becomes Ri with a length of 48 bits, where i is the round during the process. This process will be carried out as much as 16 times with the value of turnover 1 <= i <= 16 using the Expansion Table. The results of the expansion process is referred to as E(Ri-1), starting from E(R0) to E(R15). Afterwards, E(Ri-1) will be XOR processed with Ki that has been obtained from the fourth step process for each bit corresponds to running process round to produce Ai variable with a length of 48 bits and in a vector form.

6. Once obtained, Ai then broken down into eight blocks with each block consist of six bits. Each block is then distributed into eight pieces of S – Box (Substitution Box), with the first block distributed to the S – Box 1, the second block distributed to the S – Box 2 and so on. The result of the substitution process using S – Box will be collected and produce Bi variable.
7. Once Bi variable obtained, the next step is do the permutation process on each bit of Bi variable using P – Box table. The results obtained from the permutation using P – Box referred to as P(Bi), with i adapted to the round during the process, starting from P(B1) through P(B16). Thereafter, P(Bi) will be XOR processed with the Li-1obtained from the second process step in accordance with the processes running round to produce an Ri variable with a length of 32 bits and in a vector form. Ri results will then be merged with Li, which came from the Ri-1, into LiRi which is the result of the encryption process of plaintext for each round process with a length of 64 bits.
8. The eighth process step carried out when the seventh process step has obtained the L16and R16 from the 16th process round. The next step is the process of reversing positions on L16 and R16, and then combined to obtain the R16L16 form. These results are then permutated using IP-1 (Inverse Initial Permutation) table. Results obtained from the eight process step is referred to as cipher, which is a data block that has been encrypted and is ready to be sent to the recipient along with the other ciphers. A combination of several cipher is called ciphertext.

For the note, the first and second step of the process is done only once at the beginning of the DES encryption process, while the eight step of the process is done only once at the end of the DES encryption process.The third step process to the seventh process step carried out 16 times according to the number of rounds of Feistel process used by DES cryptographic method. DES encryption process block diagram is shown in Figure 1.

2.2. Triple Data Encryption Standard (3DES)

3DES cryptography method is the development of DES cryptographic method. The difference between the two methods is 3DES using triple times the DES process step used in encryption process and decryption process by using three key combination.[5] In addition, the effective length of the key used for encryption process and decryption process using 3DES cryptographic method is 168 bits (consisting of three sub - key that each have a length of 56 bits), in contrast to DES cryptographic method that uses a key with and effective length of 56 bits.[6]

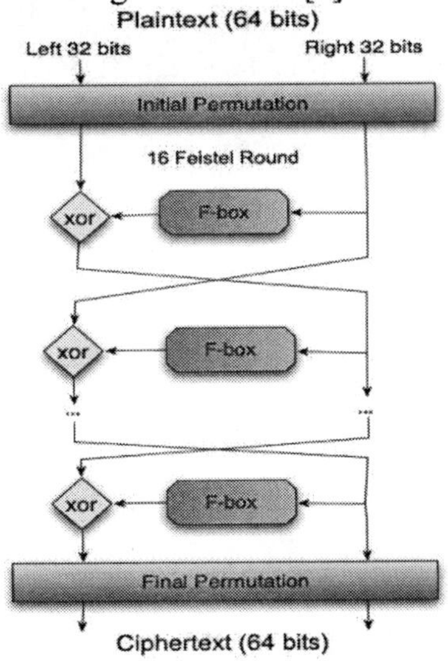

Figure. 1 DES Encryption Process Block Diagram

IOP Conf. Series: Journal of Physics: Conf. Series **954** (2018) 012009 doi:10.1088/1742-6596/954/1/012009

There are three options to use a combination of sub - key that has become standard in the encryption process and decryption process using 3DES cryptographic method,[7] that are :

1. Three sub - key have different combinations (3K3DES).
2. K1 and K2 have different combinations, whereas K1and K3have the same combinations (2K3DES).
3. Three sub - key has the same combinations..

Of the three options, the use of sub - key, the first option is the best because the three sub - key has a different combination, with an effective key length of 168 bits, so that the data is encrypted using the first option is more difficult to resolve than the use of the second and third options.[8] The second option has an effective key length of 112 bits, because the first sub - key has same combination as the third sub - key, but this option is still better than using the DES encryption process twice. The third option is the weakest compared to the previous option because the first sub - key and the second sub - key negated each other in the process so that the key used in this option has an effective length of 56 bits, the same as the length of the key used by DES cryptographic method. The second option will be used in the performance examination process of 3DES cryptographic method in this research. 3DES encryption process block diagram is shown in Figure 2.

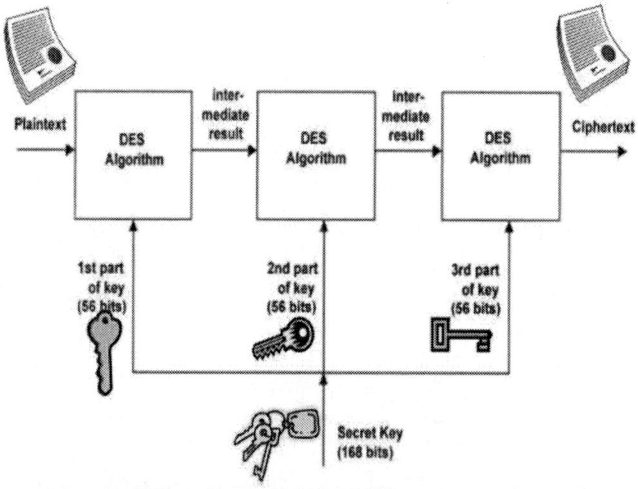

Figure. 2 3DES Encryption Process Block Diagram

2.3. *Near Field Communication (NFC)*

NFC is one of the latest communication technology and developed from RFID communication technology with working frequence of 13.56 MHz.[9] The device used in NFC based communication process is divided into two types, namely active devices and passive devices. An example of active devicee is a smart card reader, while an example of a passive device is a contactless smart card. Each NFC based active device can work in three modes of operation,[10], that are :

1. Card Emulation.
2. Card Reader / Writer.
3. Peer - to - Peer.

Card Emulation operating mode is used to change the function of active devices such as smart phones that can be used like a smart card. Card reader / writer operating mode is used to perform the writing process and reading process of the data contained in a passive device. Peer - to - peer operation mode used to perform data exchange process between two active device with similar condition, such as two smart phone devices. The second operating mode will be used in this research to examine the performance of DES and 3DES cryptography method.

IOP Conf. Series: Journal of Physics: Conf. Series **954** (2018) 012009 doi:10.1088/1742-6596/954/1/012009

3. Implementation and Testing

3.1. Program Implementation
In this research, the authors implemented DES and 3DES cryptographic methods into programs created using the Microsoft Visual Studio 2008 software and in C++ programming language. Programmes built in the form of application can be used on computers that have been based on the .NET framework. There is one application made in this research, that is the application of data writing and data reading of ACOS3 smart card using DES and 3DES cryptographic method in NFC - based systems. Contactless smart card and smart card reader that will be used in this research are ACOS3 8 KB smart card and ACS ACR1252U reader. The application display can be seen in Figure 3.

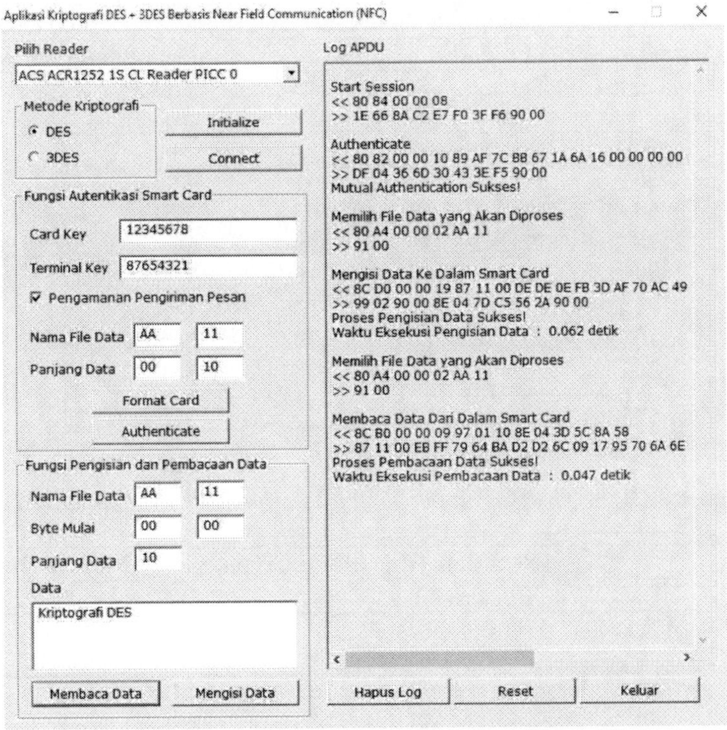

Figure. 3 Application Data Charging and Data Reading Method Using Cryptographic Smart Card ACOS3 DES and 3DES in NFC-Based Systems

The ACOS3 smart card data writing process using DES and 3DES encryption method have two inputs, that are the message that will be encrypted (plaintext / data) and key (card key and terminal key), and a single output, that is the message that has been encrypted perfectly (ciphertext) and also will be loaded into the ACOS3 smart card. The data writing process is executed by pressing the "Write Data" button. The ACOS3 smart card data reading process using DES and 3DES decryption method have two inputs, the ciphertext and key, and a single output, the message that has been perfectly decrypted (plaintext) and also will be displayed in the "Data" text box in the application. The data reading process is executed by pressing the "Reading Data" button.

3.2. DES and 3DES Performance Testing
Performance examination was conducted to examine two things, the data writing execution time and data reading execution time of ACOS3 smart card using DES and 3DES cryptographic method. There are 10 observational data for each trial based on the length of data to be processed, starting from 8 ASCII characters up to 4096 ASCII characters. Each data observation tested three times to generate

the average execution time. The performance examination results table that is carried out can be seen in Table 1 up to Table 3.

Performance examination that is done on ACOS3 smart card shows the result that the data writing execution time using DES encryption method is faster than using 3DES encryption method for each observational data. It also occurs in the data reading process using DES decryption method that is faster than using 3DES for each observational data. The execution time difference between that two cryptographic method is growing along with increasing length of the data processed.

Table 1. DES Cryptographic Method Performance Examination Results

No.	Data Length (ASCII Character)	Execution Time DES Cryptography Method (milliseconds)							
		Data Write Process				Data Read Process			
		Test 1	Test 2	Test 3	Average	Test 1	Test 2	Test 3	Average
1	8	57	57	57	57	44	41	40	42
2	16	60	59	62	61	43	42	43	43
3	32	64	63	68	65	47	46	48	47
4	64	78	77	78	78	56	56	58	57
5	128	99	99	100	100	73	72	71	72
6	256	148	149	146	148	113	113	115	114
7	512	208	206	209	208	145	145	147	146
8	1024	265	263	263	264	181	180	183	182
9	2048	321	327	326	325	217	219	220	219
10	4096	387	383	390	387	255	254	258	256

Table 2. 3DES Cryptographic Method Performance Examination Results

No.	Data Length (ASCII Character)	Execution Time 3DES Cryptography Method (milliseconds)							
		Data Write Process				Data Read Process			
		Test 1	Test 2	Test 3	Average	Test 1	Test 2	Test 3	Average
1	8	62	62	60	62	44	43	44	44
2	16	64	66	66	66	46	49	46	47
3	32	68	70	68	69	53	50	50	51
4	64	84	84	85	85	61	61	61	61
5	128	110	111	112	111	83	85	84	84
6	256	169	172	170	171	133	133	131	133
7	512	253	255	260	256	192	195	194	194
8	1024	321	325	331	326	244	248	247	247
9	2048	393	399	402	398	297	294	300	297
10	4096	472	468	473	471	365	369	372	369

Table 3 – DES and 3DES Cryptographic Method Performance Examination Average Results

No.	**Data Length (ASCII Character)**	**Execution Time 3DES Cryptography Method (milliseconds)**			
		Data Write Process		**Data Read Process**	
		DES	**3DES**	**DES**	**3DES**
1	8	57	62	42	44
2	16	61	66	43	47
3	32	65	69	47	51

No.	Data Length (ASCII Character)	Execution Time 3DES Cryptography Method (milliseconds)			
		Data Write Process		Data Read Process	
		DES	3DES	DES	3DES
4	64	78	85	57	61
5	128	100	111	72	84
6	256	148	171	114	133
7	512	208	256	146	194
8	1024	264	326	182	247
9	2048	325	398	219	297
10	4096	387	471	256	369

Through conducted performance examination, it also found that the data writing execution time of ACOS3 smart card using DES and 3DES decryption method is faster than the data writing execution time of ACOS3 smart card using DES and 3DES encryption method for each observational data. The average execution time chart of ACOS3 smart card data writing process and data reading process using DES and 3DES cryptographic method for each observational data can be seen in Figure 4 and Figure 5.

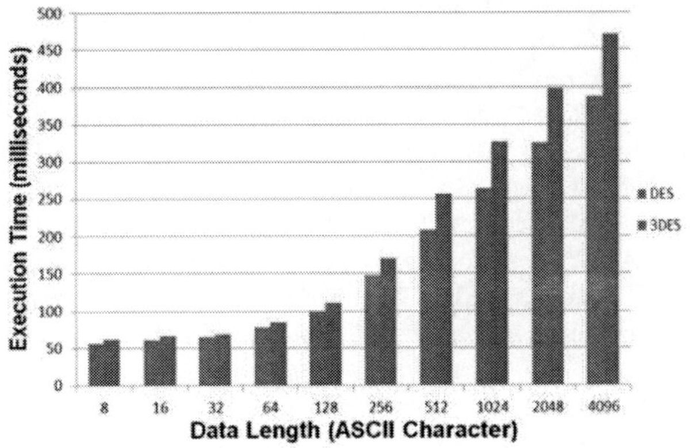

Figure. 4 Average Execution Time Chart of ACOS3 Smart Card Data Writing Process Using DES and 3DES Encryption Method

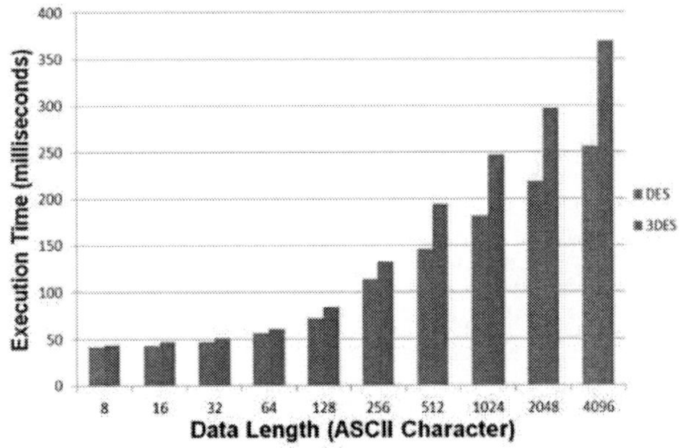

Figure 5. Average Execution Time Chart of ACOS3 Smart Card Data Reading Process Using DES and 3DES Decryption Method

4. Conclusion

Based on performance examination conducted on DES and 3DES text data cryptographic method with the help of the application of data writing and data reading of ACOS3 smart card using DES and 3DES cryptographic method in NFC -based systems, there are several things that can be concluded :

1. DES and 3DES text data cryptographic method can be implemented on application of ACOS3 smart card data writing process and data reading process in NFC - based systems.
2. The execution time of ACOS3 smart card data writing process and data reading process using DES cryptographic method is faster than 3DES cryptographic method.
3. The execution time of data reading process is faster than data writing process for ACOS3 smart card, using DES or 3DES cryptographic method.

References

[1] S.Mankotia and M. Sood. 2015 "A Critical Analysis of Some Symmetric Key Block Cipher Algorithms". *International Journal of Computer Science and Information Technologies*, Vol. 6.

[2] G. Singh and Supriya. 2013. "A Study of Encryption Algorithms (RSA, DES, 3DES and AES) for Information Security". *International Journal of Computer Applications* Vol. 67, No. 19.

[3] C. Paar and J. Pelzl. 2010. *Understanding Cryptography – A Textbook for Students and Practitioners*. Heidelberg : Springer – Verlag GmbH & Co.

[4] W. Stallings. 2014. *Cryptography and Network Security – Principles and Practice 6th Edition*. Upper Saddle River, New Jersey : Pearson Education Limited.

[5] S. Rao. 2015. "Performance Analysis of DES and Triple DES". *International Journal of Computer Applications*. Vol. 130, No. 14.

[6] M. Mathur and A. Kesarwani. 2013. "Comparison Between DES, 3DES, RC2, RC6, BLOWFISH and AES". *Proceedings of National Conference on New Horizons in IT.*

[7] P. Kumar and N.S. Rajaanadan 2016. "Data Encryption and Decryption Using By Triple DES Performance Efficiency Analysis of Cryptosystem". *International Journal of Innovative Research in Computer and Communication Engineering* Vol. 4, No. 3.

[8] A. Kahate. 2003. *Cryptography and Network Security*. Singapore : Tata McGraw – Hill Publishing Company Limited.

[9] A. Makhmali and H. M. Jani. 2013. "Comparative Study On Encryption Algorithms And Proposing A Data Management Structure". *International Journal of Scientific & Technology Research* Vol. 2, No. 6.

[10] H. A. Al-Ofeishat and M.A.A.Al Rababah. 2012. "Near Field Communication (NFC)". *International Journal of Computer Science and Network Security* Vol. 12, No. 2.

IOP Conf. Series: Journal of Physics: Conf. Series **954** (2018) 012010 doi:10.1088/1742-6596/954/1/012010

Modeling Data Containing Outliers using ARIMA Additive Outlier (ARIMA-AO)

Ansari Saleh Ahmar[1,2*], Suryo Guritno[3], Abdurakhman[3], Abdul Rahman[4], Awi[4], Alimuddin[4], Ilham Minggi[4], M Arif Tiro[1], M Kasim Aidid[1], Suwardi Annas[1], Dian Utami Sutiksno[5], Dewi S Ahmar[2], Kurniawan H Ahmar[2], A Abqary Ahmar[2], Ahmad Zaki[4], Dahlan Abdullah[6], Robbi Rahim[7], Heri Nurdiyanto[8], Rahmat Hidayat[8], Darmawan Napitupulu[9], Janner Simarmata[10], Nuning Kurniasih[11], Leon Andretti Abdillah[12], Andri Pranolo[13], Haviluddin[14], Wahyudin Albra[15], A Nurani M Arifin[2]

[1]Department of Statistics, Universitas Negeri Makassar, Indonesia
[2]AHMAR Institute, Makassar, Indonesia
[3]Department of Mathematics, Universitas Gadjah Mada, Indonesia
[4]Department of Mathematics, Universitas Negeri Makassar, Indonesia
[5]Department of Business Administration, Politeknik Negeri Ambon, Indonesia
[6]Universitas Malikussaleh, Aceh, Indonesia
[7]School of Computer and Communication Engineering, Universiti Malaysia Perlis, Malaysia
[8]STMIK Dharma Wacana, Lampung, Indonesia
9Lembaga Ilmu Pengetahuan Indonesia, Banten, Indonesia
[10]Universitas Negeri Medan, Medan, Indonesia
[11]Universitas Padjadjaran, Bandung, Indonesia
[12]Universitas Bina Darma, Palembang, Indonesia
[13]Department of Informatics, Universitas Ahmad Dahlan, Yogyakarta, Indonesia
[14]Universitas Mulawarman, Samarinda, Indonesia
[15]Faculty of Economics and Bussiness, Universitas Malikussaleh, Aceh, Indonesia

*ansarisaleh@unm.ac.id

Abstract. The aim this study is discussed on the detection and correction of data containing the additive outlier (AO) on the model ARIMA (p, d, q). The process of detection and correction of data using an iterative procedure popularized by Box, Jenkins, and Reinsel (1994). By using this method we obtained an ARIMA models were fit to the data containing AO, this model is added to the original model of ARIMA coefficients obtained from the iteration process using regression methods. In the simulation data is obtained that the data contained AO initial models are ARIMA (2,0,0) with MSE = 36,780, after the detection and correction of data obtained by the iteration of the model ARIMA (2,0,0) with the coefficients obtained from the regression $Z_t = 0{,}106 + 0{,}204Z_{t-1} + 0{,}401Z_{t-2} - 329X_1(t) + 115X_2(t) + 35{,}9X_3(t)$ and MSE = 19,365. This shows that there is an improvement of forecasting error rate data.

1. Introduction

In time series data is sometimes there's a far different data values from other data and do not reflect the characteristics of a set of data. The data value is called the outliers. In the analysis of time series data are often obtained outlier. Outlier data is a major impact on forecasting data if we use forecasting methods such as ARIMA, ARMA, and others. Forecasting is an activity to predict what will happen in the future [1]. To forecast a time series then we can use forecasting method, e.g., ARIMA Box-Jenkins method [2]. Forecasting will be far off from what we predict if the data used to predict the data contained outliers. The study on outlier detection is very important because the presence of outliers can lead to parameter estimation or forecasting becomes inappropriate. If a data outlier is not enforced properly, then it will impact the forecast does not reflect the actual data. Related to the incident such as recording and typing errors, AO is the most common type of outliers are found in the time series, so that according to the discussion above, this research will discuss the detection and correction Data containing Additive outlier (AO) in the ARIMA(p,d,q) model.

If the observations in a time series can be predicted with certainty and do not require further investigation, it is called deterministic time series and if the observations can only show the structure of the probabilistic state will come a time series, the time series is called stochastic. In the modeling of time series analysis assumes that the data are stationary. Stationary time series is said if there is no change in the average trends and change in variance. Relatively stationary time series are not increasing or decrease in the value of the extreme fluctuations in the data, or the data is at about the average value of the constant.

Stationarity can be seen by using a time series chart is a scatterplot between the value of the variable Z_t with time t. If the diagram time series fluctuates around a line parallel to the time axis (t), then the series is said stationary in average. When stationary conditions in the average unmet need differentiation or differencing process [3].

The process of first-order differencing on the difference between tth data with t–1 th, i.e., $\Delta Z_t = Z_t - Z_{t-1}$. Forms for second order is differencing $\Delta^2 Z_t = \Delta Z_t - \Delta Z_{t-1} = (Z_t - Z_{t-1}) - (Z_{t-1} - Z_{t-2})$ $= Z_t - 2Z_{t-1} + Z_{t-2}$. If these conditions are not met stationary in variance, Box and Cox (1964) introduced the power transformation $Z_t^{(\lambda)} = \dfrac{Z_t^{(\lambda)} - 1}{\lambda}$, ($\lambda$ is a parameter), known as the Box-CoX transformation. Following are some provisions to stabilize the variance is [4]:

- The transformation may be made only for the series Z_t are positive.
- The transformation is done before the differencing and modeling of time series.
- The value selected based on the sum of squares error (SSE) of the transformed series. The smallest value of SSE most constant yield variance.
- Transformation is not only stabilized the variance, but it can also normalize the distribution.

Integrate models Autoregressive Moving Average (ARIMA) has been studied by George Box and Gwilym Jenkins in 1976 [5], and their names are then frequently synonymous with ARIMA processes applied to the analysis of time series. In general ARIMA models is denoted by the notation ARIMA (p, d, q), where p denote the order of the autoregressive (AR), d expressed differentiation (differencing), and q express order of the moving average (MA). Autoregressive model (AR) was first introduced by Yule on 1926 [6] and later developed by Walker on 1931 [7], while model Moving Average (MA) was first used by Slutzky on 1937 [8]. And 1938, Wold generate the theoretical basis of the combination of ARMA and combinations is then often used [9]. Box and Jenkins have effectively managed to reach an agreement on the relevant information needed to understand and use ARIMA models for time series of the variables [10].

In data time series forecasting methods ARIMA (p, d, q) there is the so-called measures or its phases. The stages in the forecasting [11]:

a. Identification Model
Identification of the model was done to see the significance and stationary of autocorrelation data, so whether or not to do transformation or differencing process (differentiation). From this stage, the model is obtained while testing the model will be made whether or not the corresponding data.

b. Assessment and Testing Models
After identifying the model, the next step is the assessment and testing of the model. At this stage, divided into two parts, namely parameter estimation and diagnostic model.

c. Parameter Estimation
After obtaining one or more models while the next step is to find estimates for the parameters in the model.

d. Diagnostic Model
Diagnostic checking to do check whether the model estimated sufficient or adequate fit with the data. Diagnostic checking is based on the analysis of residuals. The basic assumption is that the residuals of ARIMA models are independent normally distributed random variables with zero mean constant variance.

The outlier is a problem often encountered in financial data. With the data outliers are the parties to a prediction or forecast of the data will have problems such as the problem of lack of accurate in forecasting. Gounder, et al [12] describes several types of outliers are: (1) Additive outlier (AO). Outliers of the simplest and most frequently studied in the analysis of time series are an additive outlier, also known as Type I outliers. An AO only affects a single observation, which could be worth a small or larger value compared with the expected value in the data. After this effect, the data series back to the normal track as if nothing happened. Effects of an AO is independent of ARIMA models and constrained. Outlier AO is the most problematic because it contains two consecutive residues, one before and the other after AO. An AO can have serious effects on the observed properties. This will affect the residual suspicion and supposition parameters. This can be proved in general that the AO large would encourage all autocorrelation coefficients towards zero. The influence of outliers is reduced to a large sample size.

The formula of the AO are as follows [11]:

$$Y_t = \begin{cases} Z_t & ,t \neq T \\ Z_t + \omega, & t = T \end{cases} \tag{1}$$
$$= Z_t + \omega I_t^{(T)}$$

where:
Y_t = Series observations
Z_t = array outlier-free observations
ω = magnitude outlier
$I_t^{(T)}$ = Indicator variable that indicates the time outlier

(2) innovational outlier (IO), in contrast to AO, an innovational outlier is also known as a Type II outliers (Fox, 1972) that affect some observations. An AO only affects only one residue, at the time of the occurrence of outliers. Effect of IO on the observed series consists of an initial shock that spread in subsequent observations by the weight of the moving average representation (MA) of the ARIMA models. So far the weight is often explosive, may influence the IO, in some cases, the incidence has increased and continues to increase to more and more value from that time until the events of the past, as well as in the future more unwelcome. Effect of IO depends on the particular model of the series, and series with stationary transformation effect indefinitely. For series containing trend and seasonal, the IO will affect both. IO presents some serious drawbacks that should be avoided. The problem then is that the other three types of outliers that can not explain the changes in the seasonal component.

2. Method

The method used in this research is the study of literature by analyzing the detection and correction of data AO on the ARIMA(p,d q) model. The initial step in the analysis is to stage in forecasting ARIMA (p,d,q) in order to obtain the corresponding allegations of ARIMA model. Further, testing of stationary residual. If the residual of stationary is not obtained then alleged that occurs because of the data that it contains outliers. Then do the detection and correction of data containing AO with an iterative procedure to obtain the residual of stationary and has the smallest MSE.

3. Result and Discussion

The outlier is a problem often encountered in financial data. With the data outliers are the parties to a prediction or forecast of the data will have problems such as the problem of lack of accurate in forecasting.

For a stationary process, say Z_t as the observed series and X_t as an outlier-free series. Assume that X_t follows the general model ARMA (p, q):

$$\phi_p(B)Z_t = \theta_q(B)a_t, a_t \sim WN(0,\sigma^2), \theta_i, \phi_i \in R, t \in R \tag{2}$$

with $\phi_p(B) = 1-\phi_1 B - ... - \phi_p B^p$ dan $\phi_p(B) = 1-\phi_1 B - ... - \phi_p B^p$ is a stationary and invertible operators who do not share the same factor. Additive outliers (AO) is defined as:

$$Y_t = \begin{cases} Z_t, & t \neq T \\ Z_t + \omega, t = T \end{cases} \tag{3}$$

$$Y_t = Z_t + \omega I_t^{(T)} \tag{4}$$

$$Y_t = \frac{\theta(B)}{\phi(B)} a_t + \omega I_t^{(T)}, a_t \sim WN(0,\sigma^2), \theta_i, \phi_i \in R, t \in R \tag{5}$$

where:

$$I_t^{(T)} = \begin{cases} 1, t = T, \\ 0, t \neq T, \end{cases} \tag{6}$$

is an indicator variable representing whether or not an outlier at time T.
In general, if the time series data are k additive outlier (AO), the general model of an outlier this:

$$Y_t = \sum_{j=1}^{k} \omega_j I_t^{(T_j)} + \frac{\theta(B)}{\phi(B)} a_t, a_t \sim WN(0,\sigma^2), \omega_j, \theta_i, \phi_i \in R, t \in R \tag{7}$$

3.1. Estimated Effect Outlier when Outlier Time Unknown

As a matter of motivation to detect AO procedure, we consider the simple case when T and all the parameters in (3) known. For example:

$$\pi(B) = \frac{\phi(B)}{\theta(B)} = (1 - \pi_1 B - \pi_2 B^2 - ...) \tag{8}$$

and defined:

$$e_t = \pi(B) Y_t \tag{9}$$

from (6) can be established:

$$AO: \quad e_t = \omega\pi(B) I_t^{(T)} + a_t, a_t \sim WN(0,\sigma^2), \omega_i, \pi_i \in R, t \in R \tag{10}$$

For n observations made, AO models above can be written as:

$$\begin{bmatrix} e_1 \\ \vdots \\ e_{T-1} \\ e_T \\ e_{T+1} \\ e_{T+2} \\ \vdots \\ e_n \end{bmatrix} = \omega \begin{bmatrix} 0 \\ \vdots \\ 0 \\ 1 \\ -\pi_1 \\ -\pi_2 \\ \vdots \\ -\pi_{n-T} \end{bmatrix} + \begin{bmatrix} a_1 \\ \vdots \\ a_{T-1} \\ a_T \\ a_{T+1} \\ a_{T+2} \\ \vdots \\ a_n \end{bmatrix} \tag{11}$$

Let $\hat{\omega}_{AT}$ as the least squares estimator of to model AO. Because it is white noise, then the theory of least squares, found that:

$$\begin{aligned} AO: \quad \hat{\omega}_{AT} &= \frac{e_T - \sum_{j=1}^{T} \pi_j e_{T+j}}{\sum_{j=0}^{n-T} \pi_j^2} \\ &= \frac{\pi^*(F) e_T}{\tau^2} \end{aligned} \tag{12}$$

with $\pi^*(F) = (1 - \pi_1 F - \pi_2 F^2 - \ldots - \pi_{n-T} F^{n-T})$, F is the forward shift operator so that $Fe_t = e_{t+1}$ dan $\tau^2 = \sum_{j=0}^{n-T} \pi_j^2$. The variance of the estimator is:

$$\begin{aligned} Var(\hat{\omega}_{AT}) &= Var\left(\frac{\pi^*(F) e_t}{\tau^2}\right) \\ &= \frac{1}{\tau^4} Var\left[\pi^*(F) e_t\right] \\ &= \frac{\sigma_a^2}{\tau^2} \end{aligned} \tag{13}$$

3.2. Procedure Using Iterative Outlier Detection

If T is not known, but the parameters of time series are known, then we can calculate $\lambda_{1,t}$ for each $t = 1, 2, \ldots, n$, and then make a decision based on the results of these sampling. However, in practice often be found that parameter time series ϕ_j, θ_j, π_j and σ_a^2 usually unknown and must be predictable. With the outlier then make predictions parameter would be biased. In particular, σ_a^2 will tend to be exaggerated, as indicated earlier. Therefore it is the Chang and Tiao (1983) proposed an iterative procedure to detect and deal with the situation when the number of AO is not known to exist. In an outlier detection and correction of the data it needs is called procedure. The procedure to be used in the detection procedure is iterations. The procedure is as follows (Box, Jenkins, and Reinsel, 1994):

Step 1: Download the data
Step 2: Estimated parameters of ARIMA (usually assumed that the data contains no outliers)
Step 3: Calculate the residuals of the ARIMA models are:

$$\hat{e}_t = \frac{\hat{\phi}(B)}{\hat{\theta}(B)} Y_t = \hat{\pi}(B) Y_t \tag{14}$$

with :

$$\phi_p(B) = 1 - \phi_1 B - \ldots - \phi_p B^p \text{ and } \phi_p(B) = 1 - \phi_1 B - \ldots - \phi_p B^p \tag{15}$$

and let :

$$\sigma_a^2 = \frac{1}{n} \sum_{t=1}^{n} \hat{e}_t^2 \tag{16}$$

Step 4: Calculate the test statistic presence of AO ($\hat{\lambda}_{A,T}$)

$$\hat{\lambda}_{1,T} = \frac{\tau \hat{\omega}_{AT}}{\hat{\sigma}_a} \tag{17}$$

with :

$$\hat{\omega}_{AT} = \frac{e_T - \sum_{i=1}^{T} \pi_i e_{T+i}}{\sum_{i=o}^{n-T} \pi_i^2} \tag{18}$$

$$\tau^2 = \sum_{i=0}^{n-T} \pi_i^2 \tag{19}$$

Jika $\left|\lambda_{1,T}\right| > c$ then there is AO on the time t.

c is the critical value that is calculated using the formula presented by Ljung in the SAS / ETS User's Guide.

AO effects may remove of residual by defining:

$$\hat{e}_t = \hat{e}_t - \hat{\omega}_{AT} \hat{\pi}(B) I_t^{(T)} = \hat{e}_t - \hat{\omega}_{AT} \hat{\pi}_{t-T}, t \geq T \tag{20}$$

In other cases, new estimates $\hat{\sigma}_a^2$ calculated from the modified residual $\hat{e}_t$.

If there are any outliers are identified the modified residual $\hat{e}_t$ and modification estimates $\hat{\sigma}_a^2$, but with the same parameters $\hat{\pi}(B) = \dfrac{\hat{\phi}(B)}{\hat{\theta}(B)}$ used to calculate the static of $\hat{\lambda}_{1,T}$. Steps are repeated until all outliers are identified.

Let outlier identification procedure at k point in time, $T_1, T_2, \ldots, T_k$. So the overall outlier correction model of it:

$$Y_t = \sum_{j=1}^{k} \omega_j I_t^{(T_j)} + \frac{\theta(B)}{\phi(B)} a_t \tag{21}$$

The formula above is estimated for the observation sequence Z_t for AO.
And it with a revised set of residuals:

$$\hat{e}_t = \frac{\hat{\phi}(B)}{\hat{\theta}(B)} \left[Y_t - \sum_{j=1}^{k} \hat{\omega}_j I_t^{(T_j)} \right] \tag{22}$$

and obtained from this model fit.

To modify the observation time series containing the effects of outliers can remove and correction of the data done by constructing the generalized estimating equations as follows:

$$\hat{z}_t = Z_t + \sum_{j=1}^{k} \hat{\omega}_j I_t^{(T_j)} \tag{23}$$

4. Simulation

The data used in this simulation is the data obtained from the ARIMA simulation results using the R application. The steps taken in the simulation forecast above is as follows:

4.1. Phase Identification

The first step in the modeling of time series on the identification is checked stationary data by plotting the data as shown below.

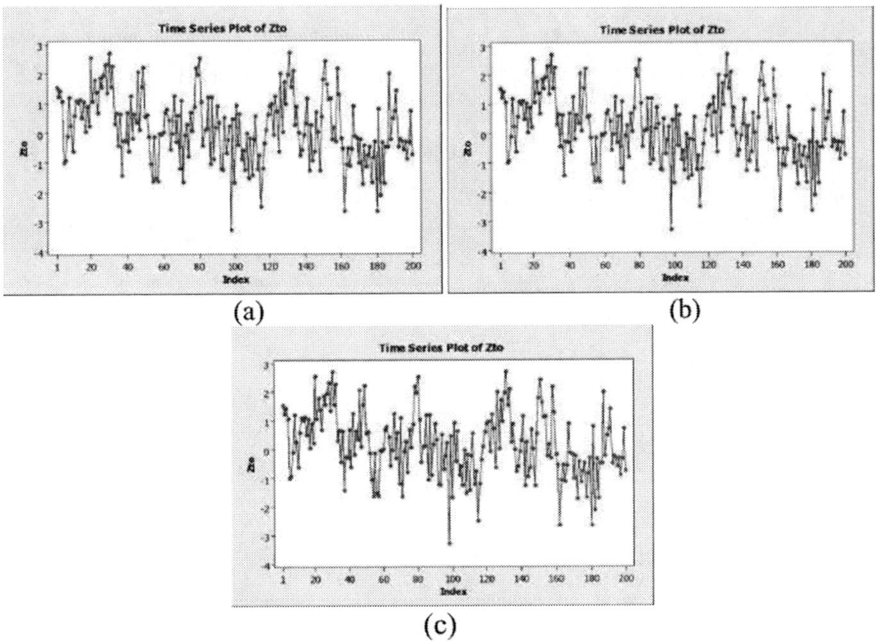

Figure 1. (a) Plot Time Series Data, (b) ACF Plot Time Series Data, (c) PACF Plot Time Series Data

From the picture looks stationary in the variance of the data and then do the ACF and PACF plots her. Based on the data diagram, the value of ACF and PACF, the data has been stationary in the average. Once the data is stationary, then the next step is to identify the model temporarily. Identification of the model done by looking at the values of the coefficient autocorrelation, the ACF and PACF plots of the data that has been stationary. From the form of the ACF and PACF are truncated form after a lag of 2 and dropped exponentially suspected that the appropriate model for these data is ARIMA (2,0,0) or AR(2) with parameters $\phi_1 = 0,2237$ and $\phi_2 = 0,4282$.

4.2. Assessment and diagnostic checks

The early stages of interpreting the results of time series analysis are to look at its significance model parameters that have been modeled. The model is the first approximation obtained AR(2), so we need to test the parameters of Autoregressive (ϕ).

Estimated parameters of the model AR(2) is significantly different from zero with 95% confidence level. This can be seen in the p-values. For parameter AR(1) that ϕ_1, p-value = 0.001 and for AR(2) is ϕ_1, p-value = 0.000. If the calculations are performed with a computer statistical package, then simply use the p-value already known. Criteria conclusion that significance testing if $p < \alpha$ and not significant when $p \geq \alpha$. Based on the results of Minitab for the data obtained p-value = 0.001 < α = 0.05, which means significant testing. After testing its significance parameters, the next step is to test the suitability of the model. To suitability models include the adequacy of the model (test whether the remaining white noise) and test the assumption of a normal distribution. In Minitab results also can be seen that the value of Ljung-Box statistics is displayed on lag 12, lag 24 and lag 36. Ljung-Box statistic value at lag 12, lag 24 and lag 36 consecutive shows p-value = 0.619, 0.520, 0.203, and 0.177 p-value is greater than α = 0.05. This means that the remaining eligible white noise.

Next, test the remaining normal distribution. The results of the test chart normal distribution for the residual shown in figure 2.

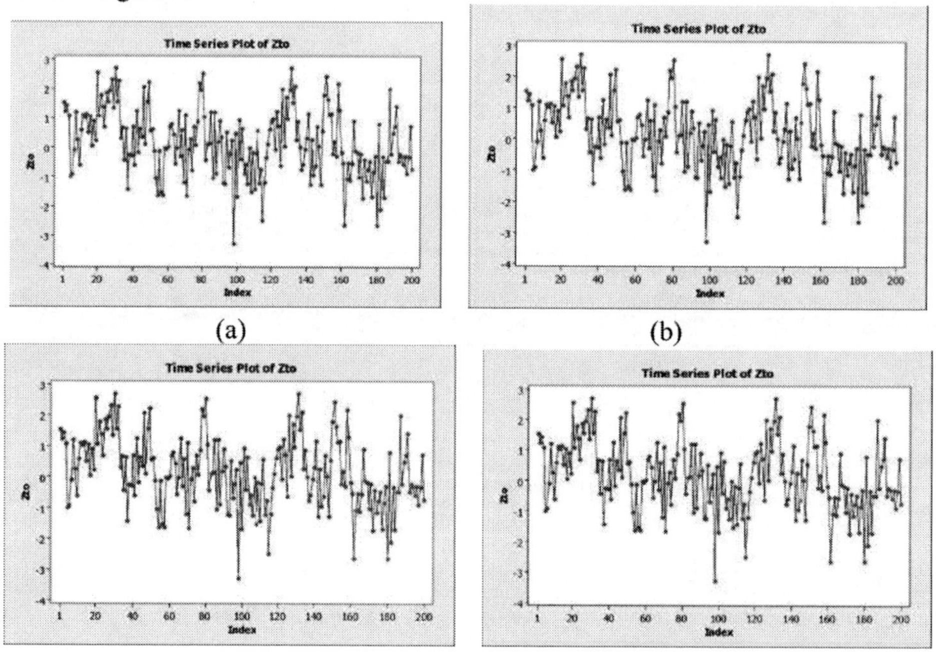

(a) (b)

(c) (d)

IOP Conf. Series: Journal of Physics: Conf. Series **954** (2018) 012010 doi:10.1088/1742-6596/954/1/012010

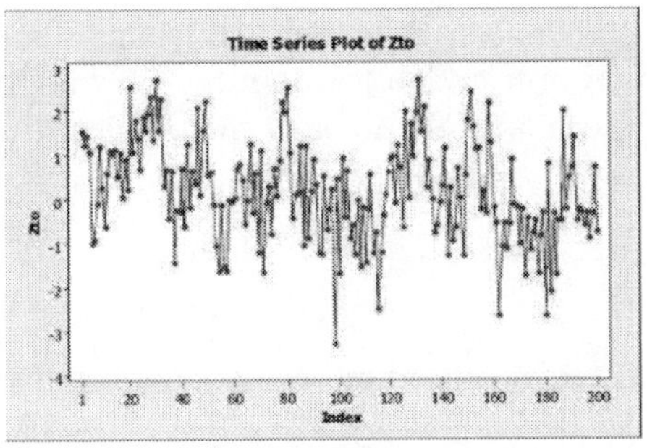

(e)

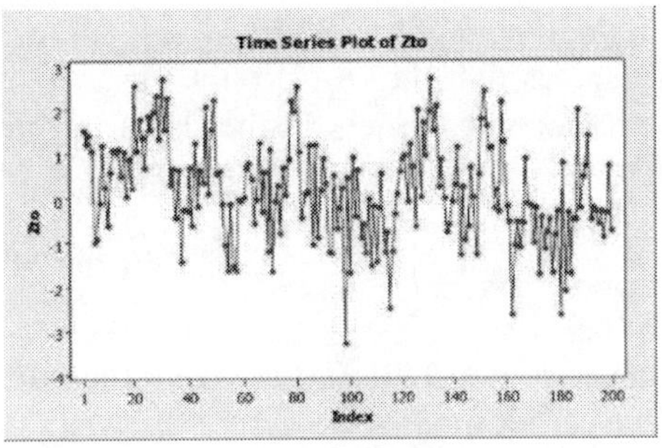

(f)

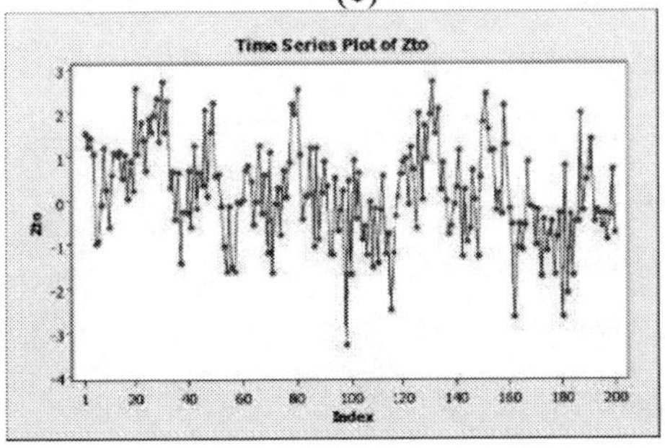

(g)

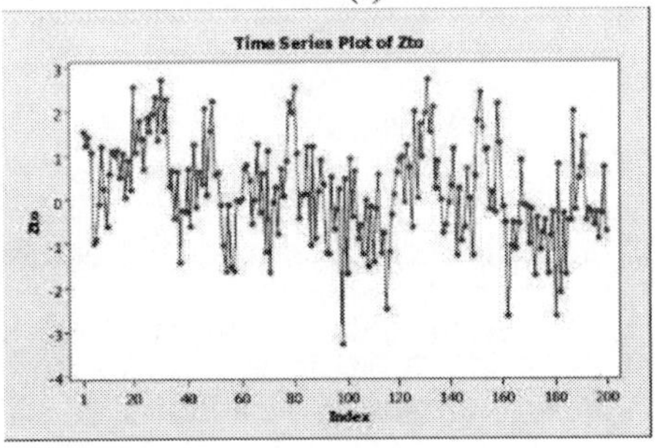

(h)

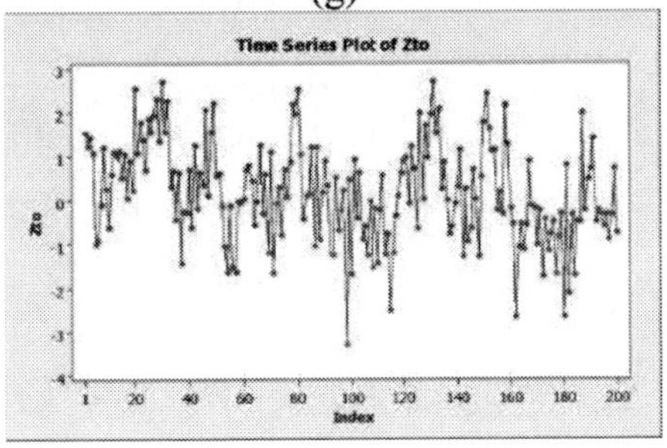

(i)

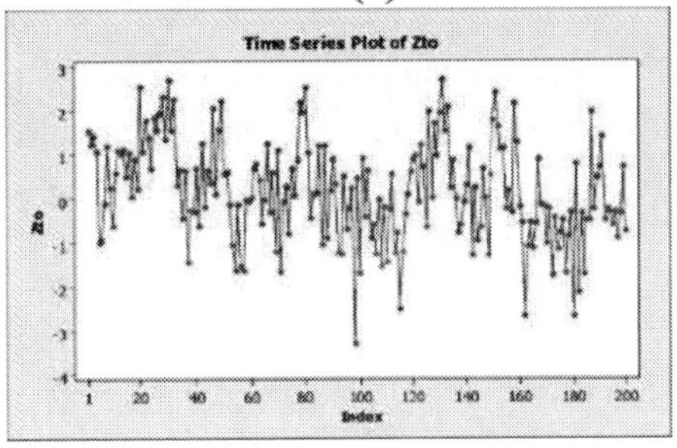

(j)

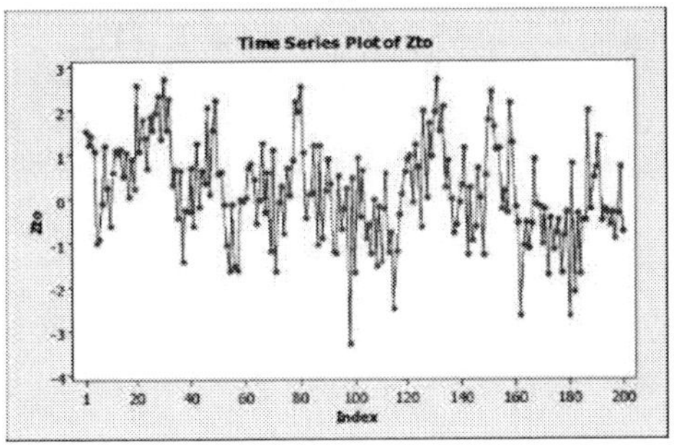

(k)

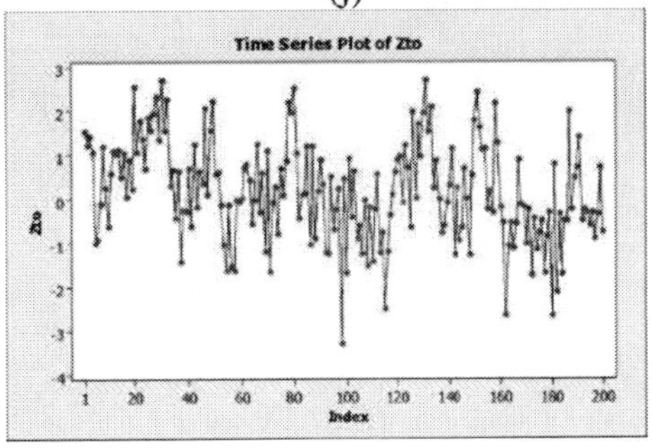

(l)

Figure 2. (a) normal distribution for the remaining test (Kolmogorov-Smirnov test); (b) Plot Residual Model AR(1); (c) Boxplot Data Residual AR(1); (d) Plot Time Series residues Addition Process Detection Outliers 1th; (e) Test the normality of Process Residues Outlier Detection Addition 1th; (f) BoxPlot Residue from Process Enhancements Outlier Detection 1th; (g) Plot Time Series Process Residues of Outlier Detection Addition 2nd; (h) Test the normality of Process Residues the addition of outlier detection 2nd; (i) residues BoxPlot outlier detection process Addition 2nd; (j) Plot Time Series Process Residues of outlier detection Addition 3rd; (k) Testing normality of Process Residues outlier detection Addition 3rd; (l) BoxPlot residues The addition of three outlier detection process.

Based on the Kolmogorov-Smirnov test (Figure 2a) obtained p-value <0.010 which is smaller than α=0.05, this shows that the rest do not meet the normal distribution assumption. This is presumably due to the presence of outliers in the data. From the results of the interim model identification, the model AR(2) written mathematically as follows:

$$Z_t = 0,2237Z_{t-1} + 0,4282Z_{t-2} + a_t \tag{24}$$

with MSE of 0.947.

To detect the presence of outliers or outliers in the data is carried out against the residual plot and boxplot of the earlier models we derive the AR(2) model.

From the figure boxplot above shows that there is an outlier at t = 98; thus obtained:

$$I_t^{(T)} = \begin{cases} 1, t = 98, \\ 0, \text{others} \end{cases} \tag{25}$$

By using the method of the least square regression equation and implementation:

$$Z_t = 0,0600 + 0,209Z_{t-1} + 0,417Z_{t-2} \tag{26}$$

with MSE = 36.780. Then with the addition of outlier detection ARIMA models will be:

$$Z_t = 0,0775 + 0,212Z_{t-1} + 0,411Z_{t-2} - 326X_1(t) \tag{27}$$

with MSE = 28.194. This step is done continuously in order to obtain a smaller MSE, normality tests are met, and there is no longer an image boxplot outlier data. Here are the results of outlier detection iteration as indicated by improved its MSE.

The addition of a second outlier, namely:

$$I_t^{(T)} = \begin{cases} 1, t = 162, \\ 0, \text{ others} \end{cases} \tag{28}$$

with the regression obtainable :

$$Z_t = 0,0922 + 0,206Z_{t-1} + 0,410Z_{t-2} - 327X_1(t) + 114X_2(t) \tag{29}$$

with MSE = 22.808.

The addition of a third outlier, namely:

$$I_t^{(T)} = \begin{cases} 1, t = 180, \\ 0, \text{ others} \end{cases} \tag{30}$$

with the regression obtainable :

$$Z_t = 0,106 + 0,204Z_{t-1} + 0,401Z_{t-2} - 329X_1(t) + 115X_2(t) + 35,9X_3(t) \tag{31}$$

with MSE = 19.365.

With the fulfillment of the residual normality test and its boxplot plot the model fit to the data is ARIMA(2,0,0) with the coefficients obtained from the regression. One of the usefulness of this detection method is that it can reduce the error rate of data prediction. This can be seen in the decline in the value of MSE of each model occurs in the first iteration to the second iteration. The following table of comparisons of each detection and correction and time series models diagram MSE impairment.

Table 1 Comparison of Model Results

MODEL	MSE
Regression ARIMA (2,0,0)	36.780
ARIMA (2,0,0) + outlier detection and correction 1th with value: $\omega_{AT} = 0,010215$. $Z_t = 0,0775 + 0,212Z_{t-1} + 0,411Z_{t-2} - 326X_1(t)$	28.194
ARIMA (2,0,0) + outlier detection and correction 2nd with value: $\omega_{AT} = -0,0227$ $Z_t = 0,0922 + 0,206Z_{t-1} + 0,410Z_{t-2} - 327X_1(t) + 144X_2(t)$	22.808
ARIMA (2,0,0) + outlier detection and correction 3rd with value: $\omega_{AT} = -0,06622$ $Z_t = 0,106 + 0,204Z_{t-1} + 0,401Z_{t-2} - 329X_1(t) + 115X_2(t) + 35,9X_3(t)$	19.365

5. Conclusion

From the discussion and simulation data on a data outlier detection and correction using iterative methods on the ARIMA model (p, d, q) can be obtained several conclusions, as follows:

(1) The process of detection and correction of data containing Additive Outlier on ARIMA (p, d, q) using iteration method and using the software Minitab 16 and Microsoft Excel 2007. The general model for equality Additive Outlier detection ARIMA (p, d, q):

$$Z_t = \sum_{j=1}^{k} \omega_j I_t^{(T_j)} + \pi(B)a_t$$

$$\text{with: } I_t^{(T)} = \begin{cases} 1, t = T \\ 0, t \neq T \end{cases} \quad and \quad \pi(B) = \frac{\theta(B)}{\phi(B)}$$

(2) From the simulation results obtained an early model for the data containing outliers i.e., Model AR(2) by the equation: $Z_t = 0,0600 + 0,209Z_{t-1} + 0,417Z_{t-2}$ by MSE of 36.780. After the process of outlier detection and correction of data obtained using an iterative method is the best model forecasting model AR (2) with a coefficient obtained from regression and end of the equation:

$Z_t = 0,106 + 0,204Z_{t-1} + 0,401Z_{t-2} - 329X_1(t) + 115X_2(t) + 35,9X_3(t)$ by MSE of 19.365.

(3) By using the detection and correction of data containing Additive Outlier on ARIMA (p,d,q) using iterative methods provide process improvements and a decrease MSE models and will certainly provide residual tends to qualify the normal distribution. In this simulation shows that there is improvement MSE value of 47.34% of the initial model.

References

[1] Ahmar, A. S., Rahman, A., Arifin, A. N. M., & Ahmar, A. A. 2017. Predicting movement of stock of "Y" using sutte indicator. *Cogent Eco. Finance,* **5**, 1 doi:10.1080/23322039.2017.1347123

[2] Rahman, A., & Ahmar, A. S. 2017. Forecasting of primary energy consumption data in the united states: A comparison between ARIMA and holter-winters models. Paper presented at the *AIP Conference Proceedings,* **1885**, doi:10.1063/1.5002357

[3] Wei, W. W. S. 2006. *Time Series Analysis : Univariate and Multivariate Methods* (2nd ed.). New York, NY: Pearson Addison Wesley.

[4] Box, G. E. P., & Cox, D. R. 1964. An analysis of transformations. *J. of Royal Stat. Soc. Series B (Method.)*, 211–252.

[5] Box, G. E. P., & Jenkins, G. M. 1976. Time series analysis, control, and forecasting. *San Francisco, CA: Holden Day*, **3226**(3228), 10.

[6] Yule, G. U. 1926. Why do we sometimes get nonsense-correlations between Time-Series: a study in sampling and the nature of time-series. *J. of Royal Stat. Soc.*, **89**(1), 1–63.

[7] Walker, G. 1931. On periodicity in series of related terms. Proceedings of the Royal Society of London. Series A, Containing Papers of a Mathematical and Physical Character, **131**(818), 518–532.

[8] Slutzky, E. 1937. The summation of random causes as the source of cyclic processes. *Econometrica: J. of Eco. Soc.*, 105–146.

[9] Wold, H. 1938. *A Study in the Analisys of Stationery: Time Series*. Almqvist & Wiksells Boktryckeri.

[10] Makridakis, S., Wheelwright, S. C., & Hyndman, R. J. 2008. *Forecasting methods and applications*. John Wiley & Sons.

[11] Wei, W.W.S., 1994 Time series Analysis:Univariate and Multivariate Methods, Addison-Wesley Publising Company, California.

[12] Gounder, M.K., Mahendran, dan Rahmatullah. 2007. Detection of Outlier in Non-linear Time series: A Review. Festschrift in honor of Distinguished Professor Mir Masoom Ali. pages 213-224

IOP Conf. Series: Journal of Physics: Conf. Series **954** (2018) 012011 doi:10.1088/1742-6596/954/1/012011

The Implementation of APIQ Creative Mathematics Game Method in the Subject Matter of Greatest Common Factor and Least Common Multiple in Elementary School

Abdul Rahman[1], Ansari Saleh Ahmar[2,3,4*], A. Nurani M. Arifin[3], Hamzah Upu[1], Usman Mulbar[1], Alimuddin[1], Nurdin Arsyad[1], Ruslan[1], Rusli[1], Djadir[1], Sutamrin[1], Hamda[1], Ilham Minggi[1], Awi[1], Ahmad Zaki[1], Asdar Ahmad[1], & Hisyam Ihsan[1]

[1]Department of Mathematics, Universitas Negeri Makassar, Indonesia
[2]Department of Statistics, Universitas Negeri Makassar, Indonesia
[3]AHMAR Institute, Makassar, Indonesia
[4]Komunitas Kolaborasi Publikasi Indonesia, Indonesia

ansarisaleh@unm.ac.id

Abstract. One of causal factors for uninterested feeling of the students in learning mathematics is a monotonous learning method, like in traditional learning method. One of the ways for motivating students to learn mathematics is by implementing APIQ (Aritmetika Plus Intelegensi Quantum) creative mathematics game method. The purposes of this research are (1) to describe students' responses toward the implementation of APIQ creative mathematics game method on the subject matter of Greatest Common Factor (GCF) and Least Common Multiple (LCM) and (2) to find out whether by implementing this method, the student's learning completeness will improve or not. Based on the results of this research, it is shown that the responses of the students toward the implementation of APIQ creative mathematics game method in the subject matters of GCF and LCM were good. It is seen in the percentage of the responses were between 76-100%. (2) The implementation of APIQ creative mathematics game method on the subject matters of GCF and LCM improved the students' learning.

1. Introduction

Mathematics is one of subjects which are important relating to the development of science and technology. Mathematics has been learned since Elementary School until University. So, it is not wonder if mathematics is considered as the king of sciences. However, that king becomes a specter that makes the students terrified. Just imagine. Students of Elementary School until Senior High School assume that mathematics is a difficult and scary subject. It is worsened by killer teacher. It causes the students not interested in learning mathematics. Education in this era, it is important to support the future. A good future will be obtained if someone have a good education and skills. To reach a higher education it is necessary that the name of learning. Mathematics learning, students need many adaptations before mastering an advance cognitive skill [1][2]. Learning styles Affect the students' learning process so that it can be used as consideration in designing learning [3][4].

This feeling has been felt since they were in elementary school until they were in Senior High School. Due to uninterested feeling has existed since Elementary School is the beginning of the stage of mathematics concept understanding, so if since elementary school this feeling is allowed, it will have negative impact on the student's future. Another factor causing uninterested feeling on the students in learning mathematics is monotonous learning method like on traditional learning method.

The reform movement of the past two decades in mathematics education has consistently emphasized problem solving as a focus of school mathematics (NCTM cited in [5]). Furthermore, Dias said recent research studies also indicates that elementary school students and even prospective elementary teachers have difficulties recognizing multiplicative relations between whole numbers [6].

Game-based learning is one of the ways to improve the interest of elementary school students to learn mathematics. This is in line with Ruseffendi that the benefit of mathematics game is for raising and improve the students' interests to learn mathematics [7]. Game-based and dramatic practices can help student to better learning and the students have a good perception about mathematiccals concept [8]. Hussain, Tan & Idris said that many researchers use digital games as a learning medium that can give positive results on motivation to learn and learning achievement of students [9]. Game-based learning can also be applied in the form of Android media. Development of Android-based learning media has been discussed by Ahmar and Rahman in 2017. The results of this study indicate that 90% of students more quickly understand if using Android-based learning media [10].

By game learning method, this research will focus on the implementation of APIQ (Aritmetika Plus Intelegensi Quantum) creative mathematics game. This method is a learning method which uses Quantum Quotient & Quantum Learning method as well as game-based creative mathematics. Based on the explanation above, the problems in this research are : (1) what are the students' responses on the implementation of APIQ creative mathematics game method in the subject matter of GCF and LCM? (2) Is there any improvement due to the implementation of APIQ creative mathematics game method on the subject matter of GCF and LCM and is there any improvement on the students' learning completeness? The purposes of this research are (1) to describe the students' responses toward the implementation of APIQ creative mathematics game method on the subject matters of GCF and LCM and (2) to find out whether by implementing this method, the students' learning completeness is improved or not [11].

2. Literature Review

According to Offenholley, game based learning is an interactive pedagogy that has as its foundation the tenet that games, by their very nature, increase learning through positive emotional experience [12]. Games do not merely entertain; rather, they can deepen connections and allow for greater learning. While educational video games are gaining currency, old-fashioned board games, card games, etc., can also o er pedagogically rich opportunities for learning mathematics. This article introduces readers to what games can do, then connects the theory to practice, providing examples of (non-computer) games in introductory college mathematics classes and how they work. The article ends with a look at some currently available middle, high school, and college video games. Furthermore, Cody, Rule, & Forsyth said that the students reached the same levels of mathematical achievement by practicing concepts through creating and playing games as with more conventional solving of practice problems [13]. Although students found the topics of the games units more difficult initially (as evidenced by pretest scores) and less interesting (as acknowledged on the attitude survey), they reported much less frustration and confusion along with more ease of learning and more fun during the units in which students practiced with games. These findings indicate that allowing students the time to create and play each other's games is at least as effective as more conventional group practice and seems to have additional benefits of less frustration and greater student clarity of concepts.

3. Method

Methode in this research is experimental research and descriptive research. Descriptive research is a type of research done to find out the value of a variable, in this case is independent variable, either one variable or more without making comparison or relating to other variables [14].

IOP Conf. Series: Journal of Physics: Conf. Series **954** (2018) 012011 doi:10.1088/1742-6596/954/1/012011

The steps done in this learning method are as follows: (1) giving pretest to the students to find out the initial ability of the students about the subject matters of GCF and LCM, (2) conveying learning indicators, (3) motivating the students, (4) delivering materials on GCF and LCM concepts by implementing APIQ creative mathematics game method, (5) giving posttest to the students, (6) helping the students in summarizing the materials, (7) giving the students response questionnaires.

The subjects of this research are 23 students in Grade IV of SDN Galesong II. The period of research implementation was in odd semester of the academic year 2011/2012. The data collecting method is by using test method to obtain the students' learning results and questionnaire method to obtain the students' responses data.

To get the students' responses data, questionnaires are used. It is aimed to find out the students' responses toward learning activity through the implementation of APIQ creative mathematics game method, and this questionnaire is given after test is taken. To find out the percentage, it can be found by applying this formula:

$$R = \frac{n}{N} \times 100\% \quad (1)$$

Description :

R = percentage of the students

n = total number of the students who give responses

N = total number of the students with Likert-scale determination.

The test is done twice, before learning (pretest) and after learning (posttest). This test is used to find out the students' learning results by referring to minimum learning completeness that is 65% and level students' responses can be shown in Table 1.

Table 1 Level Students' Responses

No	Scale	Predicate
1	76 – 100%	Excellent
2	56 – 75%	Good
3	40 – 55%	Fair
4	< 40%	Poor

4. Result and Discussion

According to (Polya, 1973), the process of mathematics problem solving aregenerally divided into four stages, (1) understanding the problem, (2) devising problem solving, (3) carrying out problem solving according to a set plan, and (4) re-examining the solutions that have been obtained. Each stage has an indicator and hierarchy implementation. However, the process of mathematics problem solving can't be separated from two important factors, namely (1) problem characteristics, and (2) cognitive maturity. Therefore, the process of mathematics problem solving posed by Polya haven't to be absolutely followed by every student in sequence.

From the results of this research, it can be found that there are some improvements in the mathematics learning results. Results of statistics relating to the initial value (Pretest) of students in learning by APIQ Creative Mathematics Game Method is presented in Table 2.

From the table 2, we can shows that pretest score of students taught using APIQ Creative Mathematics Game Method, there are 23 sample size by an average score (mean) is the score of 37,2174 of 100 and the median score of 36 with a standard of deviation of 11.78195 and variance 138,814. The minimum score obtained is 16 and the maximum score obtained is 72. If the pretest math scores of students in the experimental class is divided into five categories, the obtained frequency distribution of scores and percentages as shown in Table 3 below.

Tabel 2 Description of Pretest Score of Students Taught by APIQ Creative Mathematics Game Method

Statistic	Value
Sample Size	23
Mean	37,2174
Median	36
Mode	32
Std. Deviation	11,78195
Variance	138,814
Minimum	16
Maximum	72

Tabel 3 Distribution and Percentage of Pretest Score of Students Taught by APIQ Creative Mathematics Game Method

No.	Category	F	Persentage (%)
1	Very Good	0	0.00
2	Good	1	4.35
3	Pair	1	4.35
4	Poor	6	30,43
5	Very Poor	12	60,87
	Jumlah	**30**	**100%**

Percentage of students' mathematics scores of pretest the experimental class can further be observed in the image histogram below (figure 1).

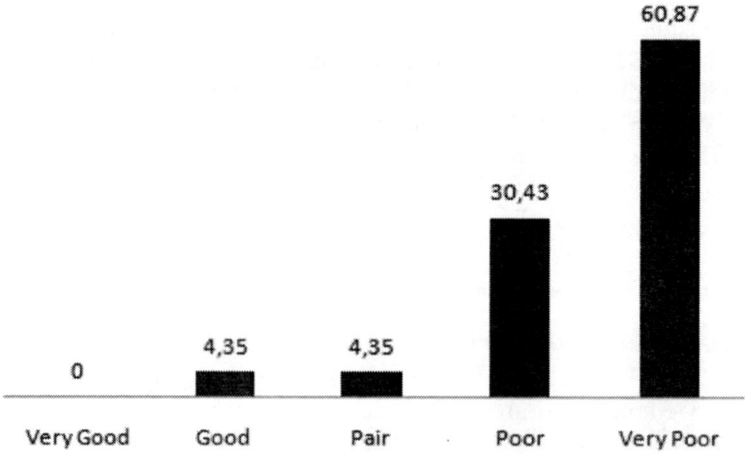

Figure 1 Scores of Pretest the Experimental Class

Based on table 3 and figure 1 above shows that students who very poor as many as 14 students (60,87%), poor category as many as 7 students (30,43%), pair and good category as many as 1 students (4,35 %). Results of statistics relating to the posttest of students in learning by APIQ Creative Mathematics Game Method is presented in Table 4 below.

Tabel 4 Description of Posttest Score of Students Taught by APIQ Creative Mathematics Game Method

Statistic	Value
Sample Size	23
Mean	80
Median	80
Mode	80
Std. Deviation	11,87817
Variance	141,091
Minimum	56
Maximum	100

Table 4 above shows that posttest score of students taught using APIQ Creative Mathematics Game Method, there are 23 sample size by an average score (mean), median, and mode is the score of 80 of 100 with a standard of deviation of 11.87817 and variance 141,091. The minimum score obtained is 16 and the maximum score obtained is 72. Percentage of students' mathematics scores of posttest the experimental class can further be observed in the image histogram below (figure 2).

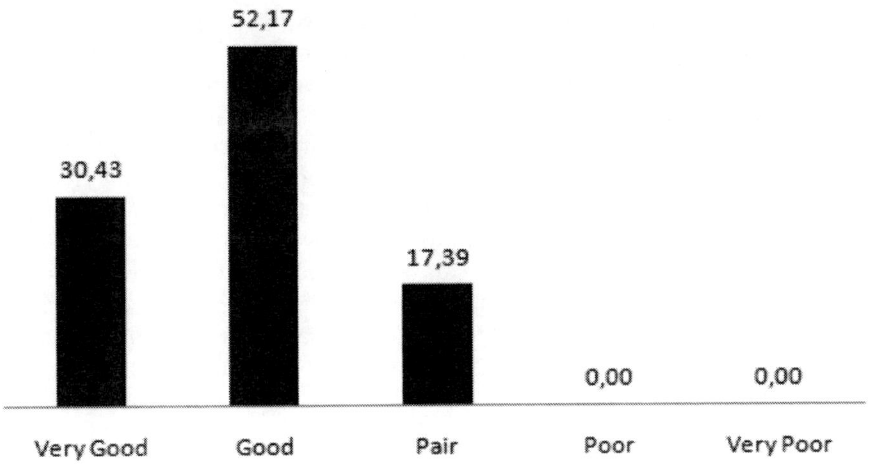

Figure 2 Scores of Posttest the Experimental Class

Based on table 4 and figure 2 above shows that students who pair as many as 4 students (17,39%), good category as many as 12 students (52,17%), and very good category as many as 7 students (30,43 %).

Table 5 Paired Samples Statistics

		Mean	N	Std. Deviation	Std. Error Mean
Pair 1	Pretest	37,2174	23	11,78195	2,45671
	Posttes	80,0000	23	11,87817	2,47677

Table 6 Paired Samples Correlations

		N	Correlation	Sig.
Pair 1	Pretest & Posttest	23	0,530	0,009

Table 7 Paired Samples Test

		Paired Differences					t	df	Sig. (2-tailed)
		Mean	Std. Deviation	Std. Error Mean	95% Confidence Interval of the Difference				
					Lower	Upper			
Pair 1	Pretest- Posttest	-42,78261	11,46915	2,39148	-47,74224	-37,82297	-17,890	22	<0,01

Based on table 5, we can see Mean, Standard of Deviation, and Standard of Error Mean from Pretest Group lower to Posttest. It shown that statistically statistics, there are enhancement form pretest to posttest. Based on table 6, there is a positive relationship between the before and after learning. This is indicated by the correlation value is 0.56. This means that learning to use APIQ method can improve students' learning ability. Value of Sig. (2-tailed) of Paired Samples Test showed results = 0.000 <0.05 (confidence interval: 95%), this means that there is difference between before and after treatment or in other words there is difference before and after the learning process by using APIQ. The difference can we see the value Mean: -42.78261 (Negative), which means there is increasing trend after the study results do use methods APIQ with the average increase is 42.78261.
The students' responses toward the implementation of APIQ creative mathematics game method in the subject matters of GCF and LCM are also found. The students who gave responses consist of (1)18 students (78.26%) who stated that they liked this kind of learning activity, (2) 19 students (82.60%) who liked learning by using APIQ creative mathematics method, (3) 22 students (95.65%) liked if this method is always used in learning. From this calculation and the data of the students' responses, it can be seen that the application of this method was good (with the scale of 76-100%). Therefore, according to Cody, Rule, & Forsyth (2015), it can be concluded that this method is good to improve the students' interests on mathematics subject. Therefore, the student may perform at higher levels with the support of the game, allowing the student to practice and gain more mathematical proficiency. The metacognitive aspects of choosing problems for games and the motivational impact of choosing a game theme that includes interesting characters or ideas related to favorite leisure activities will have a positive effect on learning.

5. Conclusion

Based on the findings of this research and the discussion, it can be concluded that: (1) The students' responses toward the application of APIQ creative mathematics game method in the subject matters of GCF and LCM were good. The percentages of the students' responses were between 76-100%. (2) The application of APIQ creative mathematics game method in the subject matters of GCF and LCM improved the students' learning. There is increasing trend after the study results do use methods APIQ with the average increase is 42,78261.

References

[1] Rahman, A., & Ahmar, A. S. 2016. Exploration of Mathematics Problem Solving Process Based on The Thinking Level of Students in Junior High Schoo. *Int. J. Environ. Sci. Educ.*, **11**, 14.
[2] Mulbar, U., Rahman, A., & Ahmar, A. S. 2017. Analysis of the ability in mathematical problem-solving based on SOLO taxonomy and cognitive style. *World Trans. Eng. Technol.*

Educ., **15**, 1, pp. 68-73. doi:10.26858/wtetev15i1y2017p6873

[3] Rahman, A., Ahmar, A. S., & Rusli. 2016. The Influence of Cooperative Learning Models on Learning Outcomes Based on Students' Learning Styles. *World Trans. Eng. Technol. Educ.*, **14**, 3.

[4] Rahman, A., & Ahmar, A. S. 2017. Relationship between learning styles and learning achievement in mathematics based on genders. *World Trans. Eng. Technol. Educ.,* **15**, 1, 74-77. doi:10.26858/wtetev15i1y2017p7477

[5] Dugdale, S., O. LeGare, J.I. Matthews, M. Ju. 1998. Mathematical Problem Solving and Computers: A Study of Learner-Initiated Application of Technology in a General Problem-Solving Context. *J. of Research on Comp. Educ.*, **30**, 3, 239–253.

[6] Dias, A. 2005. Using lattice models to determine Greatest Common Factor and Least Common Multiple. *Int. J. Math. Teach. Learn*. pp. 730–738.

[7] Ruseffendi, E. T. 1998. *Pengantar kepada Membantu Guru Mengembangkan Kompetensinya dalam Pengajaran Matematika untuk Meningkatkan CBSA* [Introduction to Helping Teachers Develop Competence in Teaching Mathematics to Improve CBSA]. Bandung, Indonesia : Tarsito.

[8] Masoum, E., Rostamy-Malkhalifeh, M., & Kalantarnia, Z. 2013. A Study on the Role of Drama in Learning Mathematics. *Math. Educ. Trends Reseach*, pp. 1-7.

[9] Hussain, S.Y.S., Tan, W.H. & Idris, M.Z. 2014. Digital Game-Based Learning for Remedial Mathematics Students: A New Teaching and Learning Approach in Malaysia. *Int. J. of Multimedia Ubiquitous Eng.*, **9**, 11, pp. 325-338.

[10] Ahmar, A. S., & Rahman, A. 2017. Development of teaching material using an Android. *Glob. J. Eng. Educ.*, **19**, 1, pp. 73-76. doi:10.26858/gjeev19i1y2017p7376

[11] Nggermanto, A. 2010. APIQ Creative Math Game : Permainan Matematika Kreatif APIQ, Petualangan Permainan dan Pembelajaran Matematika Kreatif yang Asyik dan Cerdik untuk Anak Usia Dini sampai Siswa Sekolah Menengah Atas [APIQ Creative Math Game: APIQ Creative Math Games, Adventure Games and Learning Mathematics Creative Fun and Ingenious for Early Childhood through High School Students]. Bandung, Indonesia : Nuansa.

[12] Offenholley, K. H. 2012. Gaming Your Mathematics Course: The Theory and Practice of Games for Learning. *J. of Humanistic Math.*, **2**, 2, pp. 79-92. DOI: 10.5642/jhummath.201202.07.

[13] Cody, K. J., Rule, A. C., & Forsyth, B. R. 2015. Mathematical Game Creation and Play Assists Students in Practicing Newly-Learned Challenging Concepts. *Creative Education*, **6**, pp. 1484-1495. DOI: 10.4236/ce.2015.614149

[14] Hasan, I. 2010. *Analisis Data Penelitian dengan Statistik* [Research Data Analysis with Statistics]. Jakarta, Indonesia : Bumi Aksara.

IOP Conf. Series: Journal of Physics: Conf. Series **954** (2018) 012012 doi:10.1088/1742-6596/954/1/012012

Effect of Powder Leaf Breadfruit Disposals (*Arthocarpus Altilis*) in Oil Mandar District and Polman Against Cholesterol and Glucose Mice (*Mus Musculus*)

A. Mu'nisa*, A. Asmawati , Farida A., Fressy FA., and Erni

Department of Biology, Faculty of Mathematics and Natural Sciences, Jl. Daeng Tata Raya Kampus Parangtambung, Universitas Negeri Makassar, Indonesia

*andi.munisa@unm.ac.id

Abstract. The purpose of this study was to determine the effect of powdered leaves of breadfruit (*Arthocarpus altilis*) on oil is mandated origin of the Polman glucose and cholesterol levels in mice (*Mus musculus*). This study comprised 4 treatments and each treatment consisted of 5 replicates, ie groups of mice were fed a standard (negative control); 2 groups: group of mice fed with standard and cholesterol feed (positive control); Group 3 that mice fed with standard and Selayar oil; and group 4: group of mice fed with standard and Mandar oil that has been given powdered leaves of breadfruit. Measurement of glucose and blood cholesterol levels in mice done 3 times ie 2 weeks after the adaptation period (phase 1), 2 weeks after administration of the oil (phase 2) and 2 weeks after feeding cholesterol (stage 3). Based on the analysis of data both cholesterol and glucose levels showed that in a group of 4 decreased glucose and cholesterol levels in stage 2 but at stage 3 an increase in the group of mice given only the oil while in the group of mice given the oil and powdered leaves of breadfruit indicate glucose levels and normal cholesterol. The conclusion of this study show that the addition of powdered leaves of breadfruit into cooking oil Mandar influential in glucose levels and normalize blood cholesterol levels in mice.

1. Introduction

Utilization of existing plants in the areas we need to be cultivated. Plants that live around us has many benefits. These plants need to be cultivated one plant that is very useful is the breadfruit plants (Arthocarpus altilis). Breadfruit leaves (Arthocarpus altilis) can be used for the treatment of various diseases, such as liver disease, hepatitis, enlarged spleen, heart, kidney, high blood pressure and diabetes, using breadfruit leaves are rich in compounds of flavonoids to prevent the increase in blood cholesterol levels , It also will help the community in the field of economic empowerment in particular.

Plant in Indonesia received by the public is the breadfruit plants (Arthocarpus altilis). Arthtocarpus altiltis can be utilized for human life. Breadfruit can also be used on the leaves for treatment of disease. Previous research has conducted research on the leaves of breadfruit and discovered that the leaves of breadfruit effective for treating diseases such as liver, hepatitis, enlarged spleen, heart, kidney, high blood pressure and diabetes, because it contains flavonoids [1].

The ability of breadfruit leaves in treating some chronic diseases is because the compounds contained therein. Breadfruit leaves contain several compounds that are beneficial for the body such as polyphenols, hidrosionat acid, tannin, quercetin and artoindosionin. Ortoindonesionin compounds and quercetin is a group of flavonoids derived compounds that function as antioxidants and are widely used as an active component in medicines.

Breadfruit leaves are solid green used in this study because the leaves are dark green more compounds possessed compared with light green leaves. Flavonoids are one of secondary metabolites, the possibility of its existence in the leaves is affected by the process of photosynthesis so that not too many young leaves contain flavonoids. Flavonoids are phenolic compounds contained in plant pigment, red, yellow, blue and purple.

Breadfruit leaves effectively treat diseases such as liver, hepatitis, enlarged spleen, heart, kidney, high blood pressure, diabetes and also to cure skin swelling or itching. Substances contained in its leaves could also able to overcome the inflammation in the liver, spleen enlargement, coroner of the heart, inflammation of the kidneys, high blood pressure, and diabetes [2][3][4].

Oils are triglycerides composed of fatty acids are liquid at room temperature (25oC) and mengadung more unsaturated fatty acids so susceptible to oxidation. Oil in the solid form can be called fat [5][6].

Mandar excellence origin used cooking oil because it has a distinctive aroma, is still the traditional way of making which is still maintained today, but it is also used as a cooking oil because of its non-polar solvents, as well as materials used for massage. Cooking oil is an ester that is easily oxidized resulting in the formation of free fatty acids. Damage to the quality of cooking oil can also be affected by improper storage and packaging that is not good. When oil kept too long in place where the temperature above room temperature, the oil will be susceptible to oxidation, because any increase in temperature of 15oC rate of oxidation is doubled.

Mandar origin manufacture cooking oil is treated with a technique that is still very simple, so that the level of damage during storage becomes very high. Extent of the damage can be in the form of discoloration and rancid when stored over 30 days. Damage to the oil can trigger the formation of free radicals that can cause degenerative diseases one of which is hypercholesterolemia. Treatment can be given is the addition of natural antioxidants that can prevent the oxidation of the oil.

Some research indicates that, the process of cooking oil rancidity can be inhibited by administration of antioxidants, and antioxidants both synthetic and natural antioxidants. Synthetic antioxidants are often used to prevent rancidity of oil is butylated hydroxytoluene (BHT) and butylated hydroxyanisole (BHA), but the use of excess will cause poisoning. Natural Antioxidants are chemical compounds derived from plants. This substance is able to significantly slow down or inhibit the oxidation of a substance that is easily oxidized even in low concentrations. This natural antioxidant has many double bonds are easily oxidized so it will protect fats from oxidation [7]. This study will use breadfruit leaf powder is added to the oil traditionally processed, then tested on animals who have hypercholesterolemia.

2. Method

2.1. Equipment and Materials Research

The tools used are glass tools in the form of beaker size of 250 mL and 1000 mL Erlenmeyer flasks, beakers, stir bar, basin, rang wire, eating and drinking equipment, scissors, oven, and gauges glucose and cholesterol in the form of NESCO multi check.

The main material used in this study was 70 mL Mandar oil made from coconut hybrid types, 0.9 grams of breadfruit leaves, filter paper, feed cholesterol egg yolk powder form as much as 210 grams which had been in the oven, aluminum foil, and a syringe, feed AD 1 standard form which can be from the traditional markets.

2.2. Breadfruit Leaf Powder Process

Breadfruit leaves used are old leaves that are in the growth phase stagnam. Breadfruit leaves are washed thoroughly with water and dried with aerated at room temperature 27 ° C - 28 ° C for 48

IOP Conf. Series: Journal of Physics: Conf. Series **954** (2018) 012012 doi:10.1088/1742-6596/954/1/012012

hours. Once dried breadfruit leaf, then 400 rpm in a blender until smooth and sieved using a sieve size of 60 mess. Breadfruit leaf powder put into a transparent plastic packaging to avoid contaminants and stored in a sealed container.

2.3. Breadfruit Leaf Powder Addition process into oil Mandar

A total of 70 mL Mandar oil heated in water bath at a temperature of 50 ° C - 65 ° C for ± 30 minutes. Breadfruit leaf powder as much as 0.9 grams mixed into the oil Mandar. Cooking oil as suspense for eugenol and phenol compounds contained in the powder leaves of breadfruit. Mandar oil in the filter to separate the pulp powder breadfruit leaves.

2.4. Provision of Treatment

Test animals used were mice (Mus musculus) Sprague Dawley sex male, healthy, and normal activities are kept in the Green House Biological Science UNM. Mice were 2 months old as much as 20 fish with a weight of 25 ± 5 grams. Mice were divided into 4 groups and each treatment group consisted of 5 mice in one cage.

Mice were adapted for 2 weeks with fed commercial feed in the form of standard flour secar AD1 and drinking water ad libitum fed before treatment (feed cholesterol egg yolk powder form) so that the way of life and food becomes uniform. In the third week, mice were grouped according to their group. Groups of mice were determined based on body weight were divided into 4 groups. The treatments as follows: a) Group I (negative control) is a group of male mice hanyadiberi standard feed with a dose of 1.4 grams / head / day during the trial period, b) Group II (positive control) is a group of male mice fed a standard doses 1.4 gram / head / day plus egg yolk powder as much as 1 gram / head / day, c) group III, the group of male mice fed a standard of 1.4 grams / day / head and oiled without powder and Mandar Mandar during 2 weeks at a dose of 1 ml / head / day. After 2 weeks, the mice were fed cholesterol at a dose of 1 gram / head / day for 2 weeks, and d) Group IV, which is a group of male mice fed with a standard dose of 1.4 grams / head / day and oiled Mandar plus powdered leaves of breadfruit 0.3% at a dose of 1 ml / head / day for 2 weeks. After 2 weeks fed cholesterol at a dose of 1 gram / head / day.

2.5. Period Measurement of Blood Glucose and Cholesterol

Each 14-day measurement of glucose levels. Measurement of glucose and cholesterol is carried out after a period of adaptation as glucose and cholesterol levels early and giving treatment period (after mice were given oil) and after being given a feed of cholesterol in the form of feed cholesterol.

3. Result and Discussion

3.1. Blood Glucose

The average glucose level of male mice (*Mus musculus*) after checking glucose levels stage I or prior to treatment (after a period of adaptation) and phase II (after the cooking oil) and phase III (after feeding in the form of feed cholesterol) can be seen in table 4.1 as follows:

Table 1. Average glucose levels of male mice (*Mus musculus*) before and after treatment

Treatment	The average glucose level of male mice (*Mus musculus*) mg/dL		
	Stage I	**Stage II**	**Stage III**
Negative control	74.20^{a}	81.40^{a}	97,20^{a}
Hypercholesterolemia control	99.20^{a}	97.00^{a}	104,00^{ab}
Mandar oil	90.60^{a}	119.00^{b}	123,80^{b}

Treatment	The average glucose level of male mice (*Mus musculus*) mg/dL		
	Stage I	**Stage II**	**Stage III**
Mandar oil+ breadfruit leaf powder	97.20[a]	95.00[a]	108,20[ab]

Description: The letters in the same column indicates "not significant". Different letters in one column shows "Significantly different".

In Table 1, the results of the analysis of the average glucose levels (mg / dL) male mice (*Mus musculus*) shows the average glucose levels are still normal. Gukosa levels normal limit of 50 mg / dL - 135 mg / dL. On average glucose levels between Phase I, Phase II, and Phase III, did not rise drastically despite using cooking oil Mandar origin.

3.2. Cholesterol

Data from measurements of cholesterol levels in mice (*Mus musculus*) starting from measurement before treatment (initial cholesterol levels in mice) (phase I), after administration of the oil (phase II), until after feeding cholesterol (phase III). Average cholesterol levels can be seen in Table 2.

Table 2. Average Value Cholesterol (mg / dL) of mice (*Mus musculus*) before and after treatment

Treatment	The average cholesterol level of male mice (*Mus musculus*) mg/dL		
	Stage I	**Stage II**	**Stage III**
Negative control	163,20[a]	139,40[a]	161,00[ab]
Hypercholesterolemia control	142,40[a]	121,80[a]	194,50[b]
Mandar oil	133,60[a]	146,25[a]	157,25[ab]
Mandar oil+ breadfruit leaf powder	132,60[a]	119,80[a]	150,00[a]

Description: The letters in the same column indicates "not significant". Different letters in one column shows "Significantly different".

Effect of Oil origin Mandar added powdered leaves of breadfruit (Arthocarpus altilis) to glucose and cholesterol levels in mice (Mus musculus) consists of 4 groups and each treatment group consisted of 5 replicates. Group I (negative control), the male mice (Mus musculus) were given only standard feed. Group II (positive control), the male mice (Mus musculus) were fed a standard feed supplemented with cholesterol. Group III, the male mice (Mus musculus) were given a standard feed and cooking oil given origin Mandar (1 ml / day) orally without being given powdered leaves of breadfruit (Arthocarpus altilis) for 2 weeks. Group IV, male mice (Mus musculus) are given oil mandated that has been added powdered leaves of breadfruit (Arthocarpus altilis), after 2 weeks later the male mice (Mus musculus) given the standard feed + feeding cholesterol (90 grams of feed standard and 10 grams of feed cholesterol) for 2 weeks followed by glucose and cholesterol.

The average glucose levels in tahap1, phase II and phase III, the provision of cooking oil Mandar mice to give effect to an increase in blood glucose levels. In phase I indicates the average glucose level of 90.60 mg / dL, the second phase increased to 199.00 mg / dL, and the third phase, the average glucose levels increased by 123.80 mg / dL.

Average cholesterol levels were obtained after checking cholesterol levels showed that the cooking oil origin Mandar to test animals male mice (Mus musculus) does not raise blood cholesterol levels in test animals male mice (Mus musculus) because the data obtained is still within the normal range of 40 -130 mg / dL. Likewise with the addition of powdered leaves of breadfruit (Arthocarpus altilis) into the origin Mandar cooking oil to neutralize the increase koletsrol levels in male mice (Mus musculus). Test results average cholesterol levels of male mice (Mus musculus) for 48 days with 3 times the inspection period. The average value obtained cholesterol levels showed a decrease, the use of powdered leaves of breadfruit (Arthocarpus altilis) can prevent the increase in cholesterol levels in male mice (Mus musculus).

Ketaren (1986), states that the oil is a triglyceride composed of three units of fatty acids, are liquid at room temperature (25 ° C) and contain more unsaturated fatty acids so susceptible to oxidation.

Breadfruit plants (Arthocarpus altilis) is one of the plants that can be used as a traditional medicine because it has antioxidants that are beneficial to health. The leaves of breadfruit plants are very effective for treating diseases such as inflammation of the liver (hepatitis), spleen enlargement, coronary heart disease, inflammation of the kidneys, high blood pressure, and diabetes, because it contains flavonoids [3].

The use of cooking oil as a solvent Mandar origin flavonoid compound in powdered leaves of breadfruit because it is non-polar. By using breadfruit leaves are rich in flavonoids compounds that can prevent the increase in blood glucose and cholesterol levels.

Mechanism of action of flavonoids as natural antioxidants in the hamper and slow down the oxidation process that causes the formation of free radicals is to supplement a shortage of electron free radicals that can cause oxidative stress. Flavonoid compounds work effectively because it can neutralize the fatty acid radicals and oxygen radicals. Flavonoids inhibit the formation of reactive metabolites that can bind to macromolecules network so as to prevent tissue damage. When flavonoids react with free radicals, flavonoids donate a proton and become radicals. But unpaired electrons generated didelokaslisasi by resonance, it makes radical flavonoid compounds have very low energy to be reactive radicals [8].

One source of the increased levels of cholesterol in the blood is the consumption of foods containing cholesterol and saturated fat. Sources of cholesterol comes from animal products, such as meat, spleen, brain, kidney, egg yolk and shrimp. Egg yolks contain 220-250 mg of cholesterol so that the feed containing 2.02 grams of egg yolk as already can raise cholesterol levels [9].

Excess cholesterol is a dreaded disease, which interferes with heart health. Actually, we only need a small amount of cholesterol is to create and maintain nerve cells and to synthesize the hormones in the body. If the excessive blood cholesterol levels, so most of the cholesterol it will settle. This allows for the classification or calcification, causing an increased risk of blood pressure. This situation is harmful especially when it comes to cause rupture of blood vessels. If the vessels are ruptured blood vessel in the brain that can lead to paralysis. If calcification occurs in the blood vessels going to the heart of this vital organ blood supply, so that its strength is reduced, if the flow of blood until a hiccup, cardiac infarction will occur which create irregular heartbeat or not at all strong [9].

The process of fat metabolism in the body begins with ingestion of food ingredient components in the intestine by the enzyme. Fatty acids are fused back with glycerol to form fat which is then transported by the lymph vessels. Furthermore, fat is stored in adipose (fat tissue), if needed fat to be transported to the liver in the form of lecithin is hydrolyzed by lipase into fatty acids and glycerol, glycerol is activated by ATP to glycerol phosphate and eventually undergoes oxidation. The carbon chains of fatty acids within mitochondria treated so produced acetyl coenzyme which can then be entered into the Krebs cycle. The increase in free fatty acids also increase the distribution of fatty acids in the liver. It can improve the process of gluconeogenesis, inhibits the uptake and use of glucose in the muscles. The accumulation of triglycerides in the liver and in the muscles will lead to insulin resistance. Besides the fat tissue it produces several cytokines and hormones that inhibit the action of insulin. Hormone Insulin is an important regulator in the metabolism of carbohydrates, lipids, and proteins, then any distraction in insulin action will cause degenerative diseases, including diabetes mellitus.

4. Conclusion

The conclusion of this study show that the addition of powdered leaves of breadfruit into cooking oil Mandar influential in gluokosa levels and normalize blood cholesterol levels in mice.

References

[1] Ramadhani, A. N. 2009. Uji Toksisitas Akut Ekstrak Etanol Daun Sukun (*Artrocarpus communis*) Terhadap Larva *Artemia salina* Leachdengan Metode BSLT. Skripsi. Fakultas Kedoketran Universitas Diponegoro. Semarang.

[2] Nurgraheni, S. 2000. Pengaruh Penambahan Antioksidan Terhadap Stabilitas Minyak Goreng Curah Selama Pemanasan dan Penyimpanan. Fakultas Teknologi Pertanian Institut Pertanian Bogor. Bogor.

[3] Stryer, L. 2000. *Biokimia*. EGC. Jakarta

[4] Baiyeri, K.P. and Mbah, B.N., 2006. Effects of soilless and soil-based nursery media on seedling emergence, growth and response to water stress of African breadfruit (Treculia africana Decne). *African Journal of Biotechnology*, *5*(15).

[5] Ketaren, S., (1986). *Pengantar teknologi minyak dan lemak Pangan*. Penerbit Universitas Indonesia, Jakarta.

[6] Aguilera, J.M., Michel, M. and Mayor, G., 2004. Fat migration in chocolate: diffusion or capillary flow in a particulate solid?—a hypothesis paper. *Journal of Food Science*, *69*(7), pp.167-174.

[7] Mu'nisa A., Wasmen M., Wresdiyati T., da Nastiti K., 2008. *Perbaikan aktivitas Anti Oksidan Pada Jaringan Kelinci hiperkolesterolemia Dengan Pemberian Ekstrak Daun Cengkeh.* Jurnal Veteriner (9): 182-187.

[8] Astuti, Sussi. 2008. Isoflavon Kedelai dan Potensinya Sebagai Penangkap Radikal Bebas. Jurnal Teknologi Industri dan Hasil Pertanian. Lampung.

[9] Suparman, S. 2009. *Pengaruh Pemberian Infus Daun Seledri (Apium graviolens L) Terhadap Kadar Kolesterol Serum Darah Marmut (Cavia cobaya)*. Medan: Universitas Sumatera Utara

Crime Modeling using Spatial Regression Approach

Ansari Saleh Ahmar[1,2*], Adiatma[1], M. Kasim Aidid[1]

[1]Department of Statistics, Universiats Negeri Makassar, Indonesia
[2]AHMAR Institute, Makassar, Indonesia

*ansarisaleh@unm.ac.id

Abstract. Act of criminality in Indonesia increased both variety and quantity every year. As murder, rape, assault, vandalism, theft, fraud, fencing, and other cases that make people feel unsafe. Risk of society exposed to crime is the number of reported cases in the police institution. The higher of the number of reporter to the police institution then the number of crime in the region is increasing. In this research, modeling criminality in South Sulawesi, Indonesia with the dependent variable used is the society exposed to the risk of crime. Modelling done by area approach is the using Spatial Autoregressive (SAR) and Spatial Error Model (SEM) methods. The independent variable used is the population density, the number of poor population, GDP per capita, unemployment and the human development index (HDI). Based on the analysis using spatial regression can be shown that there are no dependencies spatial both lag or errors in South Sulawesi.

1. Introduction

Criminology is a knowledge of crime or criminality that can be viewed from the perspective of biological, sociological, and others. This knowledge also provides two meanings for the term of crime or criminality, namely juridical and sociological. According to Bonger legally, crimes means the act of anti-social consciously gets a reaction from the state for the provision of the pain and then as a reaction to the legal definitions of crimes [1]. Sociologically, the crime is a human behavior that is created by the community.

Criminal acts in Indonesia increased both variety and quantity every year. As murder, rape, assault, vandalism, theft, fraud, fencing, and other cases that make people feel unsafe. According Abdulsyani that crime cases that happened was caused by various factors, both internal and external [2]. For example the level of education obtained, the salary or wages that are not sufficient, and family relations. These factors are found in juvenile detention centers in major countries such as Australia [3] and United States [4]. Additionally, Kakamu, Polasek, & Wago in their study in Japan also noted that the level of crime cases that occur are affected by unemployment, income, and level of arrest [5]. According to the research done before, it can be said that the dominant factor that influence the criminal case is an economic phenomenon.

One way that is used in defining the relationship between economic phenomena regression analysis. To determine the economic impact, especially on the level of crime cases using regression analysis, spatial linkages between regions Noteworthy because it is associated with a location / region. Spatial linkages between regions it tends often occurs due to the location adjacent to each other. To determine the effect of the relationship between the location of the existing response, then each region can not be regarded as an observation unit that stands alone and is not associated with the location area more so that the necessary spatial analysis.

Research using regression analysis with spatial approach had been done already. Kusrini, Suhartono, & Sari using regression approach spatially to model criminal cases in the city of Surabaya, the results of the study can be concluded that the measurement of the risk of a criminal act did not indicate any dependencies spatial variables significant effect is the population of junior high school education, which has negative and the level of per capita income of the population negatively [6]. Septiana & Wulandari using regression method spatially to model the dropout age from high school in East Java province, on the results of these studies it is concluded that there are dependencies in lag and error, where the model of SAR variables that have a significant effect is poor, the negative effect and the location of the house in the countryside a positive effect [7]. While SEM models, the variables are significant and positive impact is the location of the house in the countryside.

Research modeling criminality in South Sulawesi with spatial regression approach has not been done. Therefore, researchers wanted to do modeling criminality in the province of South Sulawesi with spatial regression approach that can be known or not existing spatial dependencies criminality in the province of South Sulawesi.

2. Literature

2.1. Spatial Regression Model

According to the Tobler's first law [8] that *"Everything is related to everything else, but near thing are more related than distant things"*. The law that became the reference of the regional science studies. Spatial effect is usually found on a region to another, meaning that the observations in a region depends on the observation in an area that became neighbors or adjacent areas. Spatial regression is a statistical method used to determine the relationship between the response variable and the predictor variables taking into account the linkages between regions.

Anselin developed a general model of spatial regression using *cross section data* [9]. The general model of spatial regression equation written in the form (1).

$$\mathbf{y} = \rho\mathbf{W_1}\mathbf{y} + \mathbf{X}\boldsymbol{\beta} + \mathbf{u}, \quad \text{with } \mathbf{u} = \lambda\mathbf{W_2} + \boldsymbol{\varepsilon} \tag{1}$$

$$\mathbf{y} = (\mathbf{I}_N - \rho\mathbf{W_1})^{-1}\mathbf{X}\boldsymbol{\beta} + (\mathbf{I}_N - \rho\mathbf{W_1})^{-1}(\mathbf{I}_N - \lambda\mathbf{W_2})^{-1}\boldsymbol{\varepsilon}$$

$$\boldsymbol{\varepsilon} \sim N(\mathbf{0}, \sigma^2\mathbf{I}_N)$$

From the general form of spatial regression, there are some models that can be formed in *cross-section* data, namely:

1. If $\rho = 0$ and $\lambda = 0$ then the equation becomes:

$$\boldsymbol{y} = \boldsymbol{X\beta} + \boldsymbol{\varepsilon} \tag{2}$$

   This equation is called the classical regression model by ignoring the spatial effects.

2. If $\rho \neq 0$ and $\lambda = 0$ then the equation becomes:

$$\mathbf{y} = \rho\mathbf{W_1}\mathbf{y} + \mathbf{X}\boldsymbol{\beta} + \varepsilon$$

$$\mathbf{y} = (\mathbf{I}_N - \rho\mathbf{W_1})^{-1}\mathbf{X}\boldsymbol{\beta} + (\mathbf{I}_N - \rho\mathbf{W_1})^{-1}\varepsilon \tag{3}$$

   This equation is called Spatial Autoregressive Model (SAR).

3. If $\rho = 0$ and $\lambda \neq 0$ then the equation becomes:

$$\mathbf{y} = \mathbf{X}\boldsymbol{\beta} + \mathbf{u}, \text{ with } \mathbf{u} = \lambda\mathbf{W_2} + \varepsilon$$

$$\mathbf{y} = \mathbf{X}\boldsymbol{\beta} + (\mathbf{I}_N - \lambda\mathbf{W_2})^{-1}\varepsilon \tag{4}$$

   This equation is called Spatial Error Model (SEM).

4. If $\rho \neq 0$ and $\lambda \neq 0$ then the equation becomes

$$\mathbf{y} = \rho\mathbf{W_1 y} + \mathbf{X}\boldsymbol{\beta} + \mathbf{u}\text{ , with } \mathbf{u} = \lambda\mathbf{W_2} + \varepsilon$$

$$\mathbf{y} = (\mathbf{I}_N - \rho\mathbf{W_1})^{-1}\mathbf{X}\boldsymbol{\beta} + (\mathbf{I}_N - \rho\mathbf{W_1})^{-1}\mathbf{u}$$

$$\mathbf{y} = (\mathbf{I}_N - \rho\mathbf{W_1})^{-1}\mathbf{X}\boldsymbol{\beta} + (\mathbf{I}_N - \rho\mathbf{W_1})^{-1}(\mathbf{I}_N - \lambda\mathbf{W_2})^{-1}\varepsilon \qquad (5)$$

This equation is called a General Spatial Model or Spatial Autocorrelation (SAC).

2.2. *Spatial Dependencies*

Anselin defines the spatial dependencies as their functional relationship between what happens at one point in space and what happens elsewhere [9]. To see the magnitude of the spatial dependencies can use the index Morans's I defined in equation (6).

$$I = \frac{\varepsilon^T W \varepsilon}{\varepsilon^T \varepsilon}) \qquad (6)$$

The weighting matrix that can use the index has not been standardized Moran's I defined in equation (7).

$$I = \frac{N}{S}\frac{\varepsilon^T W \varepsilon}{\varepsilon^T \varepsilon} \qquad (7)$$

To see if the magnitude of the spatial dependencies (I_j) is significant in the data can be done by testing on Moran's I index the following hypotheses.

$H_0: I_j = 0$ (There are no spatial dependencies)
$H_1: I_j \neq 0$ (There is a spatial dependencies)

Then the test statistic used in tests of significance on Moran's I index by Cliff and Ord [10] defined in equation (8).

$$Z = \frac{[I_j - E(I_j)]}{var(I_j)^{\frac{1}{2}}} \qquad (8)$$

2.3. *Spatial Weighting*

The weighting matrix/spatial weights (W) can be obtained by distance information of the adjacency *(neighborhood)* or the distance from one location to another. There are several methods for defining the intersection relations *(contiguity)* between these locations. According Lesage that the method include [11]:

1. Linear contiguity (the intersection edge);
2. Rook contiguity (intersection side);
3. Bhisop contiguity (intersection angle);
4. Double linear contiguity (the intersection of two edges);
5. Double rook contiguity (the intersection of the two sides);
6. Queen contiguity (side-angle intersection);

3. Methods

The method used in this study is *cross-sectional* spatial using data sourced from BPS-South Sulawesi province, Indonesia. The steps are as follows.

1. Exploration data on thematic maps to determine the spatial distribution patterns and dependencies on each variable and scatterplot to determine the pattern of variables and relationships.
2. Spatial regression modeling with the method of ordinary least squares (OLS), which includes estimation parameters, significant estimation models, test assumptions residual (identical, independent, and normal distribution).
3. Determine the spatial weighting (W) is using a chess queen (queen contiguity).
4. Test spatial dependency or spatial correlation between observations that are close together with the index moran or Morans' I.
   Hypotheses used to test the spatial dependencies are as follows.
   $H_0 : I = 0$ (There is no spatial dependencies)
   $H_1 : I \neq 0$ (There is a spatial dependencies)
5. The identification of the existence of spatial effect by using Lagrange Multiplier test (LM). LM Tests conducted to find out which models are in accordance with the procedure.
6. Process modeling, ie the data is modeled with Autoregresive Spatial Model (SAR), Spatial Error Model (SEM), or Autoregresive Spatial Moving Average (Sarma).
7. To examine the assumption of spatial models.
8. Summing up the results obtained.

4. Result

4.1. Administrative Map of South Sulawesi Province

Administrative map of South Sulawesi province can be seen in Figure 1, while the variable data dependent and independent variables used can be seen as follows.

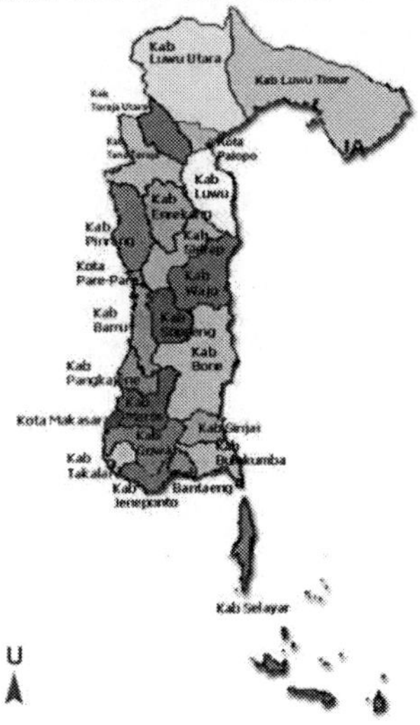

Figure 1. Administrative map of South Sulawesi Province, Indonesia

4.2. Exploration Data

Figure 2 below is a map of the affected population thematic risk of crime in the province of South Sulawesi, Indonesia, where seen no degradation of color which indicates the values of the communities affected by the risk of crime. The more intense the color of the region, the higher the risk of the communities affected by crime in the region. To determine the risk level of the communities affected by crime then categorized into five categories: very low, low, medium, high and very high.

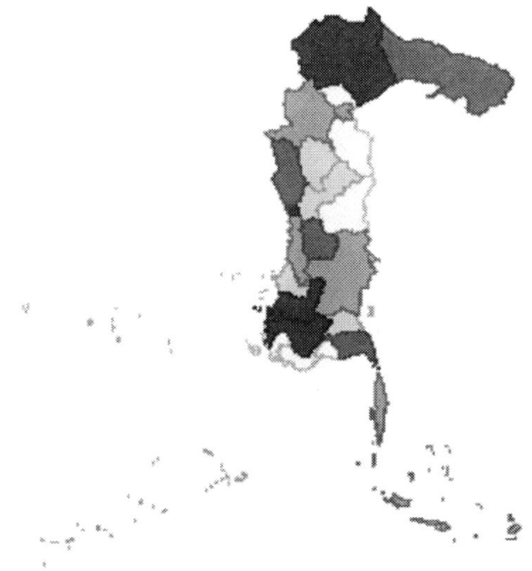

Figure 2. Map Thematic Risk Residents Affected by Crime in South Sulawesi

Risk population exposed to very low crime category (113-420 people) are in Bantaeng, Jeneponto, Luwu, Takalar and Wajo. Risk population exposed to criminality with the low category (239-295 people) are in Enrekang, Sidrap, Pangkep and Sinjai. Risk population being affected by crime category (317-449 people) are in Bone regency, Tana Toraja, Barru, Selayar and Palopo. Risk population exposed to high criminality by category (486-584 people) are in Pinrang, Soppeng, Bulukumba and East Luwu Regency. And the risk of the communities affected by crime category is very high (686-4319 inhabitants) is located in the City of Pare-pare, North Luwu, Maros, Gowa and Makassar.

4.3. Moran's I

By value $\alpha = 0{,}1$, then the value $Z\ (I) = 0{,}41851 < Z_{\alpha/2} = 1{,}645$, it means failed to reject H_0. This indicates that there is no spatial dependencies.

Table 1. Results of Moran's I Test

Variabel	Moran's I	Z(I)
Y	0,11889	0,41851
X_1	-0,01774	0,28188
X_2	0,08274	0,38236
X_3	-0,06733	0,23229
X_4	-0,03210	0,26752
X_5	0,11609	0,41571

$Z_{0{,}05} = 1{,}645\ ,\ I_0 = -0{,}0454$

Table 1 above shows that all of the dependent and independent variables generate value $Z(I) < Z_{\alpha/2} = 1{,}645$, it means failed to reject H_0. This indicates that there is no inter-regional spatial dependencies. Also seen, I_0 that all grades Moran's I value is greater than I_0 that means that all the dependent and independent variables indicating positive spatial dependency which means adjacent sites have similar values and tend to cluster. Unless the GDP per capita variable (X3) where the value of Moran's I is worth less than I_0 and negative value, the indicated that spatial dependencies have significant negative adjacent locations have different values.

4.4. Lagrange Multiplier (LM)
Selection of spatial models made by Lagrange Multiplier (LM) as the initial identification. Lagrange Multiplier (LM) is used to detect the spatial dependencies with more specific, ie with dependencies in the lag, error or both (lag and error). When LM LM lag and no significant error does occur it can be concluded either dependency on lag and error. Test conducted on a weighted spatial dependencies queen contiguity. The test results Lagrange Multiplier (LM) in Table 2. Then b erdasarkan Table 2 shows that by using a weighted queen contiguity p-value Moran's I = not significant at α = 10%. This means that there is no spatial dependencies.

Table 2. Results of Diagnostic Dependencies Spatial

Spatial Dependency Test	Value	*P-value*
Moran's I	-0.3236764	0.7461832
Lagrange Multiplier (lag)	0.9296950	0.3349422
Lagrange Multiplier (error)	0.5521177	0.4574537
Lagrange Multiplier (SARMA)	3.1437015	0.2076605

5. Conclusion
Based on the analysis and discussion drawn some conclusions about the risk of the communities affected by crime in the province of South Sulawesi showed no spatial dependencies in both the lag and error. Based on the results of this study, it presented some suggestions as follows: (1) Subsequent research include variables that are considered essential and are not included in this study. (2) Researcher further use customized methods as weighting. (3) Provincial and Local Government should be more serious in taking into pemerata an income distribution.

References
[1] Bonger, W. A. 1981. Pengantar Tentang Kriminologi. Jakarta: Ghalia Indonesia.
[2] Abdulsyani. 1987. Sosiologi Kriminalitas. Bandung: Remaja Karya.
[3] Cunneen, C. and White, R. 2011. Juvenile justice: Youth and crime in Australia. Oxford University Press.
[4] Krisberg, B. 2005. Juvenile justice: Redeeming our children. Sage.
[5] Kakamu, K., Polasek, W. and Wago, H. 2008. Spatial interaction of crime incidents in Japan. *Mathematics and Computers in Simulation*. 78(2), pp. 276–282.
[6] Kusrini, D. E., Suhartono, S. and Sari, D. M. 2013. Pemodelan Kasus Tindak Pidana di Kota Surabaya dengan Pendekatan Regresi Spasial. *J. Sains dan Seni ITS*, vol. 2, no. 2, pp. D135–D140.
[7] Septiana, L. and Wulandari, S. P. 2009. Pemodelan Remaja Putus Sekolah Usia SMA di Provinsi Jawa Timur dengan Menggunakan Metode Regresi Spasial. Institut Teknologi Sepuluh Nopember.
[8] Miller, H. J. 2004. Tobler's first law and spatial analysis. *Annals of the Association of American Geographers*. 94(2), pp. 284–289.
[9] Anselin, L. 1988. Spatial Econometrics: Methods and Models. Dordrecht: Kluwer Academic.
[10] Cliff, A. D. and Ord, J. K. 1981. Spatial processes: methods and applications. London: Pion.
[11] LeSage, J. P. 1999. The theory and practice of spatial econometrics. University of Toledo. Toledo, Ohio, 28, p. 33.

IOP Conf. Series: Journal of Physics: Conf. Series **954** (2018) 012014 doi:10.1088/1742-6596/954/1/012014

Verification Image of The Veins on The Back Palm with Modified Local Line Binary Pattern (MLLBP) and Histogram

Agus Prijono[1*], Aan Darmawan Hangkawidjaja[1], Ratnadewi[1], Ansari Saleh Ahmar[2,3]

[1]Electrical Engineering Department, Universitas Kristen Maranatha , Bandung, Indonesia
[2]Department of Statistics, Universitas Negeri Makassar, Indonesia
[3]AHMAR Institute, Makassar, Indonesia

*agprijono2006@gmail.com

Abstract. The verification to person who is used today as a fingerprint, signature, personal identification number (PIN) in the bank system, identity cards, attendance, easily copied and forged. This causes the system not secure and is vulnerable to unauthorized persons to access the system. In this research will be implemented verification system using the image of the blood vessels in the back of the palms as recognition more difficult to imitate because it is located inside the human body so it is safer to use. The blood vessels located at the back of the human hand is unique, even humans twins have a different image of the blood vessels. Besides the image of the blood vessels do not depend on a person's age, so it can be used for long term, except in the case of an accident, or disease. Because of the unique vein pattern recognition can be used in a person. In this paper, we used a modification method to perform the introduction of a person based on the image of the blood vessel that is using Modified Local Line Binary Pattern (MLLBP). The process of matching blood vessel image feature extraction using Hamming Distance. Test case of verification is done by calculating the percentage of acceptance of the same person. Rejection error occurs if a person was not matched by the system with the data itself. The 10 person with 15 image compared to 5 image vein for each person is resulted 80,67% successful Another test case of the verification is done by verified two image from different person .that is forgery, and the verification will be true if the system can rejection the image forgery. The ten different person is not verified and the result is obtained 94%.

1. Introduction

The biometric technology [1] has recently become a reliable technology to provide a high level of security in the system of recognition of a person. Among various biometric techniques and their characteristics are used to identify a person, biometric of the hand into one of the techniques most widely used and most successful. The parts of the hand that is often used for the introduction of a person's identity, among others, hand geometry, fingerprint, palm lines, the lines on the knuckles and blood vessels

Currently, the majority of a person through the hand's identifier device requires direct contact between the user's hands with the identification device to take pictures of the unique patterns of the user's hands. There are a number of problems in this regard, the first is the level of cleanliness of the

device that has been used many times by others, the problems will increase if there is a virus such as SARS can be transmitted by touching the rest of germs residing on the surface of recognition devices. The second problem is the rest of the hand imprint mold on the appliance can affect the accuracy of the system to recognize a person's identity. And the third is that biometrics can be easily forged.
These problems can be overcome by using biometrics veins on the back palm. There are several supremacy used biometric vein, that is the first image captured using an infrared camera, so there is no direct contact between the equipment and the user, the second the blood vessels inside the human body so that its are difficult to counterfeit, the third blood vessels can be used for long term.

2. Vein as Biometric

Blood vessels are part of the circulatory system that transport blood throughout the body. There are three types of blood vessels, namely: the arteries that carry blood from the heart, the capillaries that serve as the actual exchange of water and chemicals between the blood and tissues, and the blood vessels of the vein that carries blood from the capillaries back to the heart.

One biometric technology that is currently being developed is based vein biometric technology. Each individual has a unique pattern of blood vessels even in twins, though, and does not change with age but it change after experiencing the process of surgery. The advantage of vein biometric is location of vein within the skin surface. The pattern of veins which are widely used in the biometric system is the pattern of blood vessels located in the hand i.e., the back palm , the palm, and the finger. Because of it lies within the human body causes the pattern of blood vessels is impossible to forge. Therefore it is developed into vein biometric.

In the biometric system, the pattern of blood vessels can be seen clearly using infrared light. Infrared [2] is electromagnetic radiation of a wavelength longer than visible light, but shorter than radio wave radiation.
Vein biometric is adequate to the criteria of good biometric because, universal which every human being has the hands and the blood vessels in it, unique which based on the above description of each individual has a unique pattern of blood vessels even in twins, permanent which pattern of blood vessels in the hand does not change with age, can be measured quantitatively by using the imaging device near - infrared (NIR) imaging patterns so that blood vessels can be extracted and features of blood vessels can be measured.

3. Image Processing of Vein Image

Definition image processing itself is a two-dimensional image processing using a computer. Image processing is done with the aim of getting the results as expected. Based on the objective, digital image processing can be divided into several types, among others: eliminating noise in an image, improving the quality of the image by increasing or decreasing the contrast, segmentation image which divides the image into sections, analysing image by using feature extraction image for recognition and identification the objects.

Image quality improvement is one of the initial processes in image processing. Image quality improvement is needed because often the image of an object has a poor quality, for example, the image is too dark or too light, the image less sharp, and blur. There are several techniques to improve the image, such techniques gaussian low pass filter, high pass filter, and median filter.

3.1. Median Filter

At a median filter, a "window" contains a number of pixels. The window is shifted from point to point on the entire image area. On each shift created a new window. The midpoint of this window amended by the median value of the window. For example, review the window in the form of groups of pixels (square shaded) on an image in Figure 1 (a). A pixel that is being processed is an intensity of 35. Sort of pixels - the pixel

9 10 10 10 **10** 10 11 12 35

The median of the group is 10 (in bold). The midpoint of the window (35) is now replaced with the median value (10). Results of the median filter are shown in Figure 1 (b). Thus, the median filter removes the pixel values are very different from its neighboring pixels.

13	10	15	14	18
12	10	10	10	15
11	11	**35**	10	10
13	9	12	10	12
13	12	9	8	10

(a) Midpoint pixel value 35

13	10	15	14	18
12	10	10	10	15
11	11	**10**	10	10
13	9	12	10	12
13	12	9	8	10

(b) 35 replace with median 3 x 3 pixel

Figure. 1 Removal of noise with a 3x3 median filter

3.2. Feature Extraction Using Modified Local Line Binary Pattern (MLLBP)

Feature extraction is a process of making characteristic/feature of a form, in this case, is the picture/image. Feature value obtained will be analysed for further processing. Feature extraction is done by counting the number of dots or pixels that encountered in any of checks carried out in various directions tracing checks on Cartesian coordinates of the digital image is analysed, namely vertical, horizontal, diagonal right and left diagonal.

Features a unique characteristic of an object/image. Features can be divided into two, namely: features of "natural": part of the image, such as brightness and edges of objects, and features "artificial": features that are obtained with certain operations on the image, for example, gray level histogram.

The feature extraction is the process to get characteristic features that distinguish an object with other objects. Extraction feature method Local Line Binary Pattern [3] is a proposed method for the extraction of better features and generate a higher level of recognition than the previous method, namely the Local Binary Pattern (LBP). One advantage of this LLBP method is able to confirm the change of image intensity like dots, edges, and corners. LLBP Operator consists of two components, namely a horizontal component and a vertical component. The magnitude of LLBP obtained by calculating the binary code sequence from the second component.

The basic idea of the method LLBP similar to the original method is Local Binary Pattern (LBP), but the difference is: neighbourhood shape is a straight line with a length of N pixels, unlike in the LBP of the square.

Binary weight distribution starting from the left or right side that are adjacent to the central pixel (2^0) toward the tip end of the left or right pixel. Start from the centre pixel $\left(2^{\lceil N/2 \rceil -2}\right)$, for example N = 15 then $\left(2^{\lceil 15/2 \rceil -2}\right) = 2^6$ with $\lceil x \rceil$ is ceiling function, namely the function to take the smallest integer greater than or equal to x.

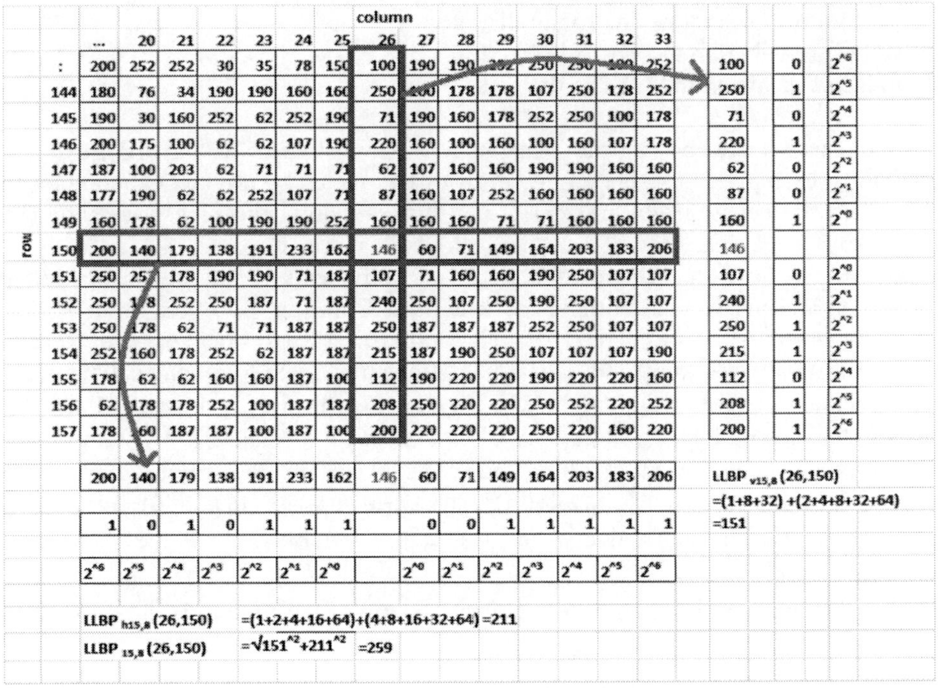

Figure 2. Illustration LLBP Operator

In LLBP algorithm, the first is to determine the binary code line in horizontal and vertical directions separately and then calculate the value of a magnitude that characterizes image intensity changes such as edges and corners. This can be expressed mathematically in equation (1) to (4).

$$s(x)=\begin{cases}1, & x\geq 0\\ 0, & x<0\end{cases} \tag{1}$$

$$LLBP_{hN,c}(x,y)=\sum_{n=1}^{n=c-1}s(h_n-h_c)2^{c-n-1}+\sum_{n=c+1}^{n=N}s(h_n-h_c)2^{n-c-1} \tag{2}$$

$$LLBP_{vN,c}(x,y)=\sum_{n=1}^{n=c-1}s(v_n-v_c)2^{c-n-1}+\sum_{n=c+1}^{n=N}s(v_n-v_c)2^{n-c-1} \tag{3}$$

$$LLBP_m=\sqrt{LLBP_h^2+LLBP_v^2} \tag{4}$$

Note:
N: is the length of the line in pixels
h_n: pixel intensity values - pixel along the horizontal row
v_n: pixel intensity values - pixel along the vertical row

$$c=\left\lceil\frac{N}{2}\right\rceil \tag{5}$$

where:
c is the position of the center pixel intensity values
h_c: pixel intensity values at the center coordinates (x_c, y_c) in horizontal row
v_c: Pixel intensity value at the center coordinates (x_c, y_c) in a vertical row
s(x): define a function threshold (barrier), this function will be worth 1 when $x\geq 0$ and will be 0 when $x<0$

$LLBP_h$: LLBP feature values in the horizontal direction
$LLBP_v$: LLBP feature values in the vertical direction
$LLBP_m$: the value of the magnitude of LLBP

Based on the equation (1) and (2), the horizontal component of LLBP ($LLBP_h$) extract the binary code of N-1 bits of each pixel. The same number of bits extracted by the vertical component of LLBP ($LLBP_v$) based on the equation (1) and (3). Thus, by combining the binary code of ($LLBP_h$) and ($LLBP_v$), a total of LLBP binary code for each - each pixel is 2 (N-1).

From Figure 2, the binary sequence of horizontal components are defined from left to right as 1010111.0011111_2 and the vertical component is defined from top to bottom as 0101001.0111011_2. Therefore, the binary code of LLBP is $1010111001111101010010111011_2$. Each binary code both horizontally and vertically are converted to decimal numbers with the following calculation:

For the horizontal component of the binary sequence to the left and right of the centre pixel: $1010111_2 = (1\times2^6)+(0\times2^5)+(1\times2^4)+(0\times2^3)+(1\times2^2)+(1\times2^1)+(1\times2^0) = 64+16+0+0+4+2+1 = 87_{10}$

For the horizontal component of the binary sequence to the right of the centre pixel: $0011111_2 = (0\times2^6)+(0\times2^5)+(1\times2^4)+(1\times2^3)+(1\times2^2)+(1\times2^1)+(1\times2^0) = 0+0+4+8+16+32+64 = 124_{10}$

By adding these two decimal values, then the values obtained LLBP the horizontal direction, namely: $87_{10}+124_{10}=211_{10}$

For the vertical component of the binary sequence the top of the centre pixel $0101001_2 = (0\times2^6)+(1\times2^5)$ $+(0\times2^4)+(1\times2^3)$ $+(0\times2^2)+(0\times2^1)+(1\times2^0) = 32+0+8+0+0+1 = 41_{10}$

For binary sequences vertical component lower part of the centre pixel:
$0111011_2 = (0\times2^6)+(1\times2^5)+(1\times2^4)+(1\times2^3)+(0\times2^2)+(1\times2^1)+(1\times2^0) = 0+2+4+8+0+32+64 = 110_{10}$

By adding these two decimal values, then the values obtained LLBP in the vertical direction, namely: $41_{10}+110_{10}=151_{10}$

From a decimal value $LLBP_v$ and $LLBP_h$ already acquired can be calculated the value of the magnitude of LLBP, namely: $LLBP_m = \sqrt{(211^2 + 151^2)} = 259$

After all the pixel value is counted with LLBP, the next process is to divided the LLBP value to be 10 x 10 block, and we have one block is 15 x 15 pixel. Each block is counted the histogram distribution and this value of histogram distribution will be counted with Hamming distance to know how similarity each block one image extracted with each block another image extracted . This process is called Modified Local Line Binary Pattern.

3.3. Hamming Distance

Hamming distance is used to determine the degree of similarity between the binary code histogram distribution image extracted with binary code histogram distribution image contained in the database. The similarity between the binary code histogram distribution of the image that will be tested with the binary code histogram distribution from the image contained in the database measured using a calculation Hamming Distance (HD). Hamming Distance is one of the testing methods to find how similar a vector to the other vector based on the value of proximity. If the value of the proximity is smaller, meaning the greater the similarity the two vectors, on the contrary, if the value is bigger means both vector similarity is smaller.

The formula for the calculation of Hamming Distance given in equation 6.

$$HD = \frac{\|(codeA \otimes codeB)\|}{codeLength} \tag{6}$$

Operator ⊗ is an Exclusive-OR Boolean operator between a couple of bits - the bits that correspond to. Code A is code sequence tested binary image, while the code B is a row of binary code from the image that was in the database. Code Length is the length of the binary code from the image to be tested. Here is an example of calculating Hamming Distance between two different codes, namely:

Hamming Distance between 11001101 and 10110010 is 7

Hamming Distance between 10111010 and 10010010 is 2

Figure 3 shows the block diagram realization equipment for verification image of the vein.

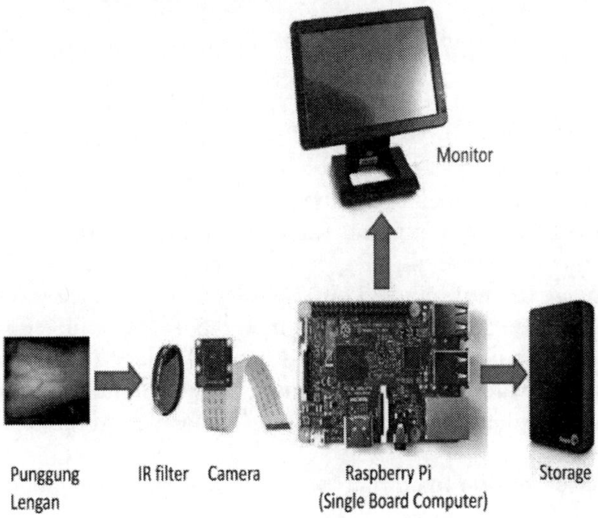

Figure 3. Block Diagram Realisation Equipment

4. Test Cases

Data collection was performed against 16 person and each person is done taking 20 pictures (V01 s / d V20), meaning a total of 320 images. In Figure 4 we look the position of the hand looks towards the camera.

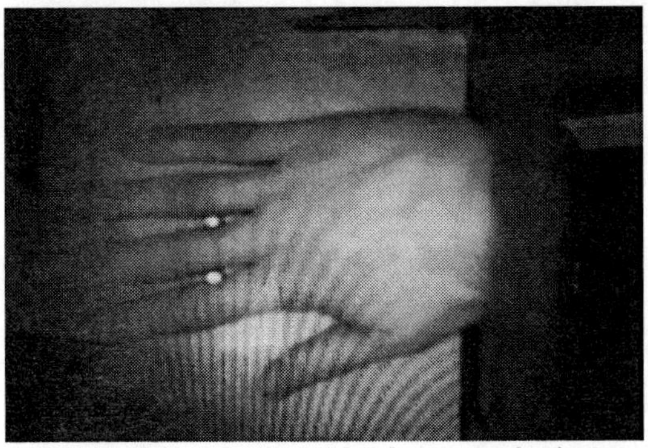

Figure 4. Position the hand looks towards the camera

After that, the captured image from the camera will be cropped into 150 X 150 pixels image as shown in Figure 5 (a). Then the image quality improvement process is carried out as shown in Figure 5 (b). Furthermore, the feature extraction process is carried out so it looks like in Figure 5 (c).

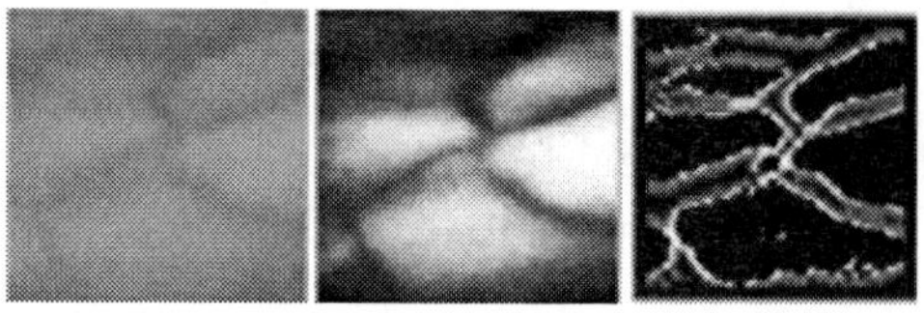

(a) (b) (c)

Figure 5. (a) the capture image from camera (b) the improvement image (c) The feature extraction image

Then the next process, the data matrix image with 150 x 150 pixels, divided into blocks numbered 10 x 10, meaning there are 100 blocks and each block have 15 x 15 pixels. From each block to get the process done histogram. Then from the histogram which will be calculated Hamming distance value.

Here is a table of determining the maximum value Hamming Distance of 5 vein image (V01 / V05) each of the 16 respondents. Suppose of first person, image vein 1 (V01) calculated the value of Hamming Distance for image 2 until image 5 (V02 until V05), then the image vein 2 (V02) calculated the value of Hamming Distance for image 3 until image 5 (V03 until V05), vein image 3 (V03) is calculated Hamming value Distance for image 4 and image 5 image (V04 and V05), and image vein 4 (V04) is calculated Hamming value Distance for image 5 (V05). Hamming value of 10 Distance then specified maximum value. Furthermore, the maximum value determination Hamming also Distance for person 2 until 16.

Hamming Distance Based on the maximum value, it will be tested image data of 6 until 20 (v06 until V20) of each person.

Figure 6 shows Percentage Successful Recognition the results of the verification data of the image of the 10 person.

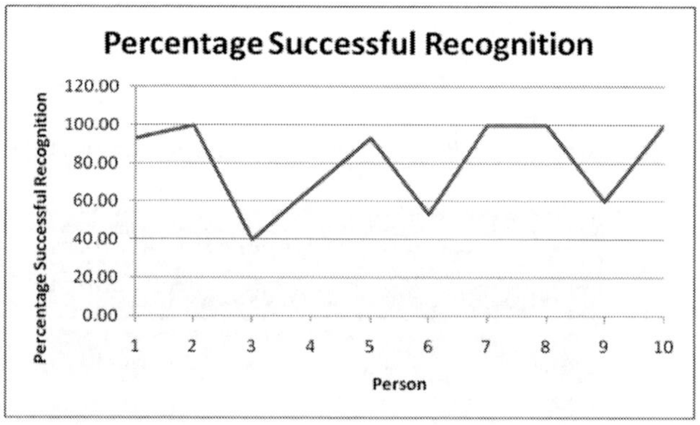

Figure 6. Percentage Successful Recognition

Figure 7 shows Percentage Successful Rejection the tested image data of different people and the obtained results are as follows:

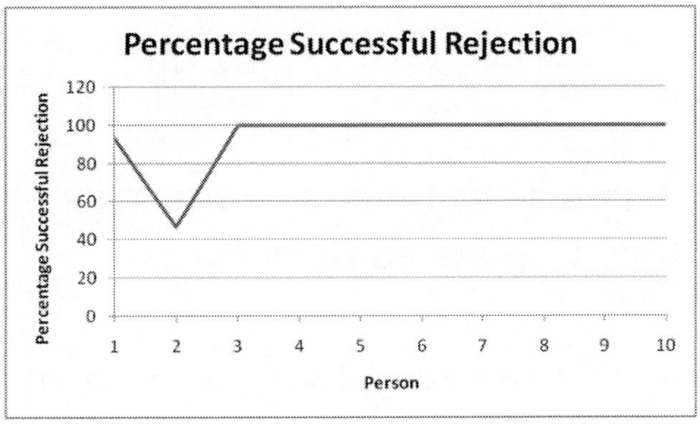

Figure 7. Percentage Successful Rejection

5. Conclusion

Test case of verification is done by calculating the percentage of acceptance of the same person. Rejection error occurs if a person was not matched by the system with the data itself. The 10 person with 15 image compared to 5 image vein for each person is resulted 80,67% successful Another test case of the verification is done by verified two image from different person .that is forgery, and the verification will be true if the system can rejection the image forgery. The ten different person is not verified and the result is obtained 94%.

References

[1] I. Sarkar, F. Alisherov, T.H. Kim, and D. Bhattacharyya, 2010 ."Palm Vein Authentication System: A Review", *Int. J. of Control Auto.*, **3**, 1, pp 27-34.

[2] R. Fuksis, M. Greitans, O. Nikisins, M. Pudzs 2010, Infrared Imaging System for Analysis of Blood Vessel Structure, *J. Electronics Elect. Eng.*, **1**, 97, pp 45-48.

[3] Rosdi, B. Affendi; C. W. Shing and S. A. Suandi 2011, Finger Vein Recognition Using Local Line Binary Pattern, *J. Sensors*, **11**, pp 11357-11371; DOI: 10.3390/s 111211357.

[4] Sec, BehavioSec, 2009, Behaviometrics, *Measuring FAR/FRR/EER in Continuous Authentication,* Technical White Paper.

IOP Conf. Series: Journal of Physics: Conf. Series **954** (2018) 012015 doi:10.1088/1742-6596/954/1/012015

The Impact of Airport Performance towards Construction and Infrastructure Expansion in Indonesia

T D Laksono[1]*, N Kurniasih[2], C Hasyim[3], M I Setiawan[4], A S Ahmar[5,6]

[1]Department of Civil Engineering, Universitas Wijayakusuma, Purwokerto, Indonesia
[2]Faculty of Communication Science, Library & Information Science Program, Universitas Padjadjaran, Bandung, Indonesia
[3]Darul Ulum University, Jombang, Indonesia
[4]Department of Civil Engineering, Narotama University, Surabaya, Indonesia
[5]AHMAR Institute, Makassar, Indonesia
[6] Department of Statistics, Universiats Negeri Makassar, Indonesia

*taufikdwilaksono@yahoo.com

Abstract. Development that is generated from airport areas includes construction and infrastructure development. This research reviews about how the implementation of material management in certain construction project and the relationship between development especially construction and infrastructure development with Airport Performance. The method that is used in this research is mixed method. The population in this research is 297 airports that are existed in Indonesia. From those 297 airports then it is chosen airports that have the most completed data about construction project and it is obtained 148 airports. Based on the coefficient correlation (R) test it is known that construction and infrastructure development has relatively strong relation with airport performance variable, but there are still other factors that influence construction and infrastructure development become bigger effect.

1. Introduction

Development is the amount of value added goods and services that are generated from all regional activities in certain area [1]. That is generated from airport areas includes construction and infrastructure development. Construction project is the activity that consists of series of works that is interdependent with each other. The construction project activity starts from the planning stage to project implementation is really influenced by the ability to set or manage project resources. There are several resources that are necessary in the construction project, such as material, workforce, cost or even equipment. The incapability to manage resources that is required in the project can cause various problems, such as inflated cost, project time finishing that is not according to plan, inaccuracy of work quality that is generated and others, which eventually can harm the project implementation as a whole.

Material is one of the important components that influence the sustainable of the construction project without any material management that can cause problem towards on-going construction project. Barne and Paulson mentioned that the lateness of certain small instrument arrival can stop a big job [2]. This statement shows that the lateness of required materials in the construction project can cause the implementation of construction project activity becomes stopped. However, the arrival of materials in the huge amount at the same time can cause problem too, such as project location that is narrow, material can be damaged when it is stored and it needs a huge amount of cost.

According to Patel and Vyas, one of the major problems in delaying construction projects is poor materials and equipment management [3]. The research that they have done shows that the construction materials and equipment may constitute more than 70% of the total cost for a typical construction project. The same thing is stated by Donyavi and Flanagan, cost for civil engineering projects showed the materials and plant component can be up to 70% of the project cost dependent upon the type of project and the extent of mechanization and plant used, whilst on commercial building and housing projects the proportion is around 45 – 50% [4]. As for Messah, Widodo and Adoe mentioned that the causing factor of construction project implementation according to contractor is the availability of workforce, while according to the owner and supervisor consultant, it is about material mobilization factor [5].

Table 1. Quarterly Construction index and Growth (percent), Quarter II/2016 - II/2017

Description	2016 Q2	2016 Q3	2016 Q4	2017 Q1	2017 Q2
Permanent Workers	116.22	117.35	118.12	117.19	118.21
Permanent Workers Growth	0.37	0.97	0.65	-0.78	0.87
Mandays	191.46	196.46	205.09	200.70	209.57
Mandays Growth	2.10	2.61	4.40	-2.14	4.42
Compensation and Wages	197.75	203.34	213.22	208.83	219.07
Compensation and Wages Growth	2.47	2.83	4.86	-2.06	4.90
Value of Construction	203.91	209.50	219.56	214.44	224.50
Value of Construction Growth	2.37	2.74	4.80	-2.33	4.69
Value of Building Construction	206.63	210.96	216.42	210.81	222.94
Value of Building Construction Growth	2.18	2.10	2.59	-2.59	5.75
Value of Civil Construction	252.15	259.73	274.30	268.29	281.62
Value of Civil Construction Growth	2.56	3.01	5.61	-2.19	4.97
Value of Especially Construction	205.43	208.91	213.28	206.35	214.81
Value of Especially Construction Growth	1.24	1.69	2.09	-3.25	4.10
Business Prospect	53.74	53.59	52.49	58.10	56.26
Business Condition	52.46	56.64	51.37	51.28	54.61
Business Problems	27.64	27.25	29.06	27.62	27.43

Table 1 shows the growth data and construction index in the second quarter of 2016-2017 and table 1 shows the Value of Construction Growth, Building Construction Growth, Civil Construction Growth and Especially Construction Growth in the last 12 months, with the Value of Especially Construction Growth has the highest score and Value of Especially Construction Growth has the lowest score. This is in accordance with the condition of the last 12 months where the government pushes the infrastructure especially in Java and outside of Java. Meanwhile, Construction Business Prospect, Condition and Problems in the last 12 months, with the Construction Business Prospect and Condition is over 50%, as well as Construction problem is below 30%. This shows that Construction Business is getting stable and the emerging problem is getting lower in Indonesia. This research reviews about the implementation of material management in certain construction project and relationship between construction and infrastructure development and Airport Performance in which it includes real estates.

2. Research Method

This research uses mixed method approach. In mixed method, we combine two research methods, which is qualitative and quantitative research [6]. This method is used for getting the comprehensive understanding about the research problem [7]. The population in the research is 296 airports that have the most completed data about construction project and it is obtained 148 airports.

In this research, quantitative data is acquired from Construction Indicator, 2nd Quarter-2017 that is published by Indonesian Central Statistics Department, while qualitative data is obtained through interview. We interview several contractors that implement certain building development project that the structure of its building uses steel material. The interview is focused to know the process that is done by contractor party in taking steel material that is needed in the project. The research analysis includes correlation analysis between Y variable (construction and infrastructure development) and X Variable (airport performance) and descriptive analysis.

3. Result and Discussion

3.1. Material management in Construction Project

In construction project that is reviewed it is known that the project mainly need steel material for at least 1,460,000 kg. The material needs that are needed by contractors are taking by using two ways, which is through supplier and directly buys from the factory that produces steel. The ordering and arriving material that are needed is done by referring to the each necessity detail of steel dimension, the available cost and schedule of each necessity of steel dimension that will be worked.

The main factor that becomes the consideration of contractor in taking this steel material is that the availability of steel dimension that is needed in the market or even the level of fluctuation of steel material price for each kilogram. The phase of material management in the construction project can be categorized into three processes that is:

1. Material Identification Process

   In the material identification process that will be used in the project, can be illustrated in the Figure 1.

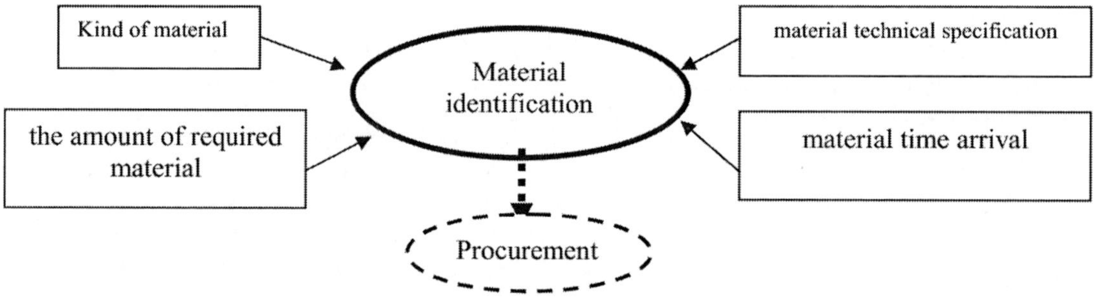

Figure 1. Material Identification Process

   In the material identification process, it is an activity to know what material that will be used, in this phase all material-related matters should be able to be identified clearly and carefully so that there is no missing matters, so the needs of material will be fulfilled. This material identification process really determines towards the fulfillment of obstacles that are faced by each project construction i.e. time, quality and cost. The incapability in identifying material that is needed accurately and carefully can cause time, quality and cost incompatibility that have been determined before.

2. Material Procurement Process

   This process of material procurement involves other parties either supplier or directly from factory. Moreover, the material procurement process can be illustrated in Figure 2.

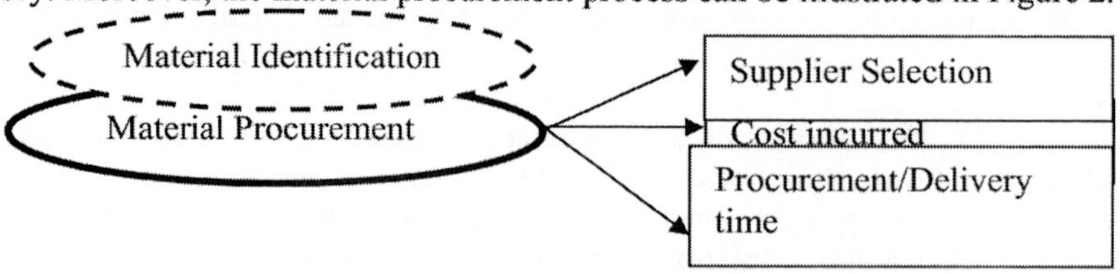

Figure 2. The diagram of Material Procurement Process

The phase of this material procurement process is the phase where material that is required should be available to be further processed. The mistake in doing material procurement can cause the impact of lateness in job implementation in the field in which not only causes the lateness of certain job, but also it can cause the lateness of job finishing entirely.

3. Material Usage Process
   In this phase, material that has been ordered and ready to be procured or sent will be processed in order to be able to be used as the part of the building as a whole. The process of material usage can be seen in Figure 4.

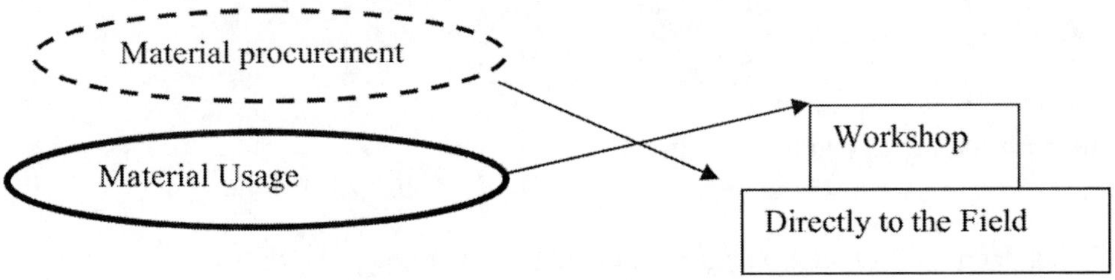

Figure 3. Material Usage Process

Material that will be used can be differentiated between workshop processed and directly to the field. When the material needs to be processed in order to be cut, base-painted and others so that material usage process is done in the workshop first, after the process in the workshop is complete, so the material is immediately sent to the field to do the following job. However, when the material sent can be directly used so that the material usage is not necessarily done in workshop but directly in the field. The evaluation towards this process relates to precision and accuracy in making the work method and project time schedule. A good teamwork is needed between workshop team and field team so that the materials that will be used in the field will not be late that is caused by the activity in the workshop. Therefore, a good coordination is needed in order to make the project activity can run well according to that has been determined.

3.2. The relationship between Y variable (construction and infrastructure development) and X Variable (Airport Performance)

Multiple linear regression analysis is used to know the amount of construction and infrastructure development influence towards Airport Performance by using multiple linear regression formula. The data that is used is 148 airports in Indonesia, where each town and district is taken 1 (one) Airport.

X Variable includes:

The result of the analysis by using SPSS version 20 is obtained regression coefficient as follows:

Table 2 The Result of Multiple Linear Regression Analysis

Model		Unstandardized Coefficients		Standardized Coefficients			Correlations
		B	Std. Error	Beta	t	Sig.	Zero-order
1	(Constant)	38.455	139.384		.276	.783	
	X1	-.492	.535	-4.468	-.919	.360	.437
	X2	.623	.512	5.571	1.217	.225	.441
	X3	.007	.002	7.227	3.088	.002	.433
	X4	-.004	.002	-4.225	-1.824	.070	.426
	X5	.000	.000	.942	1.038	.301	.433
	X6	.000	.000	-4.298	-3.304	.001	.416
	X7	-2.924E-005	.000	-.219	-1.979	.050	.249
	X8	-1.449E-005	.000	-.195	-1.421	.157	.278

Model		Correlations	
		Partial	Part
1	(Constant)		
	X1	-.078	-.063
	X2	.103	.084
	X3	.253	.212
	X4	-.153	-.125
	X5	.088	.071
	X6	-.270	-.227
	X7	-.166	-.136
	X8	-.120	-.097

By using multiple linear regression formula:

$$Y = a + b_1X_1 + b_2X_2 + b_3X_3 + b_4X_4 + b_5X_5$$

So that it is obtained multiple linear regression equation as follows:

$$Y = 38.445 - 0.492X_{11} + 0.623X_{12} + 0.007X_{13} - 0.004X_{14}$$

Based on the multiple linear regression equation above, it can be seen that X_{12} and X_{13} (aircraft departure and passenger departure) have positive improvement impact towards construction and infrastructure development, while X_{11} and X_{14} variables (aircraft arrival and passenger arrival) have negative decline effect towards construction and infrastructure development.

Tabel 3 Model Summary

Model	R	R Square	Adjusted R Square	Std. Error of the Estimate
1	.588[a]	.346	.308	1365.78117

a. Predictors: (Constant), X8, X7, X5, X2, X4, X6, X3, X1

The above table shows that the number of coefficient correlation (R) is at 0.588, which means that construction and infrastructure development variable has relatively strong relation with Airport Performance variable and 30.8% influence of Airport Performance variable towards the improvement of construction and infrastructure development, while the rest 69.2% improvement of construction and infrastructure development is influenced by factors other than Airport Performance, this can be seen from Adjusted R Square score, at 0.308.

4. Conclusion

The material management phase in construction project can be categorized into three processes i.e. identification process, procurement process, and material usage process. In the last 12 months, value of especially construction growth has the highest value and value of especially construction growth has the lowest value. This is in accordance with the condition where government pushes infrastructure in java and outside java. Meanwhile, construction activities looks getting stable and emerging problem in Indonesia is getting low. Based on the coefficient correlation (R) test it is known that construction and infrastructure development has relatively strong relation with airport performance variable, but there are still other factors that influence construction and infrastructure development become bigger effect.

References

[1] Kropinova, E.G., Zaitseva, N.A. and Moroz, M., 2015. Approaches to the assessment of the contribution of tourism into the regional surplus product: case of the Kaliningrad region.

Mediterranean Journal of Social Sciences, *6*(3 S5), p.275.

[2] Badan Pusat Statistik Subdirectorate of Statistical Publication and Compilation. 2017. Construction Indicator, 2nd Quarter-2017. Jakarta: Badan Pusat Statistik.

[3] Patel, K.V. and Vyas, C.M., 2011, May. Construction materials management on project sites. In *National Conference on Recent Trends in Engineering & Technology* (pp. 1-5).

[4] Donyavi, S. and Flanagan, R., 2009, September. The impact of effective material management on construction site performance for small and medium sized construction enterprises. In *Proceedings of the 25th Annual ARCOM Conference, Nottingham, UK* (pp. 11-20).

[5] Messah. Y., A., Widodo, T., Adoe, M., L. " Kajian Penyebab Keterlambatan Pelaksanaan Proyek Konstruksi Gedung di Kota Kupang". Jurnal Teknik Sipil, Vol II, No.2 september 2013, 157-168.

[6] Morse, J.M., 2016. *Mixed method design: Principles and procedures* (Vol. 4). Routledge.

[7] Creswell, J.W. and Clark, V.L.P., 2007. Designing and conducting mixed methods research. Australian and New Zealand Journal of Public Health, 31: 388–389. doi:10.1111/j.1753-6405.2007.00097.x

IOP Conf. Series: Journal of Physics: Conf. Series **954** (2018) 012016 doi:10.1088/1742-6596/954/1/012016

Airport Performance and Construction Enlargement Activities

Y Hanun [1*], M I Setiawan[2], N Kurniasih[3], C Hasyim[4], A S Ahmar[5,6]

[1]Department of Civil Engineering, Tarumanagara University, Jakarta, Indonesia
[2]Department of Civil Engineering, Narotama University, Surabaya, Indonesia
[3]Faculty of Communication Science, Universitas Padjadjaran, Bandung, Indonesia
[4]Darul Ulum University, Jombang, Indonesia
[5]AHMAR Institute, Makassar, Indonesia
[6] Department of Statistics, Universiats Negeri Makassar, Indonesia

yunan.hanun65@gmail.com

Abstract. The evaluation of transportation infrastructure project should consider the contribution towards infrastructure growth. This research aims to analyze the effect of Construction enlargement activities towards airport performance. This research is correlation study. The population includes 148 airports in Indonesia. By using total sampling, there were 148 sample airports . The result shows that the construction enlargement activities variable has relatively strong relationship to Airport Performance variable, while the adjusted R Square score shows the increasing construction enlargement activities that affected by the other factors aside from airport performance.

1. Introduction

The construction activities include new job, repair, addition and change, the establishment of pre-fabricated of building or structure in the project location and also construction that are temporary. The construction activity shall be carry out by general contractor, i.e. the company that has a construction job for other parties, even by special contractor, that is a business unit or individual that conducts construction activity to be used by them [1]. The result of construction activity is residence building construction, non-residence building construction, civil building construction, electric and telecommunication building construction, power plant, transmission, distribution and installation of communication network building, civil building, electricity installation, dredging, the finishing of civil construction, floor, wall, and building plafond, painting, interior finishing and decoration in finishing; exterior and gardening in the building and other civil buildings, the rent of construction tools, etc [2][3].

In the period of 1991-2010, Malaysia has strong correlation both construction sector. Considering the role of substantial sector's construction in the development of Malaysia, the government of Malaysia needs to pay attention and focus on the construction sector to fulfill the requirements of developed country [4]. Meanwhile, Indonesia is still left behind in supporting service sector to generate construction enlargement activities.

In Sumatera, the construction performance is getting slower, 4.22% (yoy) in the 2nd quarter of 2017, that lower than the previous quarter 4.48% (yoy), highway toll construction, port and the other supporting facilities of the Asian Games 2018 will support the improvement of construction performance. In Java, the construction enterprises growth is 5.52% (yoy), that higher than the 1st quarter of 2017, at 4.64% (yoy), supported by the government project development or private, such as the establishment of MRT, LRT, New Tanjung Priok, Fly Over Semanggi, PLTU Batang, Central and East Java highway construction, new airport in D.I.Yogyakarta has a potential to give a positive contribution. The increasing Index Residential Property Price also reflects the optimism of construction sector. The increasing house price is encouraged by the incline of small type house price, especially in DKI Jakarta and Banten. Fifty-five infrastructure projects in Java are still in progress in this quarter of the total of asset, at Rp 399,1 trillion or 43.01% from the whole projects in Java. In east part of Indonesia, it improves from 4.86% (yoy) to 6.24% (yoy), acceleration happens in the major part of the province in KTI, except South-East Sulawesi, North Maluku, Papua, West Papua, and NTT, the supporting factor of the improvement is the acceleration of National Strategic Project (PSN), the construction of new factory, the addition of capacity in several areas of Kalimantan, Sulawesi, as well as North Maluku and NTB, the acceleration of the growth of the new project that worth more than 0.2 millions USD is started in the 2nd quarter of 2017 [5].

In the 3rd quarter of 2017, the enterprises construction growth is estimated slightly higher than the previous quarter, supported by the acceleration of construction project realization from the local government, including the process of action from new projects in 2017, the improvement of oil refineries' capacity in Kalimantan, mineral downstream in Sulawesi and Papua, as well as Balinusra processed food industry development, the construction in supporting infrastructure in Country Border Post (PLBN), the construction of Special Zone Maloy Batuta, the construction of land and sea port, road access together with the construction of CPO storage tank, the construction of Trans-Papua road, the construction of power plant in North and West Sulawesi, the development of Buli Industrial Area in North Maluku [6]. In 2016 the realization of development in Indonesia has improved, especially in construction sector, at approximately Rp. 14.2 Trillion, restaurant and hospitality at Rp. 2.4 Trillion, transportation sector, warehousing and communication at Rp. 27.5 Trillion, as well as real estate and company service sector at Rp. 11.5 Trillion [7]. The improvement of this development has indirectly affects to airport infrastructure. On the other hand, evaluation of infrastructure project of new transportation should consider the contribution to infrastructure growth. This research aims to analyze the effect construction enlargement activities towards Airport Performance.

2. Method

This research is correlation study. The correlation study is used when the research has two variables or more [8]. In this research, airport performance becomes X variable and construction enlargement activities becomes Y Variable. The population in the research include 148 airports in Indonesia. By using total sampling, there were 148 sample airports. The data of the research is analyzed by using linear regression analysis.

3. Result and Discussion

3.1. The Comparison of Construction Enlargement Activities with Airport Performance

From 148 airports that are taken as sample in this research, there are 30 airports that have the highest Construction enlargement activities in Indonesia. However, thirty airports with this highest Construction enlargement activities has not yet supported by a good airport performance. Table 1 shows that the area with the highest Construction enlargement activities in Jakarta, Semarang, Pekanbaru, Batam, Makassar, Bandung, Sidoarjo, Medan and Balikpapan. Yet only Semarang, Batam, Makassar, Sidoarjo, Medan and Balikpapan that enter top 10 highest Airport Performance in relation to Aircraft Arrival and Passenger Departure.

IOP Conf. Series: Journal of Physics: Conf. Series **954** (2018) 012016 doi:10.1088/1742-6596/954/1/012016

Table 1 The Comparison of Construction Enlargement Activities with Airport Performance

No	Airport & City		Construction activities	Aircraft Arrival		Passenger Departure	
				rank	total	rank	total
1	*Halim Perdanakusuma (Jakarta Timur)*	*DKI Jakarta*	41,771.72	13	13,683	12	1,482,125
2	*Achmad Yani (Semarang)*	*Semarang*	36,287.62	11	17,598	9	1,754,468
3	*Sultan Syarif Kasim (Pekanbaru)*	*Pekanbaru*	25,143.48	23	8,629	15	1,260,344
4	*Hang Nadim (Batam)*	*Batam*	23,429.80	10	19,212	8	2,313,360
5	*Sultan Mahmud Badaruddin II (Palembang)*	*Palembang*	19,752.67	15	12,773	11	1,632,277
6	*Hasanuddin (Makassar)*	*Makassar*	19,585.35	3	43,807	4	3,306,534
7	*Husein Sastranegara (Bandung)*	*Bandung*	17,632.27	19	10,258	16	1,250,087
8	*Juanda (Sidoarjo)*	*Sidoarjo*	13,277.29	2	61,412	2	6,821,775
9	*Kualanamu (Medan)*	*Deli Serdang*	11,629.64	7	23,804	6	3,226,695
10	*Sepinggan (Balikpapan)*	*Balikpapan*	11,479.67	5	33,405	5	3,242,802
11	*Temindung (Samarinda)*	*Samarinda*	10,627.77	48	2,717	63	43,035
12	*Adi Sumarmo (Surakarta)*	*Surakarta*	9,410.74	28	6,858	22	738,990
13	*Abdul Rachman Saleh (Malang)*	*Malang*	9,332.77	50	2,531	36	310,307
14	*Soekarno Hatta (Tangerang)*	*Tangerang*	8,754.51	1	152,914	1	19,151,202
15	*Rogojampi/Blimbingsari (Banyuwangi)*	*Banyuwangi*	6,856.50	73	960	61	50,796
16	*Raja Haji Fisabilillah (Tanjung Pinang)*	*Tanjung Pinang*	5,839.97	27	1,270	19	125,620
17	*Tunggul Wulung (Cilacap)*	*Cilacap*	5,095.06	39	3,729	106	6,697
18	*Japura Rengat (Indragiri Hulu)*	*Indragiri Hulu*	4,897.10	97	401	150	20
19	*Supadio (Pontianak)*	*Pontianak*	4,710.99	16	11,949	14	1,316,686
20	*Noto Hadinegoro (Jember)*	*Jember*	4,148.54	107	338	87	15,402
21	*Adi Sucipto (Sleman)*	*Sleman*	3,631.13	8	23,200	7	2,992,745
22	*Ngurah Rai (Badung)*	*Badung*	3,451.42	4	36,966	3	4,191,558
23	*Raden Inten II (Lampung Selatan)*	*Lampung Selatan*	3,354.77	26	7,476	23	716,616
24	*Jeffman -DEO (Sorong)*	*Sorong*	3,148.02	29	6,599	30	359,998
25	*Mutiara(Palu)*	*Palu*	3,005.69	34	4,451	27	519,254
26	*Pinang Kampai (Dumai)*	*Dumai*	2,819.88	88	591	79	23,709
27	*Eltari (Kupang)*	*Kupang*	2,726.92	78	10,048	102	752,753
28	*Sam Ratulangi (Manado)*	*Manado*	2,609.25	20	10,162	18	1,014,885
29	*Melak (Kutai Barat)*	*Kutai Barat*	2,540.52	79	820	85	16,403
30	*Bontang (Badak Bontang)*	*Bontang*	2,528.47	90	583	84	17,993

Table 1 also shows that 30 areas with the highest Construction enlargement activities, has a very low rank in airport performance, i.e. Samarinda, Malang, Banyuwangi, Cilacap, Indragiri Hulu, Jember, Palu, Dumai, Kupang, Kutai Barat and Badak Bontang.

3.2. The Effect of Construction Enlargement Activities towards Airport Performance

The multiple linear regression analysis is used for identifying the effect of construction enlargement activities to airport performance by using multiple linear regression formula. The data analysis used data of 148 airports in Indonesia, where we took one airport each town and district.

The research variables include:

The result of the analysis by using SPSS 20 version is obtained the regression coefficient as follows:

Table 2. The Result of the Analysis of Multiple Linear Regression

Model		Unstandardized Coefficients		Standardized Coefficients	t	Sig.	Correlations		
		B	Std. Error	Beta			Zero-order	Partial	Part
1	(Constant)	673.008	487.898		1.379	.170			
	X31	-4.904	1.874	-12.096	-2.617	.010	.349	-.217	-.171
	X32	5.440	1.792	13.201	3.036	.003	.356	.249	.198
	X33	.022	.008	6.457	2.903	.004	.336	.239	.189
	X34	-.007	.007	-2.333	-1.060	.291	.328	-.090	-.069
	X35	.000	.000	1.076	1.249	.214	.340	.105	.081
	X36	-.002	.001	-5.530	-4.474	.000	.314	-.355	-.292
	X37	-8.784E-005	.000	-.179	-1.699	.092	.233	-.143	-.111
	X38	-9.897E-005	.000	-.362	-2.773	.006	.151	-.229	-.181

a. Dependent Variable: Y3

The formula of multiple regressions:

$$Y = a + b1X1 + b1X2 + b3X3 + b4X4 + b5X5$$

From the result of regression above, it is obtained the equation of multiple linear regressions as follows:

$$Y = 673.008 - 4.904X_{31} + 5.44X_{32} + 0.022X_{33} - 0.007X_{34} - 0.007X_{36}$$

Based on the equation of multiple linear regression above, it can be seen that X_{32} and X_{33} (aircraft depart and passenger depart) have positive impact towards the improvement of construction enlargement activities, while variable X_{31}, X_{34} dan X_{36} (aircraft arrival, passenger arrival and baggage loaded) have negative impact of the decline of construction enlargement activities.

Table 3. Model Summary

Model	R	R Square	Adjusted R Square	Std. Error of the Estimate
1	.640[a]	.409	.375	4780.76009

a. Predictors: (Constant), X38, X37, X35, X32, X34, X36, X33, X31

Table 3 shows that the number of coefficient correlation (R) is at 0.409, which means Construction enlargement activities variable has relatively strong relationship with Airport Performance variable, while Adjusted R Square score is 0.375, which means 37.5% effect of Airport Performance variable has an impact towards the improvement of Construction enlargement activities, the rest 62.5% of Construction enlargement activities improvement is influenced by the other factors aside from airport performance.

4. Conclusion

In 2016 the development realization has increased, especially in construction sector, restaurant and hospitality sector, transportation sector, warehousing and communication as well as real estate and company service. The result of coefficient correlation (R) test shows that construction enlargement activities variable has relatively strong relation with Airport Performance variable, while the adjusted R Square value shows that the improvement of construction enlargement activities were influenced by aside from airport performance. Therefore, the government's commitment is required for airport performance to have direct impact on construction enlargement activities.

References

[1] Hughes, W., Champion, R. and Murdoch, J., 2015. *Construction contracts: law and management*. Routledge.
[2] Allen, E., Thallon, R. and Schreyer, A.C., 2017. *Fundamentals of residential construction*. John Wiley & Sons.
[3] BPS, Produk Domestik Bruto Indonesia Triwulan 2012-2016. 2016.
[4] Khan, R.A., Liew, M.S. and Ghazali, Z.B., 2014. Malaysian construction sector and Malaysia vision 2020: Developed nation status. *Procedia-social and behavioral sciences, 109*, pp.507-513.
[5] A. Ghazali and L. Martini, "Bandung as Servic ce City in n Indonesia : Role e of Academician, Business and Commmunity," vol. 52, pp. 317–324, 2012.
[6] Bank Indonesia. "Perkembangan Properti Komersial, Triwulan II, 2017," 2017.
[7] BPS, *Statistical Yearbook of Indonesia 2017*. BPS statistics indonesia, 2017.
[8] Yin, R.K., 2013. Case study research: Design and methods. Sage publications.

Pavement Technology and Airport Infrastructure Expansion Impact

Sabib[1], M I Setiawan[2*], N Kurniasih[3], A S Ahmar[4,5], C Hasyim[6]

[1]Department of Civil Engineering, Tarumanagara University, Jakarta, Indonesia
[2]Department of Civil Engineering, Narotama University, Surabaya, Indonesia
[3]Faculty of Communication Science, Library & Information Science Program, Universitas Padjadjaran, Bandung, Indonesia
[4]AHMAR Institute, Makassar, Indonesia
[5]Department of Statistics, Universiats Negeri Makassar, Indonesia
[6]Darul Ulum University, Jombang, Indonesia

*syabibalimran1710@gmail.com

Abstract. This research aims for analyzing construction and infrastructure development activities potential contribution towards Airport Performance. This research is correlation study with variable research that includes Airport Performance as X variable and construction and infrastructure development activities as Y variable. The population in this research is 148 airports in Indonesia. The sampling technique uses total sampling, which means 148 airports that becomes the population unit then all of it become samples. The results of coefficient correlation (R) test showed that construction and infrastructure development activities variable have a relatively strong relationship with Airport Performance variable, but the value of Adjusted R Square shows that an increase in the construction and infrastructure development activities is influenced by factor other than Airport Performance.

1. Introduction

Developed countries such as Japan, USA, UK, Australia, New Zealand, and Singapore, have a high percentage of development, which attributed with services. Indonesia is still lagging in supporting the service sector to produce more development. This situation confirms how developing countries such as Indonesia still exploit natural resources rather than focus on knowledge, innovation, or service. In addition to get more development, the growth of service industries also contributed to the number of people working in an area. Which means, the government needs to accommodate the labor force, both in terms of policy, political, social and education as became the capital of the nation [1]. The policy and political will of the government can create conducive environment for the growth of the service industry. The social aspect of the service industry is preventing the social gap between the rich and the poor. And the educational process needed to create a workforce, with competence, knowledge, and skills, then becomes an added value in business [2]

Indonesia as part of the *Eastern Development Corridor* is currently the best strategic location [3]. Industrial sector greatly affects the development of the region. The area that has Industrial area, contributes to a bigger industrial impact. Java Island which has 55 Industrial Areas (75.89 percent of total area of the Indonesian archipelago) is capable of being a driving force for the national

IOP Conf. Series: Journal of Physics: Conf. Series **954** (2018) 012017 doi:10.1088/1742-6596/954/1/012017

development by contributing 57.99 percent to the national development. In addition, the industrial sector on the island of Java was able to become the leading sector with a contribution of 29.87 percent. In a Middle-term of National Development Plan 2015-2019 (RPJMN), the government sets to build and facilitate the construction of 14 industrial estates. The fourteen regions are (1) Bintuni, West Papua for oil, gas and fertilizer industrial areas; (2) Buli of East Halmahera in North Maluku for industrial area of ferronickel smelter, stainless steel and downstream of stainless steel; (3) Bitung of North Sulawesi for agro and logistics industry; (4) Palu, Central Sulawesi for industrial area of rattan, rubber, cocoa, and smelter; (5) Morowali, Central Sulawesi; (6) Konawe, Southeast Sulawesi; and (7) Bantaeng, South Sulawesi is focused on ferronickel smelter, stainless steel and downstream struck steel smelter industries. (8) Batulicin, South Kalimantan for steel industry area; (9) Jorong, South Kalimantan for bauxite industrial zone; (10) Ketapang, West Kalimantan for industrial area of alumina; (11) Landak West Kalimantan for rubber industry area, CPO; (12) Kuala Tanjung, North Sumatra for industrial area of aluminum, CPO; (13) Sei Mangke, North Sumatra for CPO processing industry; and (14) Tanggamus, Lampung to regional maritime and logistics industry [4]

In terms of transportation, there are three types of road pavements that are used in Indonesia namely, flexible pavement, rigid pavement and composite pavement. The choice of pavement type depends on several considerations, in particular the cost factor of pavement construction. Pavements of bending roads that are most widely used in Indonesia are to use almost 95% of Asphalt Concrete road surface layers. Aggregate gradation can be distinguished by uniform gradations, dense gradations and gaps gradations. Uniform graded is an aggregate gradation that has roughly the same size or also called as open-graded for fine aggregate proportion only slightly so that the air spaces between the aggregates are large enough. Dense is graded aggregate having a grain size of course to fine continuously called graded well. The type of construction that uses this gradation is Asphalt Concrete (AC). The graded gap is an aggregate gradation that has grain size on a part of a fraction that does not exist or is small and is called Hot Rolled Sheet (HRS). Gradation aggregate will determine quality road. Damage to roads during the wet state through weight puts pressure on the road surface layer will cause the cracked street surfaces. Surface holding capacity of the street will reduced or weak (stripping). For this case, sticking aggregate level with asphalt can be raised by doing with anti-stripping addition [5] [6] [7]

Airport pavement technologies that have high quality yet development in cost is needed in Airport infrastructure in Indonesia. The utilization of air transport access through airports in 2016, including 835.9 thousand domestic and international flight departures, 99.7 million domestic and international passengers transported, 75.6% *load factor* of domestic and international passengers, 715.8 thousand tons domestic and international goods transported [8]. Indonesia is currently still in 10th rank in the world in the highest amount of passengers. With the high potential number of passengers in the next 10 years (2027) Indonesia is predicted will rise to 6th world rank over Japan, Germany, Thailand and France. On the other hand, Indonesian flights frequency is currently still low in comparison to the high number of the population, so that the airport infrastructure investment would be beneficial for accelerating the regional development growth in the future [9]

The increase in investment indirectly impacts on comfort of Airport instruction. This study aims to analyze contribution potential of construction and infrastructure development towards Airport Performance.

2. Method

This research is correlation study with research variables includes Airport Performance as X Variable and construction and infrastructure development activities as Y Variable. The population in this research is 148 airports in Indonesia. The sampling technique uses total sampling, which means 148 airports are the population units then all of them become samples. Data analysis was done by using correlation coefficient (R) test. In addition, this study is also included a descriptive analysis of 30 airports with the highest construction and infrastructure development activities in Indonesia.

IOP Conf. Series: Journal of Physics: Conf. Series **954** (2018) 012017 doi:10.1088/1742-6596/954/1/012017

3. Result and Discussion

Thirty airports in Indonesia with the highest construction and infrastructure development activities are not Supported with good performance of airport. There are some areas that come at top 30 of construction and infrastructure development activities, but it is not included in 30 major Airport Performance in Indonesia in relation to Aircraft and Passengers, namely Jember, Malang, Jambi, Banyuwangi, Palu, Cilacap, Bengkulu and Bone areas. Besides that airport with the highest construction and infrastructure development activities is not an airport with the best airport performance, namely Bandung, East Jakarta, Palembang, Batam, Pekanbaru areas. The comparison analysis of construction and infrastructure development activities with Airport Performance, among 30 of 151 airports in Indonesia, can be seen on Table 1.

Table 1 Comparison Sonstruction and Infrastructure Development Activities in area with Airport Performance

No	Airport	Construction and Infrastructure Development Activities	Aircraft rank	Aircraft		Passenger (Person) Rank	Passenger (Person)		
				Arrival	Departure		Departure	Arrival	Transit
1	Husein Sastranegara (Bandung)	11,166.02	19	10,258	10,305	16	1,250,087	1,240,116	3,558
2	Hasanuddin (Makassar)	6,834.53	3	43,807	43,811	4	3,306,534	3,793,351	2,090,777
3	Achmad Yani (Semarang)	5,950.78	11	17,598	17,649	9	1,754,468	1,782,455	9,039
4	Halim Perdanakusuma (Jakarta Timur)	5,521.89	13	13,683	14,064	12	1,482,125	1,473,207	-
5	Sultan Mahmud Badaruddin II (Palembang)	5,486.29	15	12,773	12,803	11	1,632,277	1,643,238	77
6	Temindung (Samarinda)	4,219.96		-				-	
7	Hang Nadim (Batam)	4,192.43	10	19,212	19,210	8	2,313,360	2,489,168	161,301
8	Soekarno Hatta (Tangerang)	3,101.29	1	152,914	148,782	1	19,151,202	20,802,860	1,935,806
9	Sepinggan (Balikpapan)	3,003.56	5	33,405	33,405	5	3,242,802	3,301,423	757,063
10	Sultan Syarif Kasim (Pekanbaru)	2,971.44	23	8,629	8,612	15	1,260,344	1,242,567	1,755
11	Supadio (Pontianak)	2,420.46	16	11,949	11,972	14	1,316,686	1,311,463	11,414
12	Kualanamu (Medan)	2,209.62	7	23,804	25,554	6	3,226,695	3,024,679	123,523
13	Sam Ratulangi (Manado)	1,941.76	20	10,162	10,138	18	1,014,885	1,017,956	27,480
14	Juanda (Sidoarjo)	1,844.31	2	61,412	61,499	2	6,821,775	7,731,353	903,730
15	Adi Sumarmo (Surakarta)	1,326.07	28	6,858	6,908	22	738,990	720,087	-
16	Noto Hadinegoro (Jember)	1,314.94		-				-	
17	Bandara Internasional Lombok (Mataram)	1,304.84	17	11,531	11,527	17	1,112,940	1,141,037	50,320
18	Abdul Rachman Saleh (Malang)	1,267.03		-				-	
19	Sultan Thaha Syarifuddin (Jambi)	1,251.24		-		25	594,176	574,043	-
20	Eltari (Kupang)	1,217.97	22	10,048	10,044	21	752,753	752,017	18,556

No	Airport	Construction and Infrastructure Development Activities	Aircraft rank	Aircraft		Passenger (Person) Rank	Passenger (Person)		
				Arrival	Departure		Departure	Arrival	Transit
21	Adi Sucipto (Sleman)	1,115.22	8	23,200	23,192	7	2,992,745	3,019,951	5,574
22	Ngurah Rai (Badung)	1,096.99	4	36,966	36,917	3	4,191,558	4,276,388	99,276
23	Rogojampi/Blimbingsari (Banyuwangi)	1,055.22		-				-	
24	Mutiara(Palu)	970.85		-		27	519,254	514,118	4,406
25	Tunggul Wulung (Cilacap)	877.13		-				-	
26	Fatmawati Soekarno (Bengkulu)	844.22		-		29	417,412	414,110	-
27	Nusawiru (Ciamis)	818.58		-				-	
28	Tjilik Riwut (Palangkaraya)	812.12	24	8,470	8,463			-	
29	Pattimura (Ambon)	738.18	25	8,379	8,353	26	583,505	627,988	105,928
30	Bone Mappalo (Bone)	699.08		-				-	

Table 1 shows the highest construction and infrastructure development activities is in Bandung, but Airport in Bandung, only ranked 19th in Aircraft Arrival, and only ranked 16th in Passengers' Departure. Thus also with airports in Makassar, Semarang, East Jakarta, Palembang, Batam, Pekan Baru, Pontianak, Manado, Surakarta and Kupang areas, the airports in those cities have better construction and infrastructure development activities compared with the arrival and departure of aircraft passengers. The different things happen in airports in Jakarta, Batam, Balikpapan, Medan, Sidoarjo, Sleman, Badung, Palangkaraya and Ambon, where the aircraft arrival and departure have better ranks than construction and infrastructure development activities. For example, Jakarta successfully becomes the first rank in the aircraft arrival performance and departure passengers, but only gets the 8th rank in construction and infrastructure development activities. The thing that is quite unique is on the airports in Mataram city, where the airport in the city got the 17th rank both in GR construction and infrastructure development, aircraft arrival and departure passengers. Thus, the high construction and infrastructure development activities in the area are still not yet supported by airport performance. To give a positive impact on the construction and infrastructure development activities, airport performance should be increased, in particular by improving the quality of Airport infrastructure.

Multiple linear regression analysis is used to determine the amount of effect of the construction and infrastructure development activities towards Airport Performance using multiple linear regression formula. The data used are 148 airports in Indonesia, where each city and district are taken 1 (one) Airport. Variables include:

Y_2 = construction and infrastructure development

$X_{2.1}$ = Aircraft Arrival	$X_{2.5}$ = Baggage Unloaded
$X_{2.2}$ = Aircraft Departure	$X_{2.6}$ = Baggage Loaded
$X_{2.3}$ = Passenger Departure	$X_{2.7}$ = Cargo Unloaded
$X_{2.4}$ = Passenger Arrival	$X_{2.8}$ = Cargo Loaded

The results of the analysis using SPSS version 20, as follows:

Table 2. Correlations (Y_2 = construction and infrastructure development activities)

Pearson Correlation	Y_2
$X_{2.1}$	0.393
$X_{2.2}$	0.398
$X_{2.3}$	0.373

IOP Conf. Series: Journal of Physics: Conf. Series **954** (2018) 012017 doi:10.1088/1742-6596/954/1/012017

Pearson Correlation	Y_2
$X_{2.4}$	0.366
$X_{2.5}$	0.384
$X_{2.6}$	0.358
$X_{2.7}$	0.256
$X_{2.8}$	0.189

The table above shows the Pearson correlation coefficient, with the lowest correlation is $X_{2.8}$ (Cargo Loaded) of 0.189 and the highest is $X_{2.2}$ (Aircraft Departure) which means that the construction and infrastructure development activities Activities have relatively strong correlation with variable Airport Performance, unless the variable $X_{2.7}$ (Cargo Unloaded) and $X_{2.8}$ (Cargo Loaded) which has a correlation value below of 0.2

Table 3. Model Summary (Y = construction and infrastructure development activities)

Model	R	R Square	Adjusted R Square	Std. Error of the Estimate
1	.603[a]	0.363	0.326	1191.89855

a. Predictors: (Constant), X28, X27, X25, X22, X24, X26, X23, X21

The above table shows that the magnitude of the coefficient correlation (R) of 0.603 means that the construction and infrastructure development activities Activities variable has relatively strong correlation with Airport Performance variable. While the value of Adjusted R Square is 0.326, meaning that 32.6% of the effect of Airport Performance variable towards the increase of construction and infrastructure development activities, while the remaining 67.4% increase in the construction and infrastructure development activities are influenced by factors other than the Airport Performance.

Table 4. Multiple Linear Regression Equation
(Y = construction and infrastructure development activities)

Model		Unstandardized Coefficients		Standardized Coefficients	t	Sig.	Correlations
		B	Std. Error	Beta			Zero-order
1	(Constant)	118.934	121.639		0.978	0.330	
	X21	-0.336	0.467	-3.457	-0.720	0.473	0.393
	X22	0.514	0.447	5.194	1.150	0.252	0.398
	X23	0.005	0.002	5.768	2.497	0.014	0.373
	X24	-0.003	0.002	-3.641	-1.593	0.113	0.366
	X25	0.000	0.000	1.602	1.790	0.076	0.384
	X26	0.000	0.000	-4.590	-3.576	0.000	0.358
	X27	-2.885E-005	0.000	-0.245	-2.238	0.027	0.256
	X28	-2.568E-005	0.000	-0.391	-2.886	0.005	0.189

Multiple regression formula:

$$Y = a + b_1X_1 + b_2X_2 + b_3X_3 + b_4X_4 + b_5X_5$$

From the results of the above regression obtained multiple linear regression equation as follows:

$$Y_2 = 118.934 - 0.336X_{21} + 0.514X_{22} + 0.005X_{23} - 0.003X_{24}$$

Based on the multiple linear regression equation of the above, it appears that X_{22} dan X_{23} (aircraft depart and passenger depart) have a positive impact on the increase in the construction and infrastructure development activities, while the X_{31} dan X_{34} variables (aircraft arrival and passenger arrival) have an negative impact in the decline of the construction and infrastructure development activities.

4. Conclusion

The high of construction and infrastructure development activities in the area, have not been connected with Airport performances which covers aircraft performance and passengers performance. Thirty airports in Indonesia with the highest construction and infrastructure development activities, are not Supported by good airport performance. There are some areas that come in the top 30 construction and infrastructure development activities, but it does not come as the 30 major Airport Performance in Indonesia in relation to Aircraft Performance and Passengers, namely Jember, Malang, Jambi, Banyuwangi, Palu, Cilacap Bengkulu and Bone area. Besides that airport with the highest construction and infrastructure development activities is not airport with the best airport performance, namely Bandung, East Jakarta, Palembang, Batam, Pakanbaru area. The result of the coefficient correlation (R) test showed that construction and infrastructure development activities variable has a relatively strong relationship with Airport Performance variable, but the value of Adjusted R Square showed that increase in the construction and infrastructure development activities is influenced by factor other than Airport Performance.

References

[1] BPS, *Produk Domestik Bruto Indonesia Triwulan 2012-2016*. 2016.
[2] A. Ghazali and L. Martini, "Bandung as Servic ce City in n Indonesia : Role e of Academician, Business and Commmunity," vol. 52, pp. 317–324, 2012.
[3] Thailand Ministry of Industry, "Eastern Corridor Development project," Eastern Corridor Office (EECO) Ministry of Industry 75/6 Rama VI Road, Ratchathewi, Bangkok 10400, Thailand, 2017.
[4] E. S. Hartati, E. Listiyanto, A. M. Pulungan, I. Abdullah, B. Yudhistira, A. H. Firdaus, R. Abdulah, M. R. Hafiz, N. Huda, A. P. G. Talattov, D. Syafrian, M. Hanif, S. D. Nofarina, and INDEF, "Proyeksi Ekonomi Indonesia 2017: Menguji Ketangguhan Ekonomi Indonesia," 2016.
[5] Y. Li, H. Zhu, and C. Liu, "Experimental Analysis of Airport Crumb Rubber Concrete (CRC) Pavement," vol. 3, pp. 605–608, 2011.
[6] T. Willway, L. Baldachin, S. Reeves, and M. Harding, "The effects of climate change on highway pavements and how to minimise them : Technical report," 2008.
[7] The Aspalt Pavement Alliance, "Pavement Type Selection," 2010.
[8] BPS, *Statistical Yearbook of Indonesia 2017*. BPS statistics indonesia, 2017.
[9] D. Oxley, "IATA ecomonics," 2017.

Stability Analysis Susceptible, Exposed, Infected, Recovered (SEIR) Model for Spread Model for Spread of Dengue Fever in Medan

Syafruddin Side[1*], Yulita Molliq Rangkuti[2], Dian Gerhana Pane[2], Marlina Setia Sinaga[2]

[1]Department of Mathematics, Faculty of Mathematics and Natural Science, Universitas Negeri Makasar, (UNM), Makasar, South Sulawesi, Indonesia
[2]Department of Mathematics, Faculty of Mathematics and Natural Science, Universitas Negari Medan, UNIMED, 20221, Medan, North Sumatera, Indonesia

*syafruddin@unm.ac.id

Abstract. Dengue fever is endemic disease which spread through vector, Aedes Aegypty. This disease is found more than 100 countries, such as, United State, Africa as well Asia, especially in country that have tropic climate. Mathematical modeling in this paper, discusses the speed of the spread of dengue fever. The model adopting divided over four classes, such as Susceptible (S), Exposed (E), Infected (I) and Recovered (R). SEIR model further analyzed to detect the re-breeding value based on the number reported case by dengue in Medan city. Analysis of the stability of the system in this study is asymptotically stable indicating a case of endemic and unstable that show cases the endemic cases. Simulation on the mathematical model of SEIR showed that require a very long time to produce infected humans will be free of dengue virus infection. This happens because of dengue virus infection that occurs continuously between human and vector populations.

Keywords: SEIR model; Endemic; Runge-Kutta method; Stability; Dengue fever

1. Introduction

Dengue Fever Disease (DFD) is not a new disease because it happens almost every year in line with the changing seasons, from rainy season to the dry season. Based on data from the North Sumatra Health Office (DHO Sumatra), cases of DFD in 2013, there are 33 district / cities in North Sumatra Province reached 4,746 people. DFD patients in Medan are 1,270 people. From January to October 2014, the the number of dengue cases as many as 1068 cases and there are nine people died from dengue fever in the city of Medan. Based on the data, from 21 districts there are two districts that have increased the number of DFD patients in 2013 to 2014. In the Medan Helvetia sub-district, there were 101 patients and two of them died in 2013. It increases in January to October 2014, there were 104 patients and one person died. Whereas in sub-district of Medan Sunggal, there were 83 patients with DFD in 2013. Meanwhile, in January to October 2014 increased dengue fever which is reached 103 people. The high cases of dengue fever is important to investigate. Lack of awareness for Clean and Healthy Lifestyle that have not been executed so the vector is found in many areas.

A mathematical model can be applied to determine the amount of the spread of dengue fever in the city of Medan. With the development of science of mathematics, mathematical models helped provide an important role in preventing the spread of dengue fever. Meanwhile, the model is built to look for the amount of the spread of dengue disease. The mathematical model can also assist in the prediction of epidemic control in the future in order to avoid endemic. After many years, the more complex models have been derived by [1]. In some diseases, have been found for a specified period, part of the class Exposed population is infected with the disease but cannot transmit. For modeling the disease is called SEIR models used [2]. In the model of SEIR, the population is divided into four classes: class vulnerable populations (S), in which all individuals are susceptible to the disease; class exposed population (E), in which all individuals who are infected but not yet infectious; class population is infected (I), where all the people infected by the disease and can transmit; and class populations recovered (R), where individual cured [3]. Many of the researchers who discussed the model of SEIR with a variety of problems, some of them are [4] studied the stability of the global era is structured in SEIR Model with infectivity in the latency period. [5] discussed the dynamic behavior of the epidemic SEIR system with nonlinear transmission rate. [6] discussed the global stability of SEIR models with vertical transmission and saturated contact rate [7] also [8] discussed about SEIR model.

Mathematical model of the spread of dengue fever that is discussed in this paper is a dynamic model with a mathematical model, that are SIR and SEIR. SIR and SEIR model are one of the mathematical model to analyze the spread of one serotype of dengue virus between humans by vector. Then simulation model of the spread of measles to the effect of vaccination using Maple program was applied. The analysis results will be obtained parameters are most influential in the spread of measles is the rate of transmission. Thus the spread of measles epidemic can be controlled by adding a vaccination program [9]. [10,11] make a stability analysis of differential equation model. [12] and [13] conducted research on the SIR model for transmission of dengue in South Sulawesi, Indonesia and Selangor, Malaysia. [14] also conducted research on the model of SEIR for transmission of dengue with simulation. In tropical countries and become the biggest disease epidemic in Southeast Asia. Thus epidemics caused by climate change and the lack of knowledge and awareness of people about dengue so that dengue fever is endemic for a long time. In this paper, SEIR model adopted by [14] and applied for Medan city.

2. Stabilty Analysis of Model

2.1 SEIR Model of Dengue Fever

The SEIR model of dengue fever adopted by [13]. The model for the human and mosquito populations follows:

$$\frac{dS_h}{dt} = \mu_h N_h - \left(\frac{\beta_h b I_v}{N_h} + p + \mu_h\right) S_h, \tag{1}$$

$$\frac{dE_h}{dt} = \left(\frac{\beta_h b I_h}{N_h} + p\right) S_h - (\mu_h + \varphi_h) E_h, \tag{2}$$

$$\frac{dI_h}{dt} = \varphi_h E_h - (\mu_h + \gamma_h + \alpha_h) I_h, \tag{3}$$

$$\frac{dE_v}{dt} = \frac{\beta_v b I_h}{N_h} S_v - (\mu_v + \delta_v) E_v, \tag{4}$$

$$\frac{dI_v}{dt} = \delta_v E_v - \mu_v I_v. \tag{5}$$

The rate of change in human and vector populations are constant, so that:

$$N_h = S_h + E_h + I_h + R_h,$$

$$N_v = A\mu_v = S_v + E_v + I_v.$$

$$N_v - E_v - I_v = A\mu_v - E_v - I_v.$$

2.2 Equilibrium Point

To find the point of equilibrium of the system (1)-(5), we simplify the system by assuming:

$$x = \frac{dS_h}{dt}\ u = \frac{dE_h}{dt}\ y = \frac{dI_h}{dt},\quad w = \frac{dE_v}{dt}$$

$$z = \frac{dI_v}{dt} = \frac{I_v}{A/\mu_v}$$

$$N_v = \frac{A}{\mu_v} = S_v + E_v + I_v, \text{ and}$$

$$S_v = N_v - E_v - I_v = \frac{A}{\mu_v} - E_v - I_v.$$

we obtained [13]:

$$\frac{dx}{dt} = \mu_h(1-x) - (p+\alpha z)x, \tag{6}$$

$$\frac{du}{dt} = (p+\alpha z)x - (\mu_h + \varphi_h)u, \tag{7}$$

$$\frac{dy}{dt} = \varphi_h u - (\mu_h + \gamma_h + \alpha_h)y, \tag{8}$$

$$\frac{dw}{dt} = \gamma_v(1-z-w)y - (\mu_v + \delta_v)w, \tag{9}$$

$$\frac{dz}{dt} = \delta_v w - \mu_v z. \tag{10}$$

where $\gamma_v = \beta_v b$, and $\alpha = \frac{\beta_v bA}{N_h \mu_v}$. With condition $\frac{dx}{dt} = \frac{du}{dt} = \frac{dy}{dt} = \frac{dw}{dt} = \frac{dz}{dt} = 0$, then the system in Eqs. (6)-(10) is simplified by assumtion $a = \mu_h, b = a + p, c = \alpha, d = p, e = a + \varphi_h, f = \varphi_h$, $g = a + \gamma_h + \alpha_h, h = \gamma_v, j = \delta_v + \mu_v, k = \delta_v$, and $l = \mu_v$.

We have

$$a - (b+cz)x = 0 \tag{11}$$

$$(d+cz)x - eu = 0 \tag{12}$$

$$fu - gy = 0 \tag{13}$$

$$h(1(-z-w)y) - jw = 0 \tag{14}$$

$$kw - lz = 0 \tag{15}$$

Thus the system (11)-(15) gained two points of equilibrium by using MAPLE.

a. The equilibrium point for free- disease of dengue

Equilibrium point for free-disease E_0 is a condition where there are no cases of dengue disease spread between human and vector populations. Equilibrium point can be written x_0, u_0, y_0, w_0, z_0. Thus obtained free-disease equilibrium point E_0 is obtained as follows:

$$x_0 = \frac{a}{b - cm(\sqrt{q}+n)} \tag{16}$$

$$u_0 = \frac{gjlm(\sqrt{q}+n)}{-fhm(\sqrt{q}+n)(k+l) - fkl} \tag{17}$$

$$y_0 = \frac{jlm(\sqrt{q}+n)}{-hm(\sqrt{q}+n)(k+l) - kh} \tag{18}$$

$$w_0 = -\frac{m(\sqrt{q}+n)}{k} \quad (19)$$

$$z_0 = m(\sqrt{q}+n) \quad (20)$$

b. *The equilibrium point for Endemic*

Equilibrium point for endemic is the case of the spread of dengue disease between human and vector. The next set of endemic equilibrium point, symbolized $E_1 = (x_1, u_1, y_1, w_1, z_1)$, we obtained [14] follows:

$$x_1 = \frac{a}{b + cm(\sqrt{q}-n)} \quad (21)$$

$$u_1 = -\frac{gjlm(\sqrt{q}-n)}{fhm(\sqrt{q}-n)(k+l) - fi} \quad (22)$$

$$y_1 = \frac{jlm(\sqrt{q}-n)}{hm(\sqrt{q}-n)(k+l) - kh} \quad (23)$$

$$w_1 = \frac{m(\sqrt{q}-n)}{k} \quad (24)$$

$$z_1 = m(\sqrt{q}-n) \quad (25)$$

where:

$$m = \frac{1}{2c(jleg + hfka + hfla},$$

$$n = hfkda + hflda - hfkca + jlegb,$$

$$q = (hfkda)^2 + 2kl(hfda)^2 + 2dc(hfka)^2 + 2hfkdajlegb + (hflda)^2 + 2ldkc(hfa)^2 + 2hfdajegbl^2 + (hfkca)^2 - 2hfkcajlegb + (jlegb)^2 + 4hfkdaljgec.$$

2.3 *Eigenvalue for SEIR Model*

Having obtained equilibrium point, will then be analyzed stability equilibrium point. The first step there is linearized system of linear equations that arise in models of dengue epidemic.
The equations used in the linearization process is as follows [15]:

$$f(x,u,y,w,z) = \frac{dx}{dt} = a - (b+cz)x, \quad (26)$$

$$g(x,u,y,w,z) = \frac{du}{dt} = (d+cz)x - eu, \quad (27)$$

$$h(x,u,y,w,z) = \frac{dy}{dt} = fu - gy, \quad (28)$$

$$i(x,u,y,w,z) = \frac{dw}{dt} = h(1-z-w)y - jw, \quad (29)$$

$$j(x,u,y,w,z) = \frac{dz}{dt} = kw - lz. \quad (30)$$

Linearization that has been done is the Jacobian matrix J.
Jacobian matrix of the system (24)-(29) is:

IOP Conf. Series: Journal of Physics: Conf. Series **954** (2018) 012018 doi:10.1088/1742-6596/954/1/012018

$$J=\begin{bmatrix} -b-cz & 0 & 0 & 0 & -xc \\ d+cz & -e & 0 & 0 & cx \\ 0 & f & -g & 0 & 0 \\ 0 & 0 & h(1-z-w) & -hy-j & -hy \\ 0 & 0 & 0 & k & -l \end{bmatrix}$$

Eigenvalues obtained if and only if $|J-\lambda I|=0$. With J is the Jacobian matrix that linearized for endemic equilibrium point, $\boldsymbol{\lambda}$ is eigenvalues and I is the identity matrix. Then the eigenvalues for endemic equilibrium point as follows [13]:

With:

$k_1 \;=\; = C_1C_2C_3(C_4C_5-C_6),$

$k_2 \;=\; = -(C_1C_2C_3(C_4+C_5)+(C_1C_2+C_1C_3+C_2C_3)(C_4C_5-C_6)),$

$k_3 \;=\; = C_1C_2C_3+(C_4+C_5)(C_1C_2+C_1C_3+C_2C_3)+(C_1+C_2+C_3)(C_4C_5-C_6),$

$k_4 \;=\; = -(C_1C_2+C_1C_3+C_2C_3)+(C_1+C_2+C_3)(C_4+C_5)(C_4C_5-C_6),$

$k_5 \;=\; = C_1+C_2+C_3+C_4+C_5.$ and

$C_1 \;=\; = -b-bc\sqrt{q}+bn,$

$C_2 \;=\; = -e,\; C_3=-g$

$C_4 \;=\; = \frac{hljm\,(\sqrt{q}-n)}{-kh+hm(\sqrt{q}-n)+(k+l)},\; C_5=-l$

$C_6 \;=\; = \frac{hljmk\,(\sqrt{q}-n)}{-kh+hm(\sqrt{q}-n)+(k+l)}.$

The stiffness matrix of composite sandwich plate is formed by combining the stiffness matrix of the top face sheet with the upper half of honeycomb core and the bottom face sheet with the lower half of honeycomb core.

We obtained Jacobian matrix follow:

$$\begin{pmatrix} -(b+cm(\sqrt{q}-n))-\lambda & 0 & 0 & 0 & -\frac{ca}{cm(\sqrt{q}-n)+b} \\ cm(\sqrt{q}-n)+d & -e-\lambda & 0 & 0 & \frac{ca}{cm(\sqrt{q}-n)+b} \\ 0 & f & -g-\lambda & 0 & 0 \\ 0 & 0 & h\left(1-m(\sqrt{q}-n)-\frac{m(\sqrt{q}-n)}{k}\right) & -\frac{hlm(\sqrt{q}-n)}{hm(\sqrt{q}-n)(k+l)-kh}-j-\lambda & \frac{hlm(\sqrt{q}-n)}{hm(\sqrt{q}-n)(k+l)-kh} \\ 0 & 0 & 0 & k & -l-\lambda \end{pmatrix}$$

$$\left(-b-bc\sqrt{q}+bn-\lambda\right)(-e-\lambda)\left[\left(\frac{hljm(\sqrt{q}-n)}{-kh+hm(\sqrt{q}-n)(k+l)}-j-\lambda\right)(-l-\lambda)-\left(\frac{hljmk(\sqrt{q}-n)}{-kh+hm(\sqrt{q}-n)(k+l)}\right)\right]=0.$$

So that is the characteristic equation can be simplified:

$k_1+k_2\lambda+k_3\lambda^2+k_4\lambda^3+k_5\lambda^4-\lambda^5=0.$

The conclusion of the eigenvalues is,

if $k_i>0$ or $k_i<0, \nabla k_i, i=1,2\ldots,5$. Then SEIR model always give eigenvalues are Real and complex. $\lambda_1=a, \lambda_{2,3}=b\pm ci$ and $\lambda_{4,5}=b\pm ci$ with $a,b,c,d,e\in R$ and Stability analysis there are two, namely: Stable asymptotic and unstable.

3 Stability Analysis for Medan City

A parameter estimation performed in each population. Parameters and initial values mosquito populations used are the parameters of the results by [14] on Sulawesi. Because it is assumed that it has the same area that is endemic to dengue fever.

Table 1 Parameter value spread of dengue fever

Parameter Name	Notation	Value
The rate of recovery of humans infected with dengue virus	γ_h	0.328833
The rate of human deaths caused by dengue virus	α_h	0.0000002
The rate of human infection with dengue virus	φ_h	0.16667
Persentase nyamuk terinfeksi virus *dengue*	P	0.09
The rate of mosquitoes infected with dengue virus	δ_v	0.1428
The rate of human deaths	μ_h	0.000046
Mosquito mortality rate	μ_v	0.0323
The rate of contact of humans to the vector	$b\beta_h$	0.75
The rate of the vector-to-human contact	$b\beta_v$	0.375
The number of births mosquitoes	$A = 1\% x\ N_h$	2899.56

$$N_h = 289956, \alpha = \frac{\beta_h bA}{N_h \mu_v} = 0{,}232198142.$$

3.1 Equilibrium Points for Medan City

From equation (11-15) equilibrium model of SEIR the parameter values in the table above can be formulated as follows:

$$\begin{aligned} 0{,}000046 - (0{,}090046 + 0{,}232198412z)x &= 0 \\ (0{,}09 \mp 0{,}232198412z)x - 0{,}166746u &= 0 \\ 0{,}1667u - 0{,}3288792y &= 0 \\ 0{,}375(1 - z - w)y - 0{,}1751w &= 0 \\ 0{,}428w - 0{,}0323z &= 0 \end{aligned}$$

By using the software MAPLE obtained equilibrium point value. For disease-free equilibrium point is as follows: $E_0 = (199{,}0599611, -0{,}0547635, -0{,}0277582, -0{,}087716, -0{,}3877971)$
Cannot be assessed because only S_h value that is positive while the value of the other negative. This means it can be said E_0 unstable. Can be assessed by looking at its eigenvalues are:
$\lambda_1 = 0{,}6354942857,\ \lambda_2 = -0{,}1930171439 + 0{,}8220223055i,\ \lambda_3 = -0{,}0000093748,\ \lambda_4 = -1{,}023107911,\ \lambda_5 = -0{,}1930171439 - 0{,}822022306i.$

Because the equilibrium point is said to be unstable if one of the roots of the characteristic equation is real and positive, or if at least one root has a positive real part. For endemic equilibrium point $E_1 = (0{,}0005091, 0{,}0002757,\ 0{,}0001398, 0{,}0002988, 0{,}0013211)$is an asymptotically stable equilibrium point. It can be assessed by looking at the eigenvalues of E_1 namely

$$\lambda_1 = -171292 + 0{,}0091331i, \qquad \lambda_2 = -0{,}0324563,$$
$$\lambda_3 = -0{,}0893149,\ \lambda_4 = -0{,}3290751,\ \ \lambda_5 = -0{,}1731091 - 0{,}87673321i.$$

Because the equilibrium point is said to be asymptotically stable if the roots of the characteristic equation are real and negative.

3.2 Basic Reproduction Ratio

Looking for R_0 value is important because it can show how much the infection will occur to individuals as a result of the infection. If a patient is free from dengue fever, $R_0 \leq 1$ value. If the patient can transmit to other individuals and more than one, the value $R_0 \geq 1$ and DFD endemic. According to [16], R_0 is the number of secondary cases generated by a vector-borne infectious cases in the whole population vulnerable. Mathematically defined as the dominant eigenvalues of the linear equation positive (constant). To determine the value R_0 so that we can take the constant of the characteristic equation (4.5) ie:

$k_1 = 0, \quad = \quad C_1C_2C_3(C_4C_5 - C_6) = 0,$
$C_1C_2C_3C_4C_5 = C_1C_2C_3C_6$

$$\frac{C_4C_5}{C_6} = 1$$

Thus, we have the basic reproduction ratio R_0 as follow:

$$R_0 = \frac{-hjm\,(\sqrt{q} - n)}{(-kh + hm(\sqrt{q} - n) + k + l)\left(\dfrac{hjlm(\sqrt{q} - n)}{-kh + hm(\sqrt{q} - n) + k + l} - j\right)}$$

By substituting the values of the parameters, the obtained value

$$R_0 = 2.46 \geq 1,$$

That means dengue fever does not disappear from the population and cause endemic in the district are Medan Helvetia and Medan Sunggal sub-disricts. So, every patient can transmit the dengue disease to more than one new case, so may cause more extensive disease.

3.3 Numerical Simulation for Medan City

In this section, Simulation of SEIR model of spread of dengue disease using data from the district and Sunggal Medan Medan Helvetia Medan sub-districts year 2013 to 2014.

Table 2. Data of DFD epidemic in Medan Helvetia and Medan Sunggal sub districts

	2013			2014		
	Medan Helvetia	Medan Sunggal	TOTAL	Medan Helvetia	Medan Sunggal	TOTAL
S_h	145.039	144.526	289.565	145.207	148.013	293.220
E_h	5	9	14	7	24	31
I_h	96	89	185	150	147	297
R_h	99	93	192	155	176	331
N_h	145.239	144.717	289.956	145.519	148.360	293.879

From the above table can be obtained initial conditions of population in 2013:

$$S_h(0) = x(0) = \frac{289565}{289956} = 0{,}998652$$

$$E_h(0) = u(0) = \frac{14}{289956} = 0{,}000048$$

$$I_h(0) = y(0) = \frac{185}{289956} = 0{,}000638$$

and the initial value used in vector population is the result of research [12], namely:

$I_v(0) = w(0) = 0.01$
$E_v(0) = z(0) = 0.056$

IOP Conf. Series: Journal of Physics: Conf. Series **954** (2018) 012018 doi:10.1088/1742-6596/954/1/012018

Based on that description, the spread of dengue fever can be written as follows. By the time interval [0, 30] and length of integration step h = 0.001. Suppose for:

$$\frac{dx}{dt} = 0{,}000046 - (0{,}090046 + 0{,}232198412z)x = f_1(t,x,u,y,w,z),$$

$$\frac{du}{dt} = (0{,}09 + 0{,}232198412z)x - 0{,}166746u = f_2(t,x,u,y,w,z)$$

$$\frac{dy}{dt} = 0{,}1667u - 0{,}3288792y = f_3(t,x,u,y,w,z)$$

$$\frac{dw}{dt} = 0{,}375(1 - z - w)y - 0{,}1751w = f_4(t,x,u,y,w,z)$$

$$\frac{dz}{dt} = 0{,}1428w - 0{,}0323z = f_5(t,x,u,y,w,z)$$

With initial value

$x_0 = 0{,}998652$, $u_0 = 0{,}000048$,
$y_0 = 0{,}000638$, $w_0 = 0{,}01$, $z_0 = 0{,}056$.

4. Results and Discussion

Graph data for 2013-2014 on S_h population, E_h, and I_h, can be shown in figure 1, 2 and 3

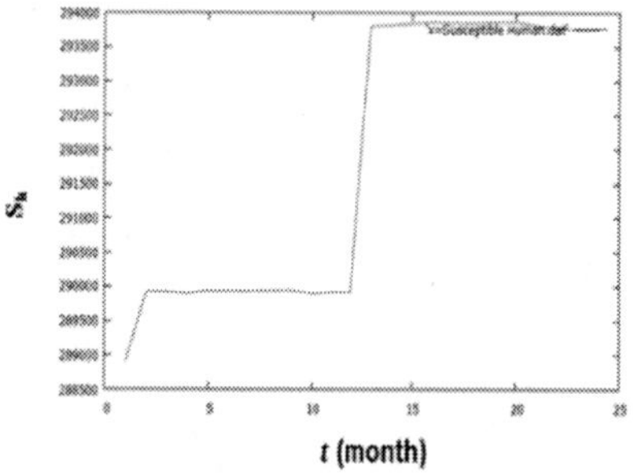

Figure 1. Graphs for Susceptible S_h Population in Medan

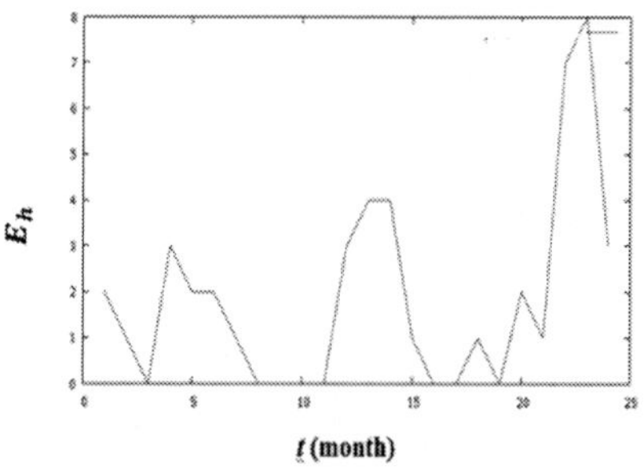

Figure 2. Graphs for Exposed E_h Population in Medan

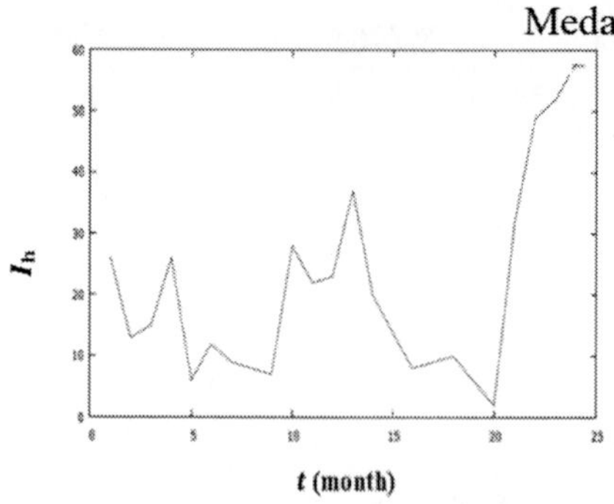

Figure 3. Graphs for Infected I_h Population in Medan

In Figure 1 it can be seen that there is an increase and decrease very significantly. In Figure 2 u_h-class growth has decreased, this is due to E_h class has moved into the class room I_h. And it can be seen in Figure 3 that an increase in the growth of human infected grade is high enough. This is because the Medan Helvetia and Medan Sunggal sub-districts the occurrence of endemic dengue disease.

IOP Conf. Series: Journal of Physics: Conf. Series **954** (2018) 012018 doi:10.1088/1742-6596/954/1/012018

Table 3. The Probability of SEIR model for DFD in year 2013-2014

Year	S_h	E_h	I_h	E_v	I_v
2013	0,9483166900	0,0486645532	0,0021747607	0,0093444069	0.0557868236
2014	0.0425493927	0.1065460550	0.0723738719	0.1244155478	0.3269441863

Graphs which were obtained by simulatiin for S_h, E_h, I_h, E_v, and I_v can be shown in figure 5, 6, 7 and 8.

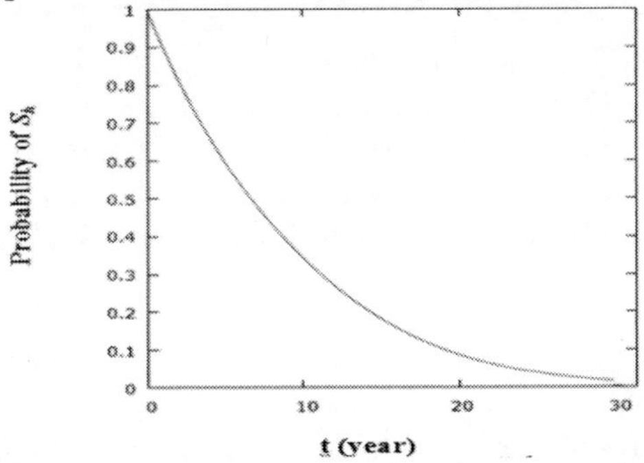

Figure 4. Probability of S_h through simulation

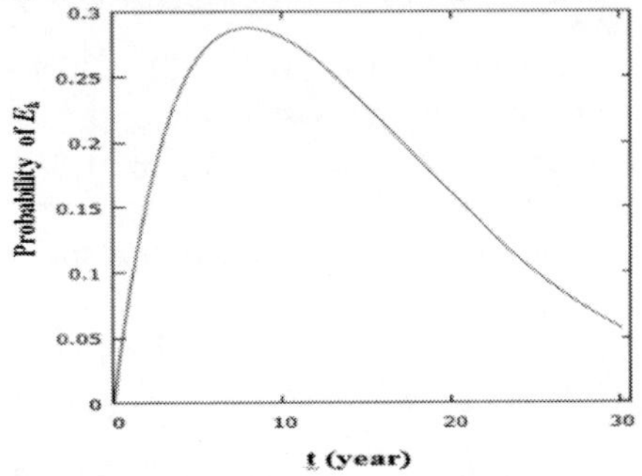

Figure 5 Probability of E_h through simulation

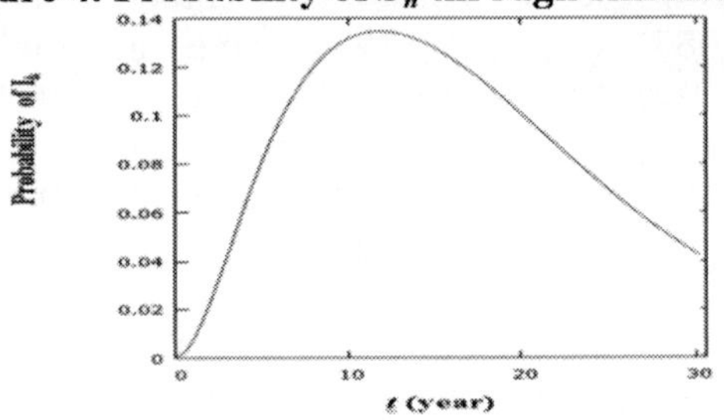

Figure 6 Probability of I_h through simulation

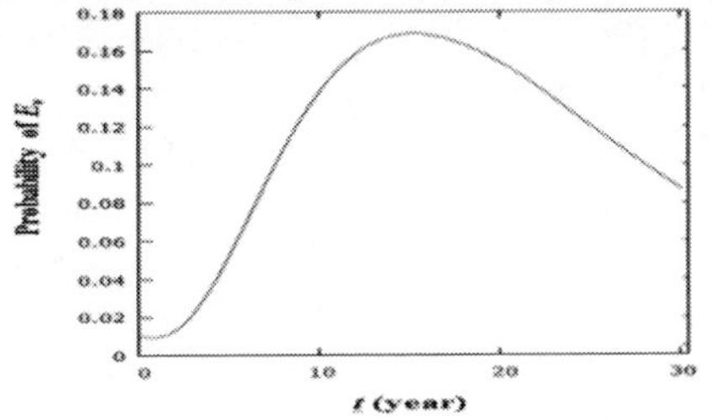

Figure 7 Probability of E_v through simulation

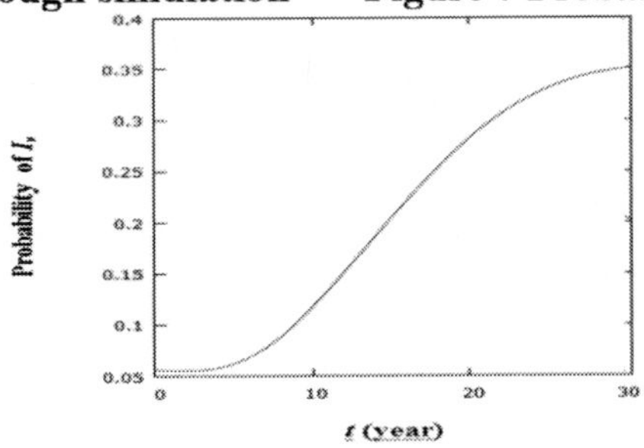

Figure 8 Probability of I_v through simulation

Based on Figure 4 shows that the rate class S_h initially fell significantly to near 0 at $t = 30$. Reduced class as this S_h due to the displacement of the class into groups and classes E_h and I_h. In the Figure 5 The growth rate is reduced E_h classes ranging from $t = 10$ due to the displacement of the class becomes class I_h and E_h. Then the growth rate was reduced I_h. Then S_h class condition, E_h, and I_h will run continuously until toward an asymptotically stable point and at that point until $t \to \infty$. This means there is no spread of the virus dengeu in each class in the human population.

Based on Figure 7 shows that the growth rate is influenced by many E_v class of vectors that bite humans infected with dengue virus. I_v class growth rate initially rose significantly to exceed 0.25 in the initial time interval until 30 months. This is because of an infected of vector will continue to transmit during his lifetime. But the rate of growth of the running time seiringnya E_v and I_v classes is reduced so that the graph slowly down until toward an asymptotically stable point and at that point until $t \to \infty$. This means there is no spread of the virus dengue in each class in the vector population. This means that, in this population had no additional classes I_v and S_v which becomes infected vector.

From Figure (4, 5, 6, 7, 8), the growth rate S_h, E_h, I_h, E_v, and I_v significant drop to near 0 in the initial time interval up to 30 months. This is because each class on the amount of human and vector populations are assumed equal. That is, every class that are susceptible will go to class exposed and infected.

1. Based on the above, the result is that dengue fever is endemic in Medan Helvetia and Medan Sunggal sub-districts. It is necessary for the eradication or termination of the chain of transmission of vector. Here are some of the efforts that can be done to break the chain of transmission of vector.
2. Protect yourself from vector bites. This can be done by installing netting on the windows, use vector repellent and insecticide in the form of a spray. Do 3M movements is shut water reservoirs, drain regular shelters and burying water cans or containers empty unused. This study gives results that dengue fever does not disappear from Medan Helvetia and Medan Sunggal sub-districts. This is consistent with the data from health center of the area where every year there is always a human infected.

5. Conclusion

Model SEIR in the spread of dengue fever has been analyzed. The analysis of SEIR model obtained: 1) The SEIR model have two points of equilibrium that are disease-free equilibrium point $E_0 = (199.0599611, -0.0547635, -0.0277582, -0.0877160, -0.3877971)$ and the endemic equilibrium point $E_1 = (0.0005091, 0.0002757, 0.0001398, 0.0002988, 0.0013211)$. Results of stability analysis E_0 unstable equilibrium point and E_1 asymptotically stable equilibrium point. 2) The value of Basic Reproduction Number R_0= 2.46 > 1, which means that the local sub-district and sub-district of Medan Sunggal Helvetia is endemic area. This means that dengue fever will not be lost on those areas. $R_0 > 1$ indicate each patient can transmit the dengue disease to more than one new cases, so that the disease can spread more widely. 3) Numerical simulation on the spread of dengue fever using Runge-Kutta Order 4. Simulation the spread of dengue disease in SEIR models indicate a case of endemic, meaning that dengue fever is still present in the human and vector populations with time $t \to \infty$. Then the rate of growth of susceptible human class, Exposed human, human Infected, vector Exposed and Infected of vector will drop significantly to near 0 in the initial time interval to 30 years. This is because each class the number of human and vector populations are assumed equal. This means that every class that is on will enter the classroom susceptible exposed and infected. And there are many other epidemic mathematical model that can be developed. One of them is the model SEIRS. In the model SEIRS, later classes who have recovered can be infected with the same disease again.

Acknowledgement

We gratefully acknowledge, the financial support received from DIPA Universitas Negeri Medan (UNIMED) Indonesia with no DIPA-000080/UN33 /KU/2015.

References

[1] Herbert, W. H, 2000, The Mathematics of Infectious Diseases, *Siam Rev.*, **42**, 4, pp. 599-653.

[2] M. M. A. El-Sheikh and S. A. A. El-Marouf, 2004, On stability and bifurcation of solutions of an SEIR epidemic model with vertical transmission, *Inter. J. of Math. & Math. Sci.* **2004**, 56, pp. 2971-2987.

[3] Al-Samadi, M., dan Gumah, G.N., 2014, On the Homotopy Analysis Method for Fractional SEIR Epidemic Model, *J. of App. Sci.*, **7**, 18, pp. 3809-3820.

[4] Xue-Zhi Li and Bin Fang, 2009, Stability of an Age-structured SEIR Epidemic Model with Infectivity in Latent Period. *Applications and Applied Math.* **4**, 1, pp. 218 – 236

[5] Na Yi, Qingling Zhang, Kun Mao, Dongmei Yang, Qin Li, 2009, Analysis and control of an SEIR epidemic system with nonlinear transmission rate. *Math. Comp. Model.* **50**, 9-10, pp. 1498-1513.

[6] Ma, Z., dan Jia, L., 2007, *Dynamical Modeling and Analysis of Epidemics,* World Scientific Publishing, Singapore.

[7] Li, J., dan Cui, N., 2013, Dynamic Analysis of an SEIR Model with Distinct Incidence for Exposed and Infectives, *Sci. World J.*, **2013**, 871393.

[8] Putra, T.R., 2011, Kestabilan Lokal Bebas Penyakit Model Epidemi SEIR Dengan Kemampuan Infeksi Pada Periode Laten, Infeksi, dan Sembuh, *Poli Rekayasa*, **7**, 1, 42-52.

[9] Kholisoh, S., Waluya, B., dan Kharis, M., 2012, Model Epidemi Pada Penyebaran Penyakit Campak dengan Pengaruh Vaksinasi, *UNNES J. of Math,* **1**, 2, 111-117

[10] M. El-Doma, 2013, Stability Analysis of a Size-Structured Population Model with Maturation Size and a Constant Inflow of Newborns. *Int. J. of Ecological Eco. Stat.***28**, 1.

[11] SANGARE Boureima, DIALLO Ouateni, SOME Longin, 2013, An Analysis of Stability and Convergence of A Finite-Difference Methods for One-Dimensional Partial Integro-Differential Equation Using A Moving Mesh. *Int. J. of Applied Math. Stat.*, **50**, 20.

[12] Side, S. and Noorani, M.S.M., 2013, A SIR model for spread of dengue fever disease (Simulation for South Sulawesi Indonesia and selangor Malaysia), *World J. of Model Simulation*, **9**, 2, pp. 96-105.

[13] Side, S. and Noorani, M.S.M., 2013, Lyapunov function of SIR and SEIR model for transmission of dengue fever disease. *Inter. J. of Simulation Proc. Model (IJSPM)*. **8**, 2/3, 177-184.

[14] Side, S. and Noorani, M.S.M., 2012, SEIR model for transmission of dengue fever, *Inter. J. of Adv. Sci. Eng. Inf. Tech.*, **9**, 1, 1-5.

[15] Rangkuti, Y.M., Sinaga M. S., Marpaung F., Side, S., 2014, A VSEIR model for transmission of Tuberculosis (TB) disease In north Sumatera, Indonesia, *AIP Conference proceedings*, **1635**, 201, doi : 10.1063/1.4903584.

[16] Diekmann O, Heesterbeek JA, & Metz JAJ, 1990. On the definition and the computation of the basic reproduction ratio R infectious diseases in heterogeneous populations. *J Math Biol.* **28**, pp. 365–382

IOP Conf. Series: Journal of Physics: Conf. Series **954** (2018) 012019 doi:10.1088/1742-6596/954/1/012019

Analysis of Student Satisfaction Toward Quality of Service Facility

D Napitupulu[1]*, R Rahim[2], D Abdullah[3], MI Setiawan[4], LA Abdillah[5], AS Ahmar[6], J Simarmata[7], R Hidayat[8], H Nurdiyanto[9], A Pranolo[10]

[1]Research Center for Quality System & Testing Technology,
Indonesian Institute of Sciences, Tangerang Selatan, Banten, Indonesia
[2]School of Computer and Communication Engineering, Universiti Malaysia Perlis, Malaysia
[3]Department of Informatics, Universitas Malikusaleh, Aceh, Indonesia
[4]Department of Civil Engineering, Narotama University, Surabaya, Indonesia
[5]Department of Information System, Universitas Bina Darma, Palembang, Indonesia
[6]Department of Statistics, Universitas Negeri Makassar, Makassar, Indonesia
[7]Department of Informatics, Universitas Malikusaleh, Aceh, Indonesia
[8]Department of Information Technology, Politeknik Negeri, Indonesia
[9]Department of Informatics Engineering, STMIK Dharma Wacana, Lampung, Indonesia
[10]Department of Informatics, Universitas Ahmad Dahlan, Indonesia

*darwan.na70@gmail.com

Abstract. The development of higher education is very rapid rise to the tight competition both public universities and private colleges. XYZ University realized to win the competition, required continuous quality improvement, including the quality of existing service facilities. Amenities quality services is believed to support the success of the learning activities and improve user satisfaction. This study aims to determine the extent to which the quality of the services effect on user satisfaction. The research method used is survey-based questionnaire that measure perception and expectation. The results showed a gap between perception and expectations of the respondents have a negative value for each item. This means XYZ service facility at the university is not currently meet the expectations of society members. Three service facility that has the lowest index is based on the perception of respondents is a laboratory (2.56), computer and multimedia (2.63) as well as wifi network (2.99). The magnitude of the correlation between satisfaction with the quality of service facilities is 0.725 which means a strong and positive relationship. The influence of the quality of service facilities to the satisfaction of the students is 0.525 meaning that the variable quality of the services facility can explain 52.5% of the variable satisfaction. The study provided recommendations for improvements to enhance the quality of services facility at the XYZ university facilities.

1. Introduction

These Quality education system of course closely related to the quality of human resources. It is based on the reality that the output of an educational system is the human resource that will be used in industry and development in an area. Efforts to improve the quality of human resources can be done by improving the educational system which can then produce a quality education system. The government is also aware of the importance of education, this can be seen by the implementation of 9-year compulsory education program for Indonesian citizens. Private universities are one of the alternatives for society to face the increasing demand of society to the education needs of today. Various advantages of private universities that will attract prospective students increasingly highlighted, ranging from degree bachelor, curriculum given to a variety of facilities that will be obtained by students will study on campus. Competition in the world of education is also getting tighter with the increase of universities, especially private universities (PTS) in Jakarta caused increasingly fierce competition. XYZ University is aware to win the competition, it needs continuous quality improvement including the quality of service facilities available to users. In addition to the continuously updated curriculum, service facilities also need to be improved for campus advancement [1].

The level of satisfaction is a function of the difference between perception and expectation [2]. The purpose of this study is to measure the gap between the perceptions and expectations of users of service facilities as a form of evaluation of the performance of the university. This research also wanted to know the relation between quality of service facility and user satisfaction. At the same time how much influence the quality of service facilities to user satisfaction. Practically, this research is expected to provide input for management related to the evaluation of condition of service facility based on user perspective so that it can formulated a comprehensive service quality improvement strategy in facing the tight competition in educational environment especially in Jakarta. Thus, in turn, it can attract high school / senior high school students to enroll in the campus environment.

Based on the background that has been described before, it can be formulated research problems is how the level of user satisfaction with the quality of existing service facilities? Then how the relationship between the quality of service facilities with user satisfaction on the XYZ university? Thus the purpose of this study is to evaluate the condition of service facilities based on user perspective so it can be known how far the level of user satisfaction of existing facilities. This can provide input for university management to make continuous improvement process.

2. Literature Review

If you Based on the literature review, similar research that has been done can be explained as follows:

- Research conducted by Sutino and Sumarno (2005) on the Effect of Product Quality (Service) and Quality of Service to Customer Satisfaction and Loyalty at PT. POS Indonesia Semarang. The results showed that the quality of service influence on satisfaction and satisfaction proved to affect loyalty [3].
- Research conducted by Samosir and Zurni (2005) on the Effect of Service Quality on Student Satisfaction Using USU Library. This research is descriptive explanatory where the method used is a survey method that uses questionnaires as a tool to collect data with the unit of analysis is USU students. The study used descriptive statistics and showed that service quality (reliability, responsiveness, assurance, empathy, and direct evidence) simultaneously had a significant effect on student satisfaction [4].
- Research conducted by Sutardji and Sri (2006) on the Analysis of Some Factors that Influence on Library User Satisfaction. The research method used is survey with questioneer as a tool to collect data. The results showed there was a positive influence between service quality variable to student satisfaction variable. Thus, the quality of service has a positive effect on student satisfaction [5].
- Research conducted by Sukandi (2010) on the Relationship Between Campus Facilities To Student Satisfaction In Facing the Competitiveness of Educational Services. This research is descriptive with survey method. The results showed that there is a positive relationship between the facilities provided by the campus on student satisfaction [6].

Based on previous literature review it can be said that there is a relationship between service quality and student satisfaction. Therefore, in this research want to know how big relation between

IOP Conf. Series: Journal of Physics: Conf. Series **954** (2018) 012019 doi:10.1088/1742-6596/954/1/012019

service quality with student satisfaction and the influence of service quality to student satisfaction especially in XYZ university.

3. Research Methodology

In this section we The sample is part of the number and characteristics possessed by that population or the delivery of something by representatives of the class, but simply represents the whole opinion. This research was conducted by questionnaire-based survey method that was distributed to students of XYZ University Computer Science Faculty. Sampling technique with purposive sampling from all study program in Computer Science Faculty (S1), Information System (S1), Computer System (S1), Computer Engineering (D3) and Information Management (D3). Here is the distribution of active student data from the Faculty of Computer Science at XYZ University as the study population presented in table 1 below:

Table 1. Population of Respondent.

No	Department	Number of Students
1.	Computer Science	213
2.	Information System	425
3.	Computer System	126
4.	Computer Engineering	25
5.	Information Management	53
	Total Population	842

In this study the number of samples is calculated using the Slovin formula [7] where the population is the total active students academic year 2015/2016 is 842 people, namely:

$$\eta = \frac{N}{1 + (N \times e^2)} \quad (1)$$

where :
η = Number of samples
N = Population
e = Inaccuracy ease due to intolerable sampling error (using 10%)

Based on the formula (1) above, the sample number is as follows:

$$\eta = \frac{842}{1 + (842 \times 0.1^2)} = 89{,}38 \quad (2)$$

Thus the minimum sample size in this study was 89 respondents.

The distributed questionnaire has been designed and tested face validity by 30 respondents. Face validity is done to ensure the respondent understands the questions or statements contained in the questionnaire. Based on the results of face test (face validity) showed that all respondents agree that each item in the questionnaire easy to read and understand as in Table 2. The questionnaire distributed using Likert scale is the scale used to calculate a perception, attitude or opinion of a person or group about a Events or social phenomena based on operational variables that have been established by researchers. The scale used has intervals of 1-5 where 1 (very unsatisfied), 2 (unsatisfied), 3 (less satisfied), 4 (satisfied) and 5 (very satisfied). A total of 89 respondents who were involved only 84

IOP Conf. Series: Journal of Physics: Conf. Series **954** (2018) 012019 doi:10.1088/1742-6596/954/1/012019

respondents who answered and filled out the research questionnaire completely. An incomplete questionnaire will not be included in the analysis phase.

4. Result and Discussions

4.1. Validity and Reliability Testing

Before the data is processed further, the validity and reliability test of the questionnaire was conducted to determine the extent to which the measuring instrument is accurate and reliable or reliable. Testing the validity and reliability of the questionnaire was done with the help of the program SPPS IBM Statistics version 22 where obtained correlation coefficient (Corrected Item-Total Correlation) above 0.3 means that all items questionnaire can be said valid [8]. While the reliability coefficient is determined based on the value of Cronbach's Alpha where obtained the overall value is 0.83, which means that the questionnaire used has been reliable because it meets the minimum requirement of 0.60 [9]. From the results of validity and reliability test can be said that the questionnaire used in this study is accurate and reliable. After validity and reliability test, based on the data that has been processed, the analysis of user satisfaction level through measurement to the perception and expectation of the users on the quality of campus service facilities that exist as presented in Table 2.

4.2. User Satisfaction Analysis

In Table 2, there are two dimensions of learning space dimension and campus environment dimension. The dimension of the classroom is related to the facilities that support learning activities in the classroom. While the dimension of the campus environment is a facility outside the classroom that supports learning activities for students. Based on Table 3 can be seen gap between the perceptions and expectations of respondents on the quality of existing service facilities. From the result of measurement of satisfaction, it seems that for each item questionnaire, perception index is smaller than user expectation index. This means that the quality of service facilities that exist today is not yet satisfactory users. When further examined, the gap between perceptions and expectations is negative (perceptions - expectations) which means perceived performance is still far from expected.

Table 2. Perception and Expectation Measurement.

No	Item	Perception	Expectation
	Class Room Dimension		
1.	Air Conditioning	3.79	4.12
2.	Facilities benches and chairs in the classroom	3.39	4.37
3.	Computer and multimedia facilities in the classroom	2.63	4.25
4.	Stationary facilities in the classroom	3.95	4.06
5.	Facilities of books - books, other reference journals	3.01	3.84
6.	Laboratory facilities on campus	2.56	4.30
	Environment Dimension		
7	Wifi network facilities on campus	2.99	4.38
8.	Library room facilities on campus	3.49	4.15
9.	Facility of Student Activity Unit on campus	3.23	4.09
10.	Sports activities facilities on campus	3.35	4.18
11.	Vehicle parking facilities on campus	3.27	4.23
12.	College canteen facility	3.24	4.21
13.	Convenience learning situations that can motivate student learning	3.69	4.45
14.	Cleaning facilities Campus toilets	3.24	4.13
	Average score	3.27	4.20

Table 3. User Satisfaction Level.

No	Item	Gap
1.	Air Conditioning	-0.33
2.	Facilities benches and chairs in the classroom	-0.98
3.	Computer and multimedia facilities in the classroom	-1.62
4.	Stationary facilities in the classroom	-0.11
5.	Facilities of books - books, other reference journals	-0.83
6.	Laboratory facilities on campus	-1.74
7.	Wifi network facilities on campus	-1.39
8.	Library room facilities on campus	-0.66
9	Facility of Student Activity Unit on campus	-0.86
10.	Sports activities facilities on campus	-0.83
11.	Vehicle parking facilities on campus	-0.96
12.	College canteen facility	-0.97
13.	Convenience learning situations that can motivate student learning	-0.76
14.	Cleaning facilities Campus toilets	-0.89
	Average Score	**-0.92**

The total user satisfaction level also shows a negative value of -0.92. This indicates that overall users feel the condition of existing service facilities is still less or have not met the expectations of users. In fact, there are 3 (three) service facilities that have the lowest perception index of laboratory facilities (2.56), computer facilities and multimedia in the classroom (2.63) and wifi network facilities (2.99). Whereas the laboratory, computer and multimedia and internet network is a very important infrastructure in developing the core competencies of computer science faculty students.

4.3. Correlation of User Satisfaction and Service Facility

In this study wanted to know the relationship between the variable quality of campus service facilities with student satisfaction. The size of the relationship between variables is expressed in numbers called the correlation coefficient (R) that is between -1 to +1 where if approaching +1 then the relationship is stronger and positive. Meanwhile, if close to -1 then there is a stronger relationship but the direction is negative. If the correlation coefficient is zero then it means there is no relationship at all between variables. The result of correlation analysis between independent and dependent variable can be presented in Table 4 below:

Table 4. Model summary.

Model	R	R Square	Adjusted R Square	Std. Error of the Estimate
1	0.725[a]	0.575	0.525	0.26961

Based on Table 4, the strength of the relationship between campus service facility quality variables and user satisfaction is 0.725. If referring to Table 5 below, the relationship or correlation between campus service facility variables and user satisfaction variables is strongly interpreted. In addition to a strong correlation, the pattern of the relationship was positive. This means that if the service quality variables increase, then the satisfaction variable will also rise and vice versa ie if the variable of campus service facilities decreased then the variable of user satisfaction will also decrease [10].

Table 5. Interpretation of Correlation Coefficient.

Coefficient Interval	Correlation
0.00 – 0.199	Very Weak
0.20 – 0.399	Weak
0.40 – 0.599	Medium
0.60 – 0.799	Strong
0.80 – 1.000	Very Strong

In Table 4, it can be shown that the variation of customer satisfaction variables can be explained by independent variables namely the quality of campus service facilities through the value of coefficient of determination R Square (R^2). The result of analysis with statistic obtained value R^2 equal to 0.575. This means the variable quality of campus service facilities as free variables contribute 57.5% to the dependent variable that is user satisfaction. Thus there are still 42.5% of the rest can be explained other variables that are not contained in the model. The results of this study reinforce previous research that the positive relationship between service quality and student satisfaction [6] and the influence of service quality on student satisfaction [3][4][5].

5. Conclusion

Based on the results of research conducted could be drawn some conclusions are follows:

1. The evaluation of the condition of the quality of the campus service facilities shows a low level of student satisfaction, which can be seen from the gap between the perceptions and expectations of users where for each item there is negative. In other words, the quality of campus service facilities is still far below user expectations.
2. Three of campus service facilities that have the lowest index based on user perception are campus: computer and multimedia laboratory in classroom and wifi network.
3. There is a strong and positive relationship (coefficient R) of 0.725 between the quality of service facilities and student satisfaction.
4. The quality of campus service facilities had an effect on student satisfaction as the user of 57.5%. It means all service facility variables in this study could explain the variation of student satisfaction by 57.5% and the rest about 42.5% would be explained with other variables.
5. Further research suggestions should be examined other variables outside this study which also affect the user satisfaction.

References

[1] Alma, B. (2007), " Strategi Pemasaran Jasa Pendidikan Tinggi", Bunga Rampai Strategic, Manajemen Bisnis berbasis hasil penelitian, Cetakan Kesatu, April 2007, Alfabeta, Bandung, p. 1 – 13.

[2] Ahmad, S., Pramono & Suratno. (2009). Faktor-Faktor Yang Mempengaruhi Keberhasilan Siswa Kelas 2 Tmo Smk Texmaco Semarang Pada Mata Diklat Service Engine Dan Komponen-Komponennya, Jurnal PTM Vol. 9, No. 2 : 57-64.

[3] Sutino & Sumarno. (2005). Pengaruh Kualitas Produk (Jasa) dan Kualitas Pelayanan terhadap Kepuasan dan Loyalitas Pelanggan di PT.POS Indonesia Semarang 50000, Majalah Ilmiah Kopertis Wilayah VI, Vol. XV, No. 23: 9 – 20.

[4] Samosir & Zurni, Z. (2005). Pengaruh Kualitas Pelayanan terhadap Kepuasan Mahasiswa Menggunakan Perpustakaan USU, Pustaha: Jurnal Studi Perpustakaan dan Informasi, Vol. 1, No. 1 : 28 – 36.

[5] Sutardji, & Sri, I.M. (2006). Analisis Beberapa Faktor yang Berpengaruh pada Kepuasan Pengguna Perpustakaan: Studi Kasus di Perpustakaan Balai Penelitian Tanaman Kacang-Kacangan dan UmbiUmbian, Jurnal Perpustakaan Pertanian, Vol. 15, No.2, 2006: 32– 37.

[6] Sukandi, P. (2010). Hubungan Antara Fasilitas Kampus Terhadap Kepuasan Mahasiswa Dalam Menghadapi Daya Saing Jasa Pendidikan, The 4th PPM National Conference on Management Research.
[7] Husein, U. (2004). Metode Penelitian untuk Skripsi dan Tesis Bisnis. Cetakan ke-6. Jakarta: PT Raja Grafindo Persada.
[8] Azwar, S. (2012). Reliabiltas dan Validitas. Edisi 4. Yogyakarta : Pustaka Pelajar.
[9] Sekaran, U. (2006). Metode Penelitian Bisnis. Jakarta : Salemba Empat.
[10] I Gusti, A., & Desak, P.N. (2008). Faktor-Faktor Penentu Kepuasan Mahasiswa Terhadap Pelayanan Fakultas Sebagai Lembaga Pendidikan (Studi Kasus Di Fmipa, Universitas Udayana) Jurnal Cakrawala Pendidikan November 2008, Th. XXVII, No. 3.

Acknowledgments
We would like to thank the research institution that has supported the research activities was being carried out properly.

Adsorption of Cd(II) Metal Ion on Adsorbent beads from Biomass *Saccharomycess cereviceae* - Chitosan

Hasri[1] & Mudasir[2]

[1]Department of Chemistry, Faculty of Mathematics and Natural Sciences, Universitas Negeri Makassar, Parangtambung, Makassar, Indonesia
[2]Department of Chemistry, Faculty of Mathematics and Natural Sciences, Universitas Gadjah Mada,Sekip Utara, Yogyakarta, Indonesia, 55281

hasriu@unm.ac.id

Abstract. The adsorbent beads that was preparation from Saccharomycess cereviceae culture strain FN CC 3012 and shrimp shells waste and its application for adsorption of Cd (II) metal ion has been studied. The study start with combination of Saccharomycess cereviceae biomass to chitosan (Sc-Chi), contact time, pH of solution and initial concentration of cations. Total Cd(II) metal ion adsorbed was calculated from the difference of metal ion concentration before and after adsorption by AAS. The results showed that optimum condition for adsorption of Cd(II) ions by Sc-Chi beads was achieved with solution pH of 4, contact time of 60 minutes and initial concentration adsorption 100mg/L. The hydroxyl (-OH) and amino ($-NH_2$) functional groups were believed to be responsible for the adsorption of Cd(II) ions.

1. Introduction

Decreased levels of heavy metal waste, especially waste of waters have become a serious concern of various parties. In line with the several methods have been made in reducing the presence of heavy metals such as waste reduction methods, precipitation, ion exchange method, coagulation, solvent extraction, reserve osmosis, adsorption or in combination with other [1][2]. However, the adsorption method is simpler easily applied and cheaper in terms of cost.

Adsorption methods made from natural materials, is now experiencing rapid development especially materials focused on material that has a high affinity, selectivity and ability to absorb metal cations in particular [3].

One of promising adsorbent material is using microorganisms as adsorbent. The use of microorganisms is very promising considering both living and dead microorganisms has the ability to accumulate metals, because the cell wall composed of polysaccharides, lipids and protein which is carboxylic group, hydroxyl, sulfhydril, phosphates and amino acids that able to binding metal ions [4]-[8].

On the other side, the selection matrix of chitosan as a supporter of microorganisms, additionally the chitosan resources that abundant especially on crab and shrimp waste, non-toxic and the results showed that the chitosan has active sites such as OH, NH_2 and -NHCO-, this group can form chelates with metal cations.

Microorganisms such as *S. cereviceae* as adsorbent has higher surface area that is suitable for absorption of cations [9]. In addition, biomass S. *cereviceae,* either dry (dead) and dead have the same

IOP Conf. Series: Journal of Physics: Conf. Series **954** (2018) 012020 doi:10.1088/1742-6596/954/1/012020

ability and easily decomposed. Technically biomass very difficult to used repeatedly for low mechanical strength. Therefore, it needs to be modified to overcome the shortage of biomass. Through the combination with chitosan, it is possible to improve the mechanical properties, chemical properties, increasing the porosity and the application can increase the ability to absorb metal cations [10]. The combination of both active sites of the adsorbent and suitable characteristics metal ion is an important factor in progress adsorption takes place [11].

2. Methods

2.1. Materials
Shrimp shell waste (*Penaeus monodon*) was obtained from P.T. Kima, Makassar. Pure culture of *S. cereviceae* strain FN CC 3012 was a collection of Microbiology Laboratory, (PAU), Universitas Gadjah Mada, Yogyakarta. Reagents in analytical grade comprising $Cd(NO_3)_2$, HCl, NaOH, CH_3COOH. Distilled water and double distilled water were acquired from the Chemistry Laboratories, UGM.

2.2. Equipments
Includes chemical glass, hot plates, analytical balance (OHAUS), Atomic Absorption Spectrometer (Perkin Elmer Model 5100 PC).

2.3. Experimental Section

2.3.1. Microorganism production. Pure culture of S. cereviceae . It was maintained on potato glucose agar at 30°C. To produce biomass powder, a 100 ml of potato glucose medium was inoculated in the 250 ml conical flasks with spore suspension and grown in submerged culture conditions. Biomass of mycelium harvested, by filtration from the growth medium and washed several time using doubly distilled water. The obtained biomass was dried at 60°C. Further processing, the powders 100 mesh.

2.3.2. Combination of S. Cereviceae. Chitosan was prepared from shrimp shell waste through deproteination followed by demineralization and deasetilation processes according to the method reported earlier (Gadd, G. M., and White, C., 1993). Amount of chitosan powder mixture of S. cereviceae (0,100 gram) was completely dissolved in 20 ml of 2% (w/w) acetic acid. The adsorben Sc-Chi were stirred slowly for 30 minute, then washed with distilled water until the solution became neutral. characterized using an FTIR (Shimadzu Prestige-21), X-ray diffraction electron, EDX (Shimadzu XRD-6000). The adsorbent after combination (Sc-Chi) were dried and stored before use.

2.3.3. Preparation of Metal Stock Solution. Metal solutions were prepared by diluting 1000 mgl^{-1} of CdI_2 solution with double distilled water to make concentration range from 10 to 200 mgl^{-1}. Each solution, the initial Cd(II) concentration and the concentration in the samples after adsorption experiment were determined using an atomic absorption spectrometer.

2.3.4. Adsorption studies. Adsorption of Cd(II) ions onto Sc-Chi from aqueous solution was investigated in batch experiments. The effects of the pH, contact time, the initial concentration of Cd(II) ions, and absorption capacity were also studied.

2.3.5. Effect of pH. On the sorption rate of Sc-Chi adsorbent were investigated in the pH from 2,0 to 9,0. Using volume 10 ml of Cd(II) ions solution 100 mgl^{-1}. The pH of the solution was adjusted using 0,1M HCl and 0,1M NaOH. 0,1gram of Sc-Chi adsorbent were transferred to this medium and the reaction mixture was stirrer for 60 minute. Uptakes of Cd(II) ions were determined from the difference of Cd(II) in the initial and final solution. The Cd(II) ions concentration, the samples was analyzed using an atomic absorption spectrophotometer (AAS)

2.4. Data Analysis

The amount of metal ion adsorbed by using adsorbent obtained by using the following expression:

$$q= (Co-Ce)V/m \qquad (1)$$

q is the amount of metal ion adsorbed onto the unit amount of the adsorbent (mgg^{-1}). Co and Ce are the initial and equilibrium concentration of Cd(II) ion (mgl^{-1}) after sorption, respectively. V is the volume of the adsorption medium (l) and m is the amount of the adsorbent (g).

The empirical Langmuir equation wich is valid for monolayer sorption onto a completely homogeneous surface is given by equation 2:

$$q_e=q_{max}KCe/(1+KCe) \qquad (2)$$

were qe is the equilibrium metal ion concentration on the adsorbent (mgg^{-1}), Ce is the equilibrium metal ion concentration in the solution (mgl^{-1}), q max is the maximum monolayer sorption capacity of the adsorbent (mgg^{-1}), and K is the Langmuir sorption constant (lmg^{-1}) related with the free energy of sorption [8][12][13].

The Freundlich model assumes a heterogeneous adsorption surface and active sites with different energy. The equation representing Freundlich model is:

$$q_e= K_fC_e^n \qquad (3)$$

Model Freundlich asumtion :

where Kf is a constant relating the sorption capacity and n is an empirical parameter relating the sorption intensity, wich varies with the heterogeneity of the material [14].

3. Result and Discussion

The weight ratio between the biomass *S. cereviceae* ratio on chitosan done with a variation of 0% to 100%, the highest adsorption was obtained at a ratio of 10%. The composition is an ideal composition, each functional group either chitosan or *S.cereviceae* means adsorbent Sc-Chi has been optimized in binding metal ions Cd (II).

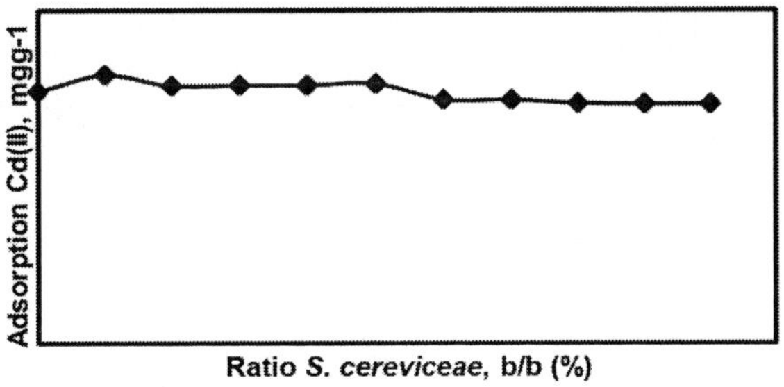

Figure 1. Ratio Chitosan with *S.cereviceae* after interaction with Cd(II) ions

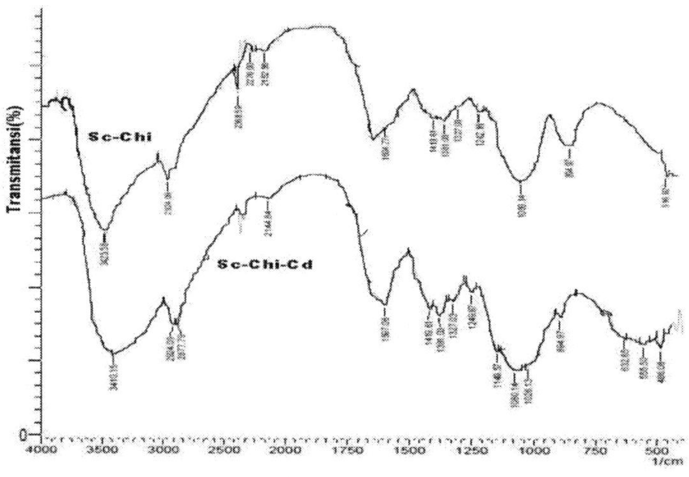

Figure 2. Infrared spectrum of adsorbent Sc-Chi before and after interaction with Cd(II) ions.

Results of FTIR spectra biomass S. (Sc) shows the absorption peak at 3387 cm-1 as the -OH group and the -NH of polysaccharides are overlapped, the absorption peak at 1651 cm-1 supported a peak at 1033 cm-1 indicates the group C = O carboxylic asymmetry and stretching vibration cm-1 indicates the group NH although not sharply (Figure 2). Based on the results of the identification FTIR spectra known that biomass *S.cereviceae* (Sc) has -OH groups, -COOH and -NH_2. It is supported by Gadd, G. M., 1990, states that imobilization cell support matrix for example; alginate, poliakrilamide, and chitosan has a high adsorption efficiency, although the diffusion restriction would be a problem Gadd and white, 1993; Tobin et. All., 1994). The study used by Sc-Chi adsorbent ratio of 10%. *S.cerevicea* XRD analysis results in Figure. 3 there are three peaks at 2θ = 18,9o; 19.8o and 18.5o. peak chitosan (Chi) at 2θ = 20,2o then biomass *Saccharomyces cerevisiae* (Sc) with 2θ = 18,9o, After modified, amended the form of diffraction peaks at 2θ = 19.8o crystal is a reflection of the (100). This reinforces the information that modification Sc-Chi has occurred,, the diffraction pattern is between the two adsorbents.

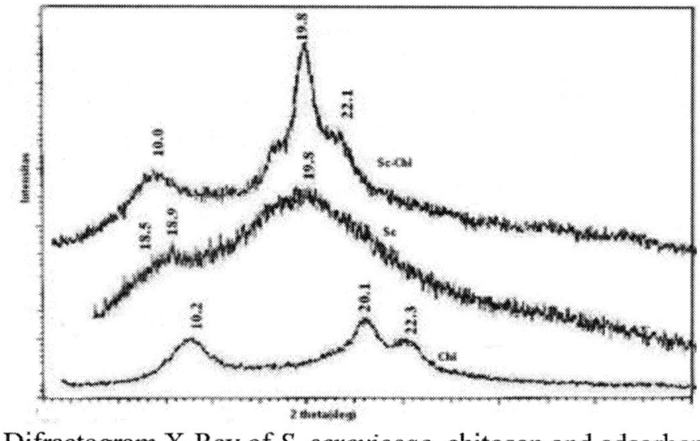

Figure 3. Difractogram X-Ray of *S. cereviceae*, chitosan and adsorbent Sc-Chi.

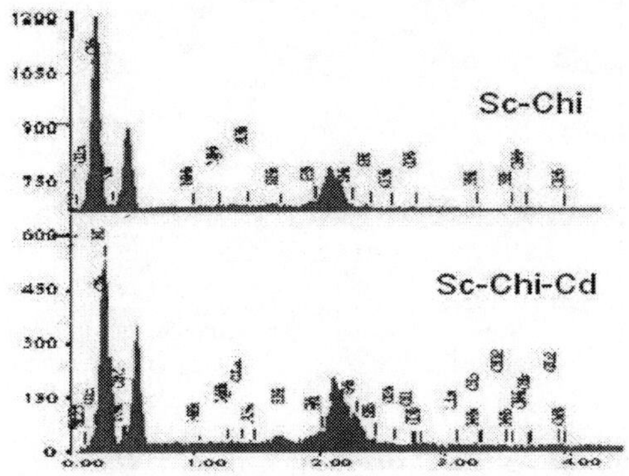

Figure 4. EDX adsorbent Sc-Chi before and after interaction with Cd(II)

The characterization results on EDX spectra (Figure 4), the adsorbent constituent elements Sc-Chi showed sharper spectra and growing constituent elements.

3.1. Effect of pH

Acidity (*p*H) represent important factor which have an in with metal ion adsorption process in condensation effect of active site or H^+ ion competition of cation for binding of active site, at the same time have an effect on to metal ion interaction with active site.

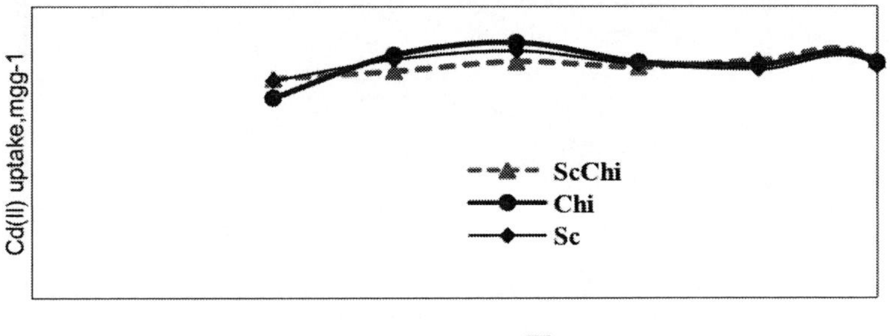

Figure 5. Effect of *p*H on Sc-Chi adsorbent, Cd(II) 100mgL^{-1}; sample volume of 10 ml; time contact 60 minute.

Figure 5. The sorption of Cd(II) increase as the *p*H increased. This is likely due to the protonation of the functional groups on the *Saccharomycess cereviceae* immobilized on chitosan structure [10]. At higher *p*H of 2,0, the sorption quantity increased up to pH of 6,0. At higher than 7,0, Pb(II) ions precipitated out because of high concentration of OH^- ions [7][8].

3.2. Effect of Contact Time

Maximum Cd(II) ions sorption with Sc-Chi adsorbent was initial period of contact time up to 60 minute. After that decreased amount of Cd(II) ions in solution

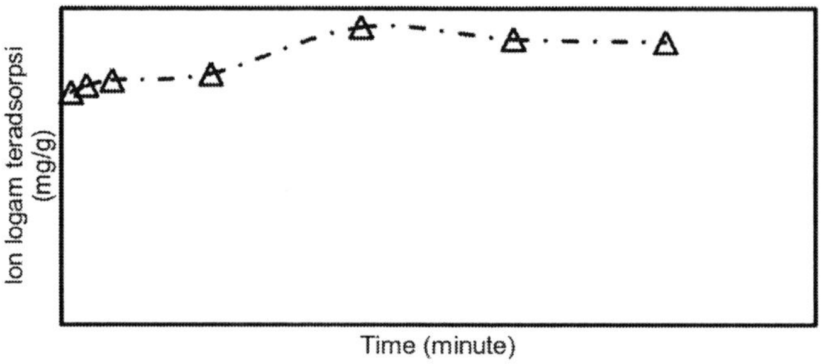

Figure 6. Effect of contact time on Sc-Chi adsorbent, Cd(II) $100 mgL^{-1}$; sample volume of 10 ml.

This rapid sorption stage indicate that surface sorption occur on Sc-Chi adsorbent. The extent of adsorption efficiency increases sharply with time and attains equilibrium at about 60 minute

3.3. Effect of Initial Cd(II) Ion Concentration

The sorption of Cd(II) ion Sc-Chi adsorbent was carried out at different initial Cd(II) ion concentration 10 - 200 mgL^{-1}.

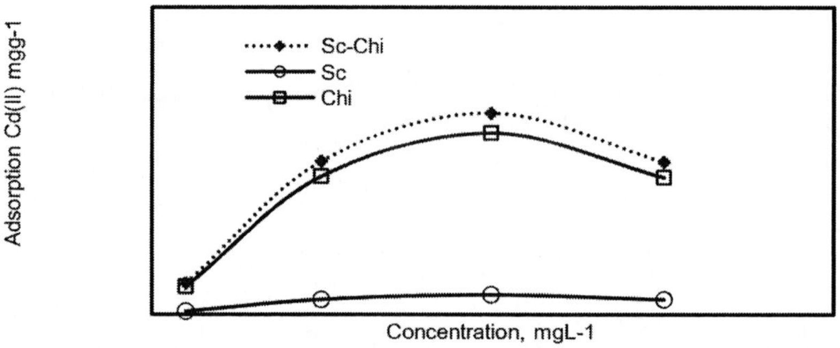

Figure 7. Effect of initial concentration on sorption of Cd(II) ions on Sc-Chi adsoorbent; volume of 10 ml; time contact 60 minute.

Adsorption of Cd(II) ions was increased as the initial concentration up to $25mgL^{-1}$, all Cd(II) ions present in solution could interact with binding site. Nearly 100% adsorption was observed when concentration increased from 10-25 mgl-1. This could be because at higher concentration, as more ions are competiting for the avaible binding sites [10].

Analysis in the of the isotherm data is important in order to develop an equation.Adsorption isotherm also describes how solutes interact with adsorbent and so is critical in optimizing the use of adsorbent. The lead uptake capacity of Sc-Chi adsorbent was evaluated using pseudo order-2 adsorption isotherm.

Table 1. The number of reaction rate constant (k) and the correlation coefficient (R2) metal ion adsorption kinetics model of Cd (II)

Kinetics Model	**Reaction rate constant (k)**	
	Sc-Chi	$\mathbf{R^2}$
Orde-1	0,006 min^{-1}	0,434

Kinetics Model	**Reaction rate constant (k)**	
	Sc-Chi	**R^2**
Orde-2	20,45 M min^{-1}	0,356
Orde pseudo-2	0,035 g $mol^{-1}min^{-1}$	0,995

Table. 1 shows that the pseudo-second-order more appropriate with the Sc-Chi adsorbent because forming the linear regression than model of kinetics first order and order-2, this indicates that the adsorption of metal ions Cd (II) following model of pseudo kinetics second order. The reaction rate constant of Sc-Chi adsorbent larger than before modification, the amount of adsorption rate due to the number of active group that acts on the adsorbent Sc-Chi unequal or crowding so that the reaction rate is slower than the biomass of Saccharomyces cereviceae Sc, but chitosan shows the adsorption rate same with an adsorbent Sc-Chi, this is likely due to the active site that play a part in adsorbing metal ions Cd (II) is not much different.

4. Conclusion

The result of research showed that optimum condition of adsorption of Cd(II) ion by Sc-Chi adsorbent with ratio of 10%, *p*H of solution of 7, contact time of 60 minutes and initial concentration adsorption 25mgl^{-1}. This adsorbent can be a good candidat for adsorption of not only Cd(II) ions but also other heavy metal ions in wasterwater stream.

Acknowledgement

The authors woud like thanks to Directore of Research and Community Services, Directorate General of Higher Education (DP2M-DIKTI), National Education Departement of Indonesia, for financial support of this research (Research Grant for Doctoral Program, 2009, with contract number: LPPM-UGM/1156/2009).

References

[1] Mahajan, S. P., 2002, Pollution in Process Industries, Tata McGraw-Hill Pub. Co. Ltd., New Delhi
[2] Kurita, K., Sannan, T., Iwakura, Y., 1998, Binding of Metal Cations, *J. Appli. Poly. Sci.*, **23**, pp. 511-515
[3] Veglio, F, Beolchini, F and Toro I., 1998, Kinetic Modelling of Copper Biosorption by Immobilized biomass, *Ind. Eng. Chem. Res.,* **37**, pp. 1107- 1111
[4] Darjito, Danar, P., and Siti Nasirotun Nisa., 2006. Study On Adsorption of Cd(II) by Chitosan-Alumina, *J. Indo. Chem.*, **6**, 3, pp. 238-244.
[5] Gadd, G. M., 1992, Microbial Control of Heavy Metal Pollution, In *Microbial Control of Polution,* Fry, J. C., Gadd, G. M., Helbert, R. A., Jones, R. W., and Watson-Craik, I.A., (eds), Society for General Microbiology Symposium, Cambridge University Press, UK., pp. 58-88
[6] Khoo, K.M.; Ting, Y.P., 2001, Biosorption of gold by immobilized fungal biomass. *Biochem. Engin. J.*, **8**, pp.51-59
[7] Tsezos, M., and Volesky, B., 1981, Biosorption of Uranium and Thorium, *Biotech. Bioeng.*, **23**, pp. 583-604
[8] Volesky B.,Holan, Z.R., 1995, *Biosorption of Heavy Metals*, Biotechnol.,**11**, pp. 234-250
[9] Gadd, G. M., and White, C., 1993, Microbial Treatment of Metal Pollution a Working Biotechnology, *Tibtech,* **11**, pp. 353-359
[10] Tobin, J. M., White, C. dan gadd, G. M., 1994, Metal Accumulation by Fungi: Application in Environment Biotechnology, *J. Ind. Microbiol.*, **13**, pp. 126-130
[11] J. acharya, et.al. 2009. *Chemical Eng. J.* **150**, pp. 25-39.
[12] Hong, N. K., Meyer, S.P., Lee, K.S., 1989, Isolation and Properties of Humus Mineral

Complexes, In : Interaction of Soil Mineral with Natural Organic and Microbes (P.M and M. Schnitzer, eds), *Soul Sci. Soc*, **VI**, pp. 103-108

[13] Holan, Z.R., an Volesky, B. 1995. Accumulation of cadmium, lead, and nickel by fungal and wood biosorbents. *Appl. Bio. Chem. Biotechnol., Prog*, **11**, pp.235-250.

[14] I. Langmuir: 1918, *J.Am.Chem.Soc.*, **40**, pp. 1361-1403

Comparative Case Studies on Indonesian Higher Education Rankings

Nuning Kurniasih[1*], C Hasyim[2], A Wulandari[3], M I Setiawan[4], A S Ahmar[5,6]

[1]Faculty of Communication Science, Library & Information Science Program, Universitas Padjadjaran, Bandung, Indonesia
[2]Darul Ulum University, Jombang, Indonesia
[3]Department of Civil Engineering, Narotama University, Surabaya, Indonesia
[4]Department of Civil Engineering, Narotama University, Surabaya, Indonesia
[5]Department of Statistics, Universiats Negeri Makassar, Indonesia
[6]AHMAR Institute, Makassar, Indonesia

nuning.kurniasih@unpad.ac.id

Abstract. The quality of the higher education is the result of a continuous process. There are many indicators that can be used to assess the quality of a higher education. The existence of different indicators makes the different result of university rankings. This research aims to find variables that can connect ranking indicators that are used by Indonesian Ministry of Research, Technology, and Higher Education with indicators that are used by international rankings by taking two kind of ranking systems i.e. Webometrics and 4icu. This research uses qualitative research method with comparative case studies approach. The result of the research shows that to bridge the indicators that are used by Indonesian Ministry or Research, Technology, and Higher Education with web-based ranking system like Webometrics and 4icu so that the Indonesian higher education institutions need to open access towards either scientific or non-scientific that are publicly used into web-based environment. One of the strategies that can be used to improve the openness and access towards scientific work of a university is by involving in open science and collaboration.
Keywords: Higher Education Rankings, comparative case studies, Indonesian Universities, Open Science

1. Introduction

In the Republic of Indonesia constitution Number 12 in 2012, it mentions that the higher education has a function to educate the nation's life, conduct the education, research and community service in order to produce academicians who are innovative, creative, skilled, competitive and cooperative, as well as able to develop science and technology by paying attention to human values [1]. To achieve these goals, the higher education must always maintain the quality. The quality of higher education is the suitability of national standards applied by universities according to the needs of the stakeholders [2]. Ministry of Research, Technology, and Higher Education of the Republic of Indonesia (Kemenristekdikti) recognizes that the quality of higher education in Indonesia is not yet in ideal condition, it can be seen from the number of universities that achieved A accreditation score (excellent) is only 12% [3]. On the other hand, higher education today are faced with massive expansion, many new higher education institutions that have sprung up with diverse profile, there are the demands on the adaptation and integration of communication technology and education, internationalization, competition, financial shape adapted to the model of governance, including increased performance, quality and accountability [4]. To achieve public accountability, there needs to be an internal quality assurance system and accreditation so as to ensure that a higher education has met the quality standards that have been determined [2].

The quality of a higher education is the result of a continuous process. There are many indicators that can be used to assess the quality of a higher education. These indicators will be depended on the purpose of the institution assessors. There are several departments in the world's universities that are used as guidelines for assessing the quality of a college. For example, 4icu takes university base rankings on web metrics where web traffic becomes the base of ranking system [5], QS World University Ranking adhering to the six metrics in raking the higher education institutions that is based on academic reputation, the reputation of the institution, the ratio of students, the number of citations per faculty, the ratio of international faculty and the ratio of international students [6], the Academic Ranking of World Universities (ARWU) uses six indicators to rank universities in the world i.e. by counting the number of staffs and alumni who won the Nobel Prizes and Fields Medals, the number of citations in Thomson Reuters, the number of articles published in journals, the number of articles indexed in Science Citation Index and the Social Sciences Citation Index, and the performance of a university [7], webometrics claim that they use a composite indicator which log webometrics normalize variables into a ratio of 1:1 before combining between activity/presence and visibility/impact of the indicator group, and several other ratings. Kemenristekdikti in 2017 had also issued a list of 100 best universities in Indonesia. The list is divided into 5 clusters i.e. cluster 1, cluster 2, cluster 3, cluster 4 and cluster 5. These clusters serve as guidelines for coaching purposes. The university ranking of Kemenristekdikti version is based on data from the database of Higher Education (Dikti) and Kemenristekdikti assessment results data that are not listed in the database. These indicators are the quality of human resources, quality of management, quality of student activities, the quality of research and publications [8]. The existence of different indicators makes different ranking results. As an example, University of Muhammadiyah Yogyakarta (UMY) questioning its position in the ranking of 71 based on Higher Education (Dikti) version, while the international rating of webometrics and 4icu, they are in 24th place in comparison to any other universities in Indonesia [9].

This study aimed to compare the ranking indicators used by Kemenristekdikti with the indicators used by international ratings by taking the example of two of rating systems that are Webometrics and 4icu. The research questions include universities that are included in the top 14 of Kemenristekdikti, Webometrics and 4icu version, what are the similarities and differences among the three ranking indicators universities? What can bridge the ranking indicators of universities in Indonesia by the trend of international rankings? The results of this study are expected to provide contribution in bridging the ranking indicators based on national needs and international trends.

2. Research Methods

This research uses qualitative research method with comparative case studies approach. Comparative study aims to understand, explain and interpret variety of processes and the outcome of the units of analysis [10]. Comparative case studies include the analysis and synthesis of the similarities, differences and patterns between two or more cases that have the same focus or target [11]. Analysis units were selected based on the purpose of the study i.e. analyze the ranking indicators of university. From a variety of higher education ranking version, we choose three ranking versions, namely Kemenristekdikti, Webometrics and 4icu versions. Kemenristekdikti version is chosen for this version that is used to determine the ranking of universities in Indonesia, while the Webometrics and 4icu selected as ranking indicators of both because those ranking systems have more emphasis on the use of technology that becomes a trend change in the digital era. The stages of analysis in this research covers identification, evaluation and interpretation of indicators of each rating system then compared to generate indicators that are related to the national interest and international trend.

3. Results and Discussion

3.1. The Comparison of Higher Education Rating System of Kemenristekdikti, Webometrics and 4icu version

In August 2017, Kemenristekdikti released the 100 best universities for non-polytechnics category and 25 best universities for polytechnic categories. This study only examines the ranking of non-polytechnic universities. Of the 100 best non-polytechnic universities there are 14 universities that come into the category of Cluster 1 (Excellent), as it is shown in Table 1.

Table 1 Universities Ranking in Indonesia, Cluster 1 Kemenristekdikti Version 2017

Rank	Kemenristekdikti version	Webometrics version	4icu Version
1	Gadjah Mada University	University of Indonesia	Gadjah Mada University
2	Institut Teknologi Bandung	Universitas Gadjah Mada	Universitas Indonesia
3	Institut Pertanian Bogor	Institut Teknologi Bandung	Institut Pertanian Bogor
4	Universitas Indonesia	Institut Pertanian Bogor	Universitas Diponegoro
5	Institut Teknologi Sepuluh Nopember	Universitas Brawijaya	Universitas Sebelas Maret
6	Universitas Diponegoro	Universitas Padjadjaran	Universitas Brawijaya
7	Universitas Airlangga	Universitas Diponegoro	Institut Teknologi Bandung
8	Universitas Brawijaya	Universitas Udayana	Universitas Negeri Yogyakarta
9	Universitas Hasanuddin	Universitas Sebelas Maret	Universitas Padjadjaran
10	Universitas Negeri Yogyakarta	Universitas Riau	Universitas Pendidikan Indonesia
11	Universitas Sebelas Maret	Universitas Syiah Kuala	Universitas Islam Indonesia
12	Universitas Andalas	Universitas Jenderal Soedirman	Universitas Airlangga
13	Universitas Pendidikan Indonesia	Institut Teknologi Sepuluh Nopember Surabaya	Universitas Islam Negeri Syarif Hidayatullah Jakarta
14	Universitas Padjadjaran	Universitas Hasanuddin	Universitas Udayana

Source: [8, 12, 5]

Table 1 shows the different ranking positions from the three ranking systems. Of the 14 universities included in Cluster 1 of Kemenristek Dikti version, there are 10 universities that go into the top 14 versions of Webometrics and 9 universities entered into the top 14 of 4icu version with different positions. One university is not available in the top 14 of Kemenristekdikti version, but included in the top 14 of Webometrics version and 4icu version namely Udayana University. Airlangga University, Yogyakarta State University and University of Indonesia are in the top 14 of Kemenristekdikti and 4ICU version, but not in the top 14 of Webometrics version. On the other hand, Riau University, University of Syiah Kuala and University of Jendral Soedirman, ranked in the top 14 of Webometrics version, but not in the top 14 of Kemenristekdikti and 4icu versions. Meanwhile, the Indonesian Islamic University and the Islamic University of Syarif Hidayatullah Jakarta are in the top 14 of 4icu version, but not included in the top 14 of Kemenristekdikti and Webometrics versions. Thus, the degree of similarity of the number of higher education institutions based on Kemenristekdikti, Webometrics and 4icu versions on the top 14 are at 9-10 institutions or 64-71% only.

3.2. The Rating Indicators in Kemenristekdikti, Webometrics and 4icu Versions
As noted in the introduction part, different indicators make different ranking results. In Table 1 we have seen the similarities and differences of university rankings from 3 rating departments. Table 2 below shows the indicators that are used to determine the ranking:

Table 2 Ranking Indicators

Kemenristekdikti version	**Webometrics version**	**4icu Version**
Quality of human resources Quality of management Quality of student activities Quality of research and publications	Presence Rank Impact Rank Openness Rank Excellence Rank	Using the algorithm by extracting data from: Moz Domain Authority, Alexa Global Rank, Similar Web Global Rank, Majestic Referring Domains and Majestic Trusts Flow.

Source: [8, 12, 13]

Table 2 shows that the indicators of Kemenristekdikti are more extensively on the institutional aspects, while the webometrics version see how a university featuring activities accurately in the website [14], especially activities related to publications indexed in Scimago and Google Scholar, the number of web pages indexed by Google and the number of online documents in the website. On the other hand, 4icu assesses the popularity of a website based on traffic, confidence and quality of links. It is intended to help academicians and international students to know the universities that are popular in a country [13].

In addition to the differences, these three ratings have the same similarity i.e. on the quality of the research, publications and other activities that have impact so that all of them are accessible by the public.

3.3. Variables that Bridging Rating Indicators between Kemenristekdikti Rating System and Webometrics as well as 4icu versions
Web-based rankings such as Webometrics or 4icu emphasize how academicians and universities globally reflect their quality to the whole world [14]. Rating indicators of Kemenristekdikti version has accommodated aspects of an institution of higher education in Indonesia which emphasizes on three main activities, such as education, research and community service. Nevertheless, the trend of the use of the website by the public should be able to encourage colleges-universities in Indonesia to boost its presence in the web. The presence of universities in the web is not only for scientific publications but also for all web-based public communication processes. In Webometrics, 5% aspects of assessment is derived from the presence of the institution in the web that includes the number of pages to be the main domain and sub-domain of a Google-indexed university; 50% comes from the number of external network (subnet) which comes from an institution by using backlinks checker tools, namely Ahrefs Majestic; 10% rated on the transparency/openness to look at the number of citations indexed by Google scholar; 35% rated on excellence / scholar by seeing the amount of paper that is most often cited based on Scimago data [15].

In order to accommodate this, Kemenristekdikti need to encourage universities to increase the presence and accessibility of scientific publications and other information through the website. The quality of research and publications itself are available in the indicators of Kemenristekdikti version. The matter that is needed to be realized is that publication trend in the digital era is through digital media so that the impact of a publication through the website is no longer just viewed from a quote in a scientific work, but also can be seen from how much a scientific paper is clicked, shared, stored or downloaded, discussed, adapted and are reviewed in an online forum [16]. Thus, Kemenristekdikti need to emphasize the quality of the publication which is accompanied by ease of accessibility and openness. Kemenristekdikri needs to encourage universities to have an open access policy in the development of its repository. Open science development is believed to have a positive impact on science development and dissemination of scientific work without limits. Open science grant to openness, integrity and productivity in the development of science. Open science can also provide contribution for innovation by presenting open data and collaboration [17]. By promoting open science, it is expected that the level of attendance of Indonesian universities and scientists in the web can increase. The increasing level of presence on the web can increase the rank of the universities on the web-based ranking system.

4. Conclusion

The differences in the indicators used by the higher education ranking departments lead to differences in each department such as Kemenristekdikti, Webometrics and 4icu. Web-based rating emphasizes the attendance of a university with a variety of attributes appear in the website environment. To bridge the indicators used by Kemenristekdikti which assess all aspects of institutional ranking of a web-based trend, then Kemenristekdikti need to encourage colleges and universities in Indonesia to open access to scientific or non-scientific information that is publicly used in a web environment. One of the strategies that can be used to improve the openness and access to the scientific work of a university is by engaging in open science and collaboration. Open science makes open science and can be disseminated quickly and ultimately expected to increase the impact of the scientific work produced. Improving the impact of a scientific work can help improve the university rankings.

References

[1] Presiden Republik Indonesia, Undang-undang Republik Indonesia Nomor 12 Tahun 2012 tentang Pendidikan Tinggi, Jakarta, Jakarta: Republik Indonesia, 2012.
[2] Kemendikbud, Bidang Pendayagunaan dan Pelayanan Data dan Statistik Pendidikan, 2013. Analisis kualitas program studi perguruan tinggi dalam mendukung pencapaian koridor ekonomi indonesia, Jakarta.
[3] Kemenristekdikti. 5 May 2017. Mutu perguruan tinggi menentukan kompetensi lulusan. Available: https://ristekdikti.go.id/mutu-perguruan-tinggi-menentukan-kompetensi-lulusan/.
[4] K. Tremblay, D. Lalancette and D. Roseveare, "Assessment of higher education learning outcomes," OECD, France, 2012.
[5] uniRank™, 2017. 2017 Indonesian university ranking updated. Available: http://www.4icu.org/id/.
[6] QS World University Ranking. 8 June 2017. QS world university ranking 2018: methodology. Available: https://www.topuniversities.com/qs-world-university-rankings/methodology.
[7] ShanghaiRanking Consultancy, 2017. About academic ranking of world universities. Available: http://www.shanghairanking.com/aboutarwu.html.
[8] Kemenristekdikti. 21 August 2017. 100 Besar perguruan tinggi non politeknik dan 25 besar perguruan tinggi politeknik di Indonesia tahun 2017. Available: https://ristekdikti.go.id/100-besar-perguruan-tinggi-non-politeknik-dan-25-besar-perguruan-tinggi-politeknik-di-indonesia-tahun-2017/.
[9] UMY. 25 August 2017. UMY pertanyakan hasil pemeringkatan Kemenristekdikti. Available: http://www.umy.ac.id/umy-pertanyakan-hasil-pemeringkatan-kemenristekdikti.html.
[10] C. C. Ragin. 1987. The comparative method: moving beyond qualitative and quantitative, Berkeley (CA): University of California Press.
[11] D. Goodrick/ 2014. Comparative case studies, methodological briefs:iImpact evaluation 9, Florence, Italy: United Nations Children's Fund (UNICEF).
[12] Cybermetrics Lab. 2017. Indonesia. Available: http://www.webometrics.info/en/Asia/Indonesia%20.
[13] uniRank™, 2017. University ranking components. Available:

http://www.4icu.org/about/index.htm#ranking.

[14] M. Kaya, E. Cetin and A. Sözeri. 2010. Introduction to webometrics: quantitative web research for the ranking of world universities; research centers and hospitals, in *Conference: Explorations in e-Government and e-Governance (ICEGEG-2010)*, Antalya-Turkey.

[15] Cybermetrics Lab. 2017. 2017 Ranking web of universities. July new edition. Available: http://www.webometrics.info/en/node/178.

[16] N. Kurniasih. 2016. The implementation of Altmetrics in library as an alternative measurement method for scholarly article impact in social web ecosystem, *in International Conference on Science Mapping and the Development of Science*, Yogyakarta.

[17] S. Jong and K. Slavova. 2014. When publications lead to products: The open science conundrum in new product development. *Research Policy* , **43**, 4, pp. 645-654.

IOP Conf. Series: Journal of Physics: Conf. Series **954** (2018) 012022 doi:10.1088/1742-6596/954/1/012022

Beyond effective teaching: Enhancing students' metacognitive skill through guided inquiry

Adnan & Arsad Bahri*
Department of Biology, Faculty of Mathematics and Natural Sciences, Universitas Negeri Makassar, Indonesia.

E-mail: arsad.bahri@unm.ac.id

Abstract. This research was quasi experimental with pretest posttes non-equivalent control group design. This research aimed to compare metacognitive skill of students between tought by guided inquiry and traditional teaching. Sample of this research was the students at even semester at the first year, Department of Biology, Faculty of Mathematics and Natural Sciences, Universitas Negeri Makassar, Indonesia. The data of students' metacognitive skill was measured by essay test. The data was analyzed by inferential statistic of ANCOVA test. The result of research showed that there was the effect of teaching model towards metacognitive skill of students. Students were tought by guided inquiry had higher metacognitive skill than tought by traditional teaching. The lecturer can use the guided inquiry model in others courses with considering the course materials and also student characteristics.

Keywords: Guided inquiry, learning model, metacognitive skill, biology teaching, effective teaching

1. Introduction

One of many factors which determine the quality of education is successful education in university. Starategy, model, or lecturers pattern are important aspects in education process beside materials to gain the achievement. Learning experiences are gained as long as the lectures ran, really determine on ability and quality from lecturers. So, the lecturers have responsibility to create learning experiences of their students by the implementation of appropriate model or learning strategy.

Science learning is expected to enhance students on how to fullfill 21 centuries skill and one of the skills is learning skill and inovation. Integrated science included Biology can be implemented by scientific inquiry to grow up the thinking skills, and could change from passive learning to active learning. So, the process of Animal Structure course as a part of Biology learning in university, emphasize on learning experiences directly through implementation and development of skill of thinking included metacognitive.

The result of observation showed that empowerment of student metacognitive skill was still unsufficient. Metacognitive skill should be trained so student so they could perform their affection and skills in terms of manage their learning [1]. Teaching method which implemented, less of giving chances and learning experiences to construct the concepts which is learned through thinking process. Metacognitive skill needs a metacognitive strategy to be taught. Metacognitive strategy can be described as routine activity, which represented managing the mental process, as a part of complexity process and apllied of getting goals [2]. Metacognitive related also with thinking level, cognitive style, and learning style of students. Thinking level, cognitive style, and learning style of students have an influence on metacognitive of students [3]-[6].

Metacognitive skill was very important to be applied because student who knew the effective strategy to learn, tend to have good learning achievement [7]. Metacognition was the strong variable to affect academic succesfull [8][9]. The research showed that there was positive contribution of metacognitive skill towards learning outcomes of students [10][11].

Based on that problem, need to apply the learning model that can empower the metacognitive skills of students in learning process. One of learning model based on constructivistic for instance guided inquiry. Implementation of guided inquiry trained the students to do many activities, for example obsevation, investigation, experiments, and compare one experiments with other experiments, ask a question, and find answer of self question. Guided inquiry learning aims to give chance for students to build their intelectual behaviour and motivated to give ideas when teacher give a problem.

2. Methods

This research was a quasy experimental designed to compare effects of guided inquiry and traditional teaching towards metacognitive skill of students in the first year, even semester in Department of Biology, Faculty of Mathematics and Natural Sciences, Universitas Negeri Makassar, Indonesia. Research design was Pretest-Posttest Nonequivalent Control Group Design [12] as showed in Table 1.

Table 1. Quasy experimental design research

Pretest	Group	Posttest
T_1	X_1	T_2
T_3	X_2	T_4

T_1& T_3: Pretest
T_2& T_4: Posttest
X_1: Guided Inquiry
X_2: Traditional Teaching

The sample of research randomly choosed, consists of 125 students in 4 classes which take Animal Structure course. Each teaching model would represent by 2 classe. The sample of research in each group consists of students with equal academic ability, based on grouping test. Before giving a treatment, 2 groups of students with classes given essay test which developed by researcher to measure metacognitive skill of students. Instrument was validated before implemetation. During 2 months, 4 classes taught with different learning model. Then, they were given final test. Data analyzed by statistical inferencial as ANCOVA test using SPSS for windows 22.

3. Results and Discussion

The results of normality and homogenity test shows that the data were normally distributed and homogeny. Normality test showed p value of pretest > 0.05 (p = 0.302) and p value for posttest = 0.312. Homogenity test showed p value > 0.05, pretest p value = 0,117 and posttest p value = 0.428.

Hipothesys test with ANCOVA test showed that the learning model give affect towards metacognitive skill of students (p < sig. 0,05). The summary of ANCOVA test of metacognitive skill of student showed on Table 2.

Tabel 2. Summary of ANCOVA test of metacognitive skill of student.

Source	Type III Sum of Squares	df	Mean Square	F	Sig.
Corrected Model	8765.919[a]	2	4382.959	34.104	<0.001
Intercept	9207.351	1	9207.351	71.643	<0.001
XmetacogSkill	7111.269	1	7111.269	55.334	<0.001
Model	1556.508	1	1556.508	12.111	0.001
Error	15678.993	122	128.516		

Total	467750.000	125
Corrected Total	24444.912	124

The result of research showed that metacognitive skill of students successfully improved through guided inquiry model higher than conventional teaching. It is stated that result gained by students with this inquiry model could enhance their metacognitive skill. Guided inquiry is one of learning model whereas lecturer provided materials, subject, and problem for investigation [13]. Student plan their procedural to overcome the problem. Steps in inquiry learning model were observation to found problem, create core of problem, make a hypothesis, plan the problem solving, do the experiments, make observation and collect data, data analysis, and make inferencial [14]. Those activity will train skill of student on how plan, manage, and evaluate their learning. Result of this research in line with reference [15] found that guided inquiry affect on metacognitive skill of students.

Metacognitive skills refers to three essential skill were planning skills, monitoring skills, and evaluating skills [16]. Steps in guided inquiry learning model had reflected metacognitive skill aspects. Founding problem phase, creating core of problem, make a hypothesis, and plan the problem solving (design experiments) in guided inquiry are planning aspects in metacognitive skills. Doing experiments phase, observation and data collection, and data analysis in guided inquiry learning are monitoring aspects in metacognitive skill. Making inferencial phase in inquiry are evaluation aspects in metacognitive skill. This shows that guided inquiry have integrated with metacognitive strategy and can be used to empower metacognitive skill of student.

At the planning phase, the students explore their thoughts to find out what previous knowledge they have, that will help them to complete the task, the students know what needs to be done first in order to help in completing the task, and the students plans the time management in planning the task [17]. When the students working on a sheet of metacognitive skills, students have used their prior knowledge to help in completing task. The prior knowledge is the knowledge, which students have gained in the material they have learned. Furthermore, in doing the problem of metacognitive skills, students know and formulate what things need to be done first in completing the task. Students have also arranged their time in working out the metacognitive skills question that can be seen from the timeliness of students in collecting answers about metacognitive skills.

Students also perform planning skills in completing tasks. As long as the students are planning (planning skills), students will ask themselves what information should be known based on the questions provided and the time needed to solve the problem [18]. If the student can know exactly the information provided in the question, then the student can determine what strategy is used to solve the problem.

In monitoring activity (monitoring skills), students take action to solve the problems, remember important information, and check whether it is on the right track. As long as the student performs the monitoring, the student will ask himself or herself what related information is important to remember and what to do to solve the problem. Furthermore, based on the information known in the question, the student can determine the further steps that must be done so that the problem can be solved.

In evaluating activity (evaluation skills), the student will check the conformity between what is known and the steps used to solve the problem. Students can ask themselves how well they have solved the problem. It is indicated by the expression of the students' reasons in determining the steps to solve the problem based on prior knowledge.

In Biology learning, guided inquiry is a very closely related to the laboratory activities. The inquiry presents students' involvement in laboratory using their own thinking process. These learning phases include focusing on students and explaining inquiry processes, presenting problems or phenomena, helping students formulate hypotheses to explain that problems or phenomena, encouraging students to collect data, formulate explanations, reflect problem situations and thought processes [13]. These activities require students to find their own concepts in solving problems related to animal structure materials. It can empower students' metacognitive skills in thinking how they learn and understand the course material.

Laboratory activities require the existence of an active role of students to prove the hypothesis and analyze the results of practice in accordance with existing theories so that students can build their own understanding. It was stated that a learning model was required to help students involve in the lab activity [19]. Through guided inquiry, lecturer helps students to investigate the problem and discover and construct the concept by themselves. It can build metacognitive skills of students. In the process of discovery of the concept, students are encouraged to conduct monitoring of each stage carried out in guided inquiry. Because the inquiry stages of the students occur over and over again, the self-learning monitoring process also occurs more than once. Such conditions can stimulate the students' metacognition skills empowerment. It can causes the application of guided inquiry can improve students' metacognitive skills.

In line with this result of research, reference [20] concluded that the application of guided inquiry in chemistry learning enhanced metacognitive skills, particularly the planning skills. Associated with metacognitive awareness as measured by MAI (metcognitive awareness inventory), planning skills show the most dominant performance. To further improve students' thinking skills, metacognitive skills need to be trained periodically on subsequent lessons.

Similarly, research found that the metacognitive skills of students experience improvement in each face to face when tought by guided inquiry [19]. Research found that metacognitive skills have a positive relationship to learning outcomes. It means that the improvement of students' metacognitive skills will followed by the improvement of cognitive learning outcomes [10][21]. Its found that after implementation of guided inquiry, there was an improvement from pretest score to posttest of self-regulation skills, especially aspects of monitoring, followed by evaluation and planning [22].

Guided inquiry makes students construct concepts and discuss in concept discovery so that students gain more meaningful knowledge in the long term. Based on the results of the research, it suggested that metacognitive skills should be trained continuously so that students are accustomed to control the students' thinking, including self-planning, self-monitoring, and self-evaluation. Thus the students will be trained to become self-regulated learners. This is in line with arguements that metacognitive skills were useful for making the students become self-regulated learners, encouraging them to become their own managers as well as assessing their own thoughts and learning [23]. Therefore, lecturers need to apply learning strategies that not only emphasize cognitive development but more metacognitive aspects are also important to be empowered during learning process.

4. Conclussion

Based on the results of research and discussion, it can be concluded that there was the effect of implementation of guided inquiry toward students' metacognitive skills. The implementation of guided inquiry proved able to empower students' metacognitive skills higher than traditional learning. Lecturers can apply guided inquiry in other courses by considering the course material and students' characteristics.

References

[1] Corebima A D 2009 *Proc. National Seminar of Bology Education* (Indonesia: Department of Biology, Faculty of Mathematics and Natural Sciences, Universitas Negeri Makassar)

[2] Hacker D J 1998 *Definitions and Empirical Foundations. In Hacker D J, Dunlosky, and Graesser A (Eds.) Metacognition in Educational Theory and Practice* (Mahwah, NJ: Erlbaum) p 1–23

[3] Rahman, A. & Ahmar, A.S. 2016. *Int. J. of Env. & Sci. Edu., **11***(14), 7278-7285.

[4] Mulbar, U., Rahman, A., & Ahmar, A. S. 2017. *World Transc. on Eng. and Tech. Edu., 15*(1), 68-73. doi:10.26858/wtetev15i1y2017p6873

[5] Rahman, A., Ahmar, A.S, & Rusli. 2016. *World Transc. on Eng. and Tech. Edu.*, **14**(3), 425-430.

[6] Rahman, A., & Ahmar, A. S. 2017. Relationship between learning styles and learning achievement in mathematics based on genders. *World Transc. on Eng. and Tech. Edu., 15*(1), 74-77. doi:10.26858/wtetev15i1y2017p7477
[7] Dunning D, Johnson K, Ehrlinger J and Kruger J 2003 *Current Directions in Psychological Science* **12** 88-87
[8] Magno C 2010 *Learning, memory and cognition* **36** 255-262
[9] Kruger J and Dunning D 1999 *J. Pers. Soc. Psychol.* **77** 1121-1134
[10] Bahri A and Corebima A D 2015 *J. of Baltic Sci. Edu.* **14** 487-500
[11] Bahri A 2016 *Proc. Int. Conf. on Mathematics, Science, Technology, Education, and their Applications* (Indonesia: Faculty of Mathematics and Natural Sciences, Universitas Negeri Makassar) p 156-162
[12] Borg W L and Gall M D 1983 *Educational Research, An Introduction. 4th Edition* (New York and London: Longman Inc.)
[13] Arends R I 2009 *Learning to Teach* (New York: Mc Graw Hill)
[14] Suryanti 2008 *Model-Model Pembelajaran Inovatif* (Surabaya, Indonesia: Universitas Negeri Surabaya Press)
[15] Irawati F, Kurniawan H C, Primandiri P R and Santoso A M 2015 *Proc. National Seminar of Biologi Education XII* (Indonesia: Faculty of Teacher Training and Educational Science, Universitas Negeri Sebelas Maret, Solo) p 483-487
[16] Woolfolk A 2009. *Educational Psychology: Active Learning Edition. 10th Edition. Part Two.* (Boston: Pearson Education, Inc.)
[17] North Central Regional Educational Laboratory 1995 *Metacognition.* (online) http://www.ncrel.org/sdrs/areas/issues/students/learning/lr1metn.htm.
[18] The Teaching Excellence in Adult Literacy 2012 *Metacognitive Processes* (American Institutes for Research)
[19] Mu'minin S K F and Azizah U 2014 *UNESA J. of Chemical Education* **3** 67-74
[20] Aprilia F and Sugiarto B 2013 *UNESA J. of Chemical Education* **2** 36-41
[21] Kristiani N, Susilo H, Roham F and Aloysius D C 2015 *Int. J. of Educational Policy Research and Review* **2** 113-120
[22] Hidayah R 2014 *Proc. of Int. Conf. on Research, Implementation and Education of Mathematics and Sciences* (Indonesia: Universitas Negeri Yogyakarta) p 91-96
[23] Peters M 2000 *J. of Nursing Education* **39** 166-170

Acknowledgment
This project was funded by the Universitas Negeri Makassar, Indonesia (Contract of PNBP Research No. 256/UN36.9/PL/2017).

The Profile of Creativity and Proposing Statistical Problem Quality Level Reviewed From Cognitive Style

Awi[1], A S Ahmar[2,3*], A Rahman[1], I Minggi[1], U Mulbar[1], Asdar[1], Ruslan[1], H Upu[1], Alimuddin[1], Hamda[1], Rosidah[1], Sutamrin[1], M A Tiro[2], and Rusli[1]

[1]Department of Mathematics, Universiats Negeri Makassar, Indonesia
[2]Department of Statistics, Universiats Negeri Makassar, Indonesia
[3]AHMAR Institute, Makassar, Indonesia

ansarisaleh@unm.ac.id

Abstract. This research aims to reveal the profile about the level of creativity and the ability to propose statistical problem of students at Mathematics Education 2014 Batch in the State University of Makassar in terms of their cognitive style. This research uses explorative qualitative method by giving meta-cognitive scaffolding at the time of research. The hypothesis of research is that students who have field independent (FI) cognitive style in statistics problem posing from the provided information already able to propose the statistical problem that can be solved and create new data and the problem is already been included as a high quality statistical problem, while students who have dependent cognitive field (FD) commonly are still limited in statistics problem posing that can be finished and do not load new data and the problem is included as medium quality statistical problem.

Keywords: scaffolding, cognitive style, field independent, field dependent

1. Introduction

Problem posing in learning is important to be owned by students or even college students. This is supported by the result of the research from many experts. Some of them are the research that was conducted by Hashimoto in [1] which showed that learning by using problem posing approach can cause positive impact towards students' ability in solving problem. In addition, Leung, Silver, English (in [2]) reported that:

> *Problem posing has a positive influence on students' ability to solve word problems, and provided a chance to gain insight into students' understanding of mathematical concepts and processes (p.150).*

Problem prosing, according to this research, has positive influence towards students' ability in solving math word problem and give chance to the teacher to know the students' understanding about the concept and mathematical process. In line with that, Mestre stated that problem posing can be used to investigate concept transfer cross context and identify knowledge, reasoning, as well as concept development that is owned by students [3]. Meanwhile, in mathematics learning, the implementation of problem posing is the approach that is really recommended by The National Council of Teachers of Mathematics (NCTM), because problem posing has an advantage in developing knowledge and children's understanding towards important concept in school mathematics [4].

Many college students, on the other hand, still experience difficulty in problem posing during teaching and learning activity. This is supported by the findings of Romagnano (in [5]) research that there are three main dilemmas in teaching and learning activity of mathematics, namely: (1) "*ask them or tell them*" *dilemma*, (2) "*good problems*" *dilemma*, and (3) "*grading*" *dilemma*. From those three dilemmas, dilemma number (2) is related to the difficulty in problem posing during teaching and learning activity.

Kilpatrick (in [2]) stated that problem quality that is posed by students is independent variable that can be predictor towards students' ability in solving problem. Then, Kontorovich et.al., mentioned that problem posing is the special type of problem solving [6]. This means that problem posing and problem solving is inseparable. College students pose problem to be then solved again by themselves.

Before doing problem posing based on the situation that is given by a teacher, college students need to pay attention the kind of information and language structure of a problem that is going to be posed. This is important to be considered to avoid the question or problem that has no solution. The result of the research from Mayer et. al., (in [1]) found that students experience difficulties in solving mathematics problem, because it is difficult to understand the question's language. Besides, the question that has relationship proposition and imagery proposition is more difficult to be finished by the students in comparison to the question that has tasking proposition. College students' creativity is really needed in this case to overcome the difficulties in solving mathematics question. Mathematics question has been done by Haylock in [7], and one of the ways to see the posing problem ability is by creative ability. This research focuses more on the result aspect of problem posing by using creativity criteria, i.e. fluency, flexibility and originality, and do not stuck on the aspect of creative process which emphasizes more on cognitive side of the students when posing problem, whether the problem that is posed of the students fulfill the creative thinking criteria or not. Moreover, Silver said that the problem solving and problem posing can increase creative ability through creative dimension, i.e. namely, fluency, flexibility, and novelty [8].

Then, in the research that had been done by Rahman for the students of XII grade IPA-1 Makassar State High School 11, it is found that students who have cognitive style field-independent ask more statistics question that has its completeness (53.21%) in comparison to the students who have field-dependent cognitive style (19,23%) [9]. This is because not all the students have the same way in receiving and processing data in the information given. In other words, there are cognitive style differences between a student and another [10][11].

From the above description, it is known that statistics problem posing in one side is important to be owned by student. On the other hand, there are still students who do not have this ability yet. Furthermore, when students have already posed problem, it turns out that the students have its own ways in receiving and processing data from the given information. This is the way that can be stated as creativity. This fact that encourages the writer to reveal the profile in the level of creativity and ability posing problem quality of statistics problem reviewed from their cognitive style [12]-[15].

2. Research Methods

This research is qualitative research. The focus of the research is to know the creativity and quality of problem posing statistics of students that have different cognitive style, by revealing response image (statement, non-statistical question, and statistical question), students that are proposed based on the given information. Response that is in the form of statistical questions is continued by language analysis that focuses on the semantic and syntaxes relationship.

The subject of this research is the students of 2014 batch Mathematics Education Study Program of Mathematics and Natural Science Faculty in the State University of Makassar. The determination of subject in this research is done by following steps as follows:

a. Guided to cognitive style test result. Based on the cognitive style test, the students are grouped into two groups, namely: students who have field-independent (FI) cognitive style and the grup of students that have field-dependent (FD) cognitive style.

b. By taking the minimum of two field-dependent students, two field-independent students minimum, and minimum of two field-dependent approaching field-independent or vice versa.
c. Thus, the number of subjects in this research is 6 students.

The data collection in this research uses main instrument, i.e. the researcher himself, besides the main instrument, supporting instrument is also used, namely: Instrument to know the students' cognitive style: Group Embedded Figures Test (GEFT) and Statistics problem posing instrument.

After developing instrument and deciding the subject of the research, so the data collection is done as follows:

a. The researchers prepare the place that is representative to conduct statistics problem posing test.
b. In the first meeting, the first subject is called, i.e. a student that have file-independent cognitive style, then test of statistics problem posing is given to be worked without giving meta-cognitive *scaffolding* with the time that has been set.
c. The researcher checks the result of the work of the first subject and decide what parts that have already correct and what parts that needs to be fixed.
d. The part that still needs to be fixed, designed type and meta-cognitive scaffolding shape that will be given at the second meeting of the first subject.
e. The second meeting of the first subject i.e. given again the test of problem posing to be worked by giving meta-cognitive scaffolding.
f. The researcher that is helped by the research helper collects data when the subject works on the test by giving meta-cognitive scaffolding in the form of interview.
g. The data that will be collected at this time is the data that is related to:
   - Subject ability in problem posing after being given meta-cognitive scaffolding.
   - Kinds of meta-cognitive *scaffolding* that are needed by the subject in posing statistics problem.
   - The students' difficulties in posing statistics problem
h. The result of the work of the first subject by giving meta-cognitive scaffolding is analyzed and recorded its profile image. When the data is considered not enough, so that additional interview will be done until it is adequate to be drawn a conclusion.

After data collecting for the first subject that has file-independent cognitive style, in the next meeting it is continued by calling the second subject namely other field-independents with the same steps like the first subject. So it goes on until all the subjects' data are taken. The implementation of this data collection is done for each subject by using 2 days minimum time.

According to Moleong, the data analysis process is started with analyzing all data that are available from various resources [16]. In this research the available data is the ability of Statistics problem posing in this case response image that is posed by students with consider kind of response, syntaxes, and semantic relationship, the result of the interview, and meta-cognitive scaffolding kinds that are given. The data of this research is analyzed by paying attention the ability of statistics problem posing, meta-cognitive scaffolding types that are given in posing statistics problem that is done by the students.

Data analysis steps that are done in this research i.e.: First: reducting data, in this case separate between data that is appropriate to this research and the other one is not appropriate, *Second:* spreading data that are gotten from all data resources by paying more attention between steps and the result of problem posing with the result of the interview with meta-cognitive scaffolding, the obstacles that are experienced by students, kinds of meta-cognitive scaffolding that are given, as well as students' response. In the data explanation of this research, labeling is also done. Meta-cognitive scaffolding type on each is also given. *Third:* the data explanation is continued by grouping data according to the level of statistics problem posing and meta-cognitive scaffolding type that is given. *Fourth,* to determine the authenticity of the data in this research, it is done resource triangulation namely, each couple two or more subjects on each cognitive style group is matched with creative equality, problem quality that is posed as well as meta-cognitive scaffolding type that is given. If

IOP Conf. Series: Journal of Physics: Conf. Series **954** (2018) 012023 doi:10.1088/1742-6596/954/1/012023

creativity, problem quality that is posed as well as meta-cognitive scaffolding type that is given is the same, so that those data are considered as valid, and if the opposite happens, the data is considered as invalid. *Fifth*, because the purpose in this research describes creativity level profile and problem posing quality and meta-cognitive scaffolding type that is given to the students when they pose statistics problem, so the valid data will be further analyzed and the invalid data will be reduced or analyzed as side findings. Reduction is done to separate invalid data and do not give a new information.

3. Result and Discussion

3.1. The Description of Students' Cognitive Style

The data of students' cognitive style in this research is obtained from the test of cognitive test on 20 students of 2014 batch in mathematics education study program FMIPA UNM. Based on the result of the test, the cognitive style itself achieves the data as seen the following table.

Table 1 Description of Students' Cognitive Style

Respondents	**Cognitive Style**		**Sum**
	FI	**FD**	
Angk. 2014	7	13	20
Percentage	35,0	65,0	100

Note: FI=field-independent; FD=field-dependent

Based on Table 6, it is obtained that 20 students who followed cognitive style test, there are 6 (35.0%) students that are in the *field-independent* cognitive style (GK-FI), and 14 (65,0%) students that are in the *field-dependent* cognitive style (GK-FD), from this grouping election of research subject is done.

3.2. The Level of Creativity and Quality of Statistics Problem Posing

GK-FI students from average in posing statistics problem from the information given pose more statistics problems than GK-FD students (1.88). And the GK-FD students pose more statistics problem in the form of Non-Statistical Question (PNs) than GK-FI students.

Table 2 The comparison of Creative Statistics Problem Posing and The level of Creative Thinking (TKBK) between GK-FI Group and GK-FD Group

	Flexibility			Sum	Fluency	Sum	Total
TKBK	1	2	3		1		
FD	1	0	0	1	12	12	13
FI	0	5	1	6	1	1	7
Total	1	5	1	7	13	13	20

The comparison of statistics problem posing creativity and the level of creative thinking (TKBK) between the GK-FI group and GK-FD group can be seen that problem posing creativity of GK-FI is more dominant on flexibility part and GK-FD more dominant on fluency which means GK-FI better than GK-FD. Meanwhile from the level of creative thinking ability (TKBK) GK-FI group has higher level (Level 2 and Level 3) if it is compared to GK-FD (Level 1). This is in accordance with Septiadi (2016) opinion which stated that field independent cognitive style subject have a tendency to finish problem by using reasoning in connecting ideas/information while field dependent cognitive style tend to finish problem by using reasoning, in other words, the level of thinking GK-FI is better than GK-FD.

Table 3 Creativity and Statistics Problem Posing Quality in Statistics Group GK-FI and GK-FD

Quality	Flexibility		Sum	Fluency		Sum	Total
	Low	Medium		Low	Medium		
FD	1	0	1	10	2	12	13
FI	0	6	6	1	0	1	7
Total	1	6	7	11	2	13	20

In relation to the comparison of statistics problem posing creativity and the creative thinking ability (TKBK), the comparison of creativity and problem posing quality of Statistics Group GK-FI and GK-FD can be seen that problem posing creativity GK-FI is more dominant in the flexibility part and GK-FD is more dominant in fluency which means that GK-FI is better than GK-FD. While from the level of problem posing quality, GK-FI has more dominance and problem posing quality in medium level while GK-FD are dominant in the quality of problem posing side and even problem posing creativity. This is in accordance with Sasongko & Siswono which states that from creative side FI subject is more creative than FD subject [17].

4. Conclusion

Based on the result of the data analysis of the research from the previous chapter, several conclusions of this research are: (1) the level of problem posing creativity GK-FI subject in flexibility level and subject GK-FD on fluency level. (2) The level of creative thinking ability (TKBK) subject GK-FI are on the level 2 and 3 and subject GK-FD is in the level 1. (3) The level of problem posing quality subject GK-FI is in the medium level and subject GK-FD is in the low level. (4) From creativity level, the creative thinking ability (TKBK) and statistics problem posing quality of the students from mathematics education 2014 batch FMIPA UNM can be stated that GK-FI subject is better than GK-FD subject.

Acknowledgment

The authors would like to express their thanks to Universitas Negeri Makassar for the opportunity provided to the authors to carry out this research, and to the management of the Research Institute of Universitas Negeri Makassar for facilitation of this research. This work was supported by funds provided by the DIPA Universitas Negeri Makassar.

References

[1] Silver, E.A. & Cai, Jinfa 1996. An analysis of aritmetic problem posing by middle school students, *Journal for Research in Mathematics Education*, 27 (5), 521-539.

[2] Christou, Constantinos., Nicholas Mousoulides, Marios Pittalis, Demetra Pitta-Pantazi, Bharath Sriraman., 2005. *An Empirical Taxonomy of Problem Posing Processes*. In ZDM 2005, 37 (3). 149-158.

[3] Mestre, P. J. 2002. Probing adults' conceptual understanding and transfer of learning via problem posing. In: *Applied Developmental Psychology*. 23, 9-50.

[4] English, Lyn D. 1998. Children's problem posing within formal and informal Contexts, *Journal for Research in Mathematics Education*, 29 (1), 83-106.

[5] Jaeng, M, 2004. Pengembangan Model Pembelajaran Matematika Sekolah dengan Cara Perseorangan Dan Kelompok Kecil. Surabaya: Disertasi. PPs. UNESA Surabaya.

[6] Kontorovich, I., Koichu, B., Leikin, R., & Berman, A. (2012). An exploratory framework for handling the complexity of mathematical problem posing in small groups. *The Journal of*

Mathematical Behavior, *31*(1), 149-161.

[7] Leung, Shuk-kwan S. 1997. *On the role of creative thinking in problem solving.* http://www.emis.de/journals/ZDM/zdm973i.html. Zentralblatt für Didaktik der Mathematik (ZDM) Volume 29 (June 1997) Number 3. Electronic Edition ISSN 1615-679X. didownload tanggal 14 Januari 2016.

[8] Silver, E. A. 1997. *Fostering Creativity through Instruction Rich in Mathematical Problem Solving and Thinking in Problem Posing.* http://www.emis.de/journals/ZDM/zdm973i.html. Zentralblatt für Didaktik der Mathematik (ZDM) Volume 29 (June 1997) Number 3. Electronic Edition ISSN 1615-679X. didownload tanggal 14 Januari 2016.

[9] Rahman, Abdul. 2006. Deskripsi pengajuan masalah statistika Berdasarkan Gaya Kognitif Siswa Kelas XII(IA)-1 SMA Negeri 11 Makassar. Surabaya: Makalah Seminar, PPs UNESA Surabaya.

[10] Mulbar, U., Rahman, A., & Ahmar, A. S. (2017). Analysis of the ability in mathematical problem-solving based on SOLO taxonomy and cognitive style. *World Transactions on Engineering and Technology Education, 15*(1), 68-73. doi:10.26858/wtetev15i1y2017p6873

[11] Rahman, A., Ahmar, A.S, & Rusli. 2016. The influence of cooperative learning models on learning outcomes based on students' learning styles. *World Transactions on Engineering and Technology Education*, 14(3), 425-430.

[12] Rahman, A., & Ahmar, A. S. (2017). Relationship between learning styles and learning achievement in mathematics based on genders. *World Transactions on Engineering and Technology Education, 15*(1), 74-77. doi:10.26858/wtetev15i1y2017p7477

[13] Rahman, A. & Ahmar, A.S. 2016. Exploration of Mathematics Problem Solving Process Based on The Thinking Level of Students in Junior High School. *International Journal of Environmental and Science Education, 11*(14), 7278-7285.

[14] Arsyad, N., Rahman, A., & Ahmar, A. S. (2017). Developing a self-learning model based on open-ended questions to increase the students' creativity in calculus. *Global Journal of Engineering Education*, *9*(2), 143-147.

[15] Ahmar, A. S., & Rahman, A. (2017). Development of teaching material using an android. *Global Journal of Engineering Education, 19*(1), 72-76. doi:10.26858/gjeev19i1y2017p7376

[16] Moleong, L. J. (2005). *Metodologi Penelitian Kualitatif.* Jakarta: Remaja Rosdakarya.

[17] Sasongko, D. F., & Siswono, T. Y. E. (2013). Kreativitas Siswa dalam Pengajuan Soal Matematika ditinjau dari Gaya Kognitif Field-Independent (FI) dan Field-Dependent (FD). Jurnal Mahasiswa Teknologi Pendidikan, 2(1).

Business Centre Development Model of Airport Area in Supporting Airport Sustainability in Indonesia

MI Setiawan[1*], S Surjokusumo[2], DM Ma'soem[3], J Johan[2], C Hasyim[4], N Kurniasih[5], A Sukoco[6], I Dhaniarti[7], J Suyono[6], IN Sudapet[6], RD Nasihien[1], SW Mudjanarko[1], A Wulandari[6], Ansari S Ahmar[8], MBN Wajdi[9]

[1]Department of Civil Engineering, Narotama University, Surabaya, Jawa Timur 60117, Indonesia
[2]Department of Civil Engineering, Tarumanagara University, Jakarta Barat DKI Jakarta 11470, Indonesia
[3]Department of Civil Engineering, Trisakti University, Jakarta Barat DKI Jakarta 11440, Indonesia
[4]Universitas Darul Ulum, Jombang Jawa Timur 61419, Indonesia
[5]Faculty of Communication Science Library & Information Science, Universitas Padjadjaran, Bandung, Jawa Barat 45363, Indonesia
[6]Department of Management, Narotama University, Surabaya Jawa Timur 60117, Indonesia
[7]Department of Law, Narotama University, Surabaya Jawa Timur 60117, Indonesia
[8]Department of Statistics, Universitas Negeri Makassar, Makassar, Indonesia
[9]STAI Miftahul Ula, Nglawak Kertosono Nganjuk, Jawa Timur 64351, Indonesia

*ikhsan.setiawan@narotama.ac.id

Abstract. Airport is expected to play the role in enhancing the economic level of the region, especially the local people around the airport. The Aero City concept in developing an airport might also develop a city centreed in the airport that combining airport oriented business development, business actors and local people around the airport area. This study aims to generate development model of business centre at the airports in Indonesia. This is a mixed method based study. The population includes 296 airports under government management, government subsidiary and military. By using stratified random sampling, there were 151 sample airports. The results show that business centre development in the airport area will be related with the airport management and the commercial property (business centre) growth at the airport. Aero City in Indonesia can be developed by partnership system between government and private sector that consists of construction, development, and implementation of commercial property such as hotel, apartment, retail, office, etc. Based on the result of T-Value test, Airport Performance variable predicted to have significant influence on Gross Regional Domestic Product Central Business District performance.
Keywords: aero city, airport business model, airport commercial property, airport business area

1. Introduction

Airport holds strategic role as the transportation network point, as the gate of economic activities to actualize development equity, growth, economic stability also national development congruity, activities centre for interconnection of transportation mode, transportation to meet the demand to improve integrated and sustainable service quality depicted as the centre for air transportation mode shifting to another transportation mode or otherwise, to support industrial activities, trade and/or tourism activities as the effort to promote national development dynamics, also integrate with other development sectors, open the access to isolate areas, develop Indonesia border areas in the islands and/or lands, disaster handling, and infrastructure to strengthen national insights and nation sovereignty [1].

Airport development in Indonesia is quite fast, even some airports successfully ranked in the global best flight ranking. For instance, in 2017 Soekarno-Hatta airport has been ranked 10th in the Top 20 Fastest Growing World Airports Hub Connectivity 2017 [2]. Meanwhile, in 2016 Juanda Airport ranked first category big airport in The Top 20 Airports [3]. However, airports development for now and in the future is not necessarily to build a luxurious airport but also play the key role in the society economic development. The Aero City concept in developing an airport might also develop a city centreed in the airport that combining airport oriented business development, business actors and local people around the airport area [4]. Therefore, airport does not only serve as the flight destination but also connecting the industry and the market, the product and the consumer from various countries.

There are many successful examples of Aero City concept, for instance Changi airport in Singapore that achieved 7,2% growth in 2016 reached S$ 1,1. This income growth generated from the concession supported by strong retail sells that grows up to 9,9% and reached highest record with S$ 2,2 billion. Changi airport is one of three top airports in the world with highest concession sells. Changi airport keeps launch innovative retail throughout the year, for instance the opening of Zara and The Shilla Beauty Loft outlet. The big profit of Changi airport's non-aeronutic business enables Changi to substitute and maintain the competitive flying costs [5]. Other example is Malaysia airport that recorded the income escalation on duty free and non-dutiable goods (contribute by 17,7%), agriculture and horticulture income (contribute by 0,8%) and hotel income (contribute by 2,2%). Those non-aeronutic incomes generally generated from business commercial on airport operational and non-airport operational business. Commercial business on airport operational includes income from commercial room leasing, duty free and non-duty free shop, food and drinks outlet, airport parking facility, ads business, hotel and free commercial zone in KLIA. In 2016, non-aeronutic reached RM 1.110 million, duty free and non-dutiable good income reached RM 740 million, airport operational income reached RM 259 million, hotel income reached RM 92 million, agriculture and horticulture income reached RM 34 million, also improvement and maintenance project income reached RM 132 million [6].

In Indonesia to date, the Aero City concept developed in the form of strategic development master plan of International Airport Kualanamu. By employing this concept the airport might have the strategic function to produce and accelerate the region economic activities. It is important to analyze the space structure, land function, and city design in accordance with the development vision of the region in order to prepare the airport as an Aero City [7].

This study aims to generate development model of business centre at the airports in Indonesia. Problems identification includes airport management, commercial properties growth (business centre) and Aero City development model in Indonesia. The study results might contribute to the Aero City development in Indonesia.

2. Research Method

This is a mixed method based study. Mixed method requires researcher to collect, analyze and combine both qualitative and quantitative data in order to gain comprehensive understanding on the research problems [8]. The population includes 296 airports under government management, government subsidiary and military. By using stratified random sampling, there were 151 sample airports then we chose 7 airports of 7 big cities that have the most complete data on commercial property development, that is Jabodetabek, Banten, Bandung, Makasar, Medan, Semarang, and

Surabaya. Primary data derived from observation activities of 7 airports area in 7 different cities, while quantitative data derived from the research data released by 151 sample airports.

Research variable includes X Variable (airport performance) and Y Variable (Gross regional domestic product Central Business District (GRDP CBD) performance), and the hypothesis as below:

H_0 : Airport Performance has no significant influence on the GRDP CBD performance

H_1 : Airport Performance has significant influence on GRDP CBD performance

Data analysis technique includes two stages, that is descriptive analysis and influence test of Airport Performance variable (X) and GRDP CBD Performance (Y) using Partial Least Square Path Modelling Structural Equation Modelling (PLS-SEM) analysis, which including convergence validity test, discriminant test (Cross Loading value), discriminant test result (AVE root value), construct reliability test, R Square test, and T-Value test.

3. Results and Discussion

3.1. Airport Management

Indonesia have 296 official airports under government management include 13 airports managed by PT Angkasa Pura I, 13 airports managed by PT Angkasa Pura II, 53 airports under Indonesian National Army (TNI) management, 178 airports managed by Transportation Ministry technical operation unit and 39 airports managed by local government technical operation unit [9]. Airport management by government differs from the private sector management. To build an Aero City requires cooperation of all parties include the government, private sector, and also the society. There are several collaborations done by state and local government with Angkasa Pura – as the government subsidiary – and private sector. The collaborations include (1) direct buy and sell agreement of goods asset such as buildings and lands complete with the facilities within; (2) collaboration agreement to establish and to utilize half of the air field physic facility such as Apron, Taxiway, GSE yard, Inspection Road, and Helipad that also include the basic design preparation and submission, technical specification, and design detail, budgeting and result submission; (3) annual contribution submission from PT Angkasa Pura I (Persero) to the local government for maintaining and operating the half air field physic facility; (4) exchanging plan of PT Angkasa Pura (Persero) land and local government's land; (5) investment inclusion cooperation on the entity association and inclusion on the private sector enterprises also other government's subsidiaries; (6) concession collaboration with private sector enterprises, government's subsidiaries, and Directorate General of Air Transportation Hub to provide airport services including aircraft service, passengers, goods, and post consists of providing and/or developing landing facility service, take off, maneuver, parking, and aircraft retention, terminal facility to serve the passengers, cargo, post, electronic facility, electricity, water, and waste disposal unit, lands for buildings, fields, and industries, also buildings to support the air transportation smoothness.

Transportation Ministry of Indonesia Regulation Number PM.129 year 2015 arranges provision that an airport requires to provide minimum 70% of operational service facility and maximum 30% of commercial service facility from the total of passengers terminal space minus 20% of the circulation room and utility building [10]. This regulation affects the financial income of Angkasa Pura airports as one of the airport managerial in Indonesia where there is income decreasing of tenant leasing in 13 airports managed by Angkasa Pura.

3.2. Commercial Property (Business Centre) Growth

Based on the result of Central Bank of Indonesia (Bank Indonesia) data analysis, commercial property growth consists of 5 commercial property types, that is offices, retails, apartments, hotel, and industrial land [11]. Demand property growth in Jabodetabek and Surabaya shows positive trend and pretty promising, while commercial property demand in other areas is quite low such as in Banten the retail is quite low, offices is low in Makasar, in Bandung office and hotel is low, apartment and hotel is low in Semarang while in Medan offices, retail, and hotel is quite low. Commercial property supply growth in Jabodetabek (specifically apartment and hotel) and Surabaya (specifically offices) are both pretty positive, while hotel growth in Makasar has positive trend, apartment in Medan is pretty good, both in Semarang and in Bandung apartment is quite low while in Banten both retail an hotel are quite

IOP Conf. Series: Journal of Physics: Conf. Series **954** (2018) 012024 doi:10.1088/1742-6596/954/1/012024

low. Commercial property price growth shows a very promising commercial property price growth. Only few indicate price decreasing tendency such as in Jabodetabek only apartment's price is decreasing, in Banten retail's price is decreasing, in Bandung, Makasar and Semarang only hotel's price is decreasing, in Surabaya only offices is decreasing, and in Medan only apartment and hotel price are decreasing [11, 12].

The data of 151 airports in Indonesia from Statistics Indonesia (*Badan Pusat Statistik*) shows that the highest growth of real estate at airport area is in Tangerang and the lowest is in Halmahera Tengah (in the middle of Halmahera) [13]. Reliability test used in this research was Cronbach alpha and composite reliability or known as "Dillon Goldstein". Composite Reliability to measure internal consistency uses Formula 1 as follow [14]:

$$\rho c = \frac{\left(\sum \lambda_i\right)^2}{\left(\sum \lambda_i\right)^2 + \sum_i \operatorname{var}(\varepsilon_i)} \tag{1}$$

λ_i : factor loading
ε_i : error variance

While to calculate Cronbach's Alpha uses Formula 2 as follow [14]:

$$\alpha = \frac{k}{k-1}\left(1 - \frac{\sum \sigma_b^2}{\sum \sigma_t^2}\right) \tag{2}$$

α : alpha reliability coefficient
k : amount of question items
σ_b^2 : score variant of each question item
σ_t^2 : score variant of question items total

The result of convergence validity test can be seen in Table 1 as follow:

Table 1. Convergence Validity Test upon GRDP CBD

Variable	Indicator	Outer Loading	AVE	Communality	Status
Airport Performance (X1)	Aircraft Arrival (x1.1)	0.995	0.881	0.881	VALID
	Aircraft Departure (x1.2)	0.994			VALID
	Passenger Departure (x1.3)	0.989			VALID
	Passenger Arrival (x1.4)	0.989			VALID
	Baggage Unloaded (x1.5)	0.988			VALID
	Baggage Loaded (x1.6)	0.990			VALID
	Cargo Unloaded (x1.7)	0.685			VALID
	Cargo Loaded (x1.8)	0.828			VALID
APBD Performance (Y1)	Revenue (Y1.1)	0.997	0.982	0.982	VALID
	District Own Source Revenue (PAD, Y1.2)	0.996			VALID
	The result of separated regional wealth management (Y1.3)	0.979			VALID
	Budget (Y1.4)	0.996			VALID
	Capital Expenditure (Y1.5)	0.986			VALID
	Regional Investment participation (Y1.6)	0.991			VALID

IOP Conf. Series: Journal of Physics: Conf. Series **954** (2018) 012024 doi:10.1088/1742-6596/954/1/012024

As can be seen on Table 1, the result of convergence validity test shows that outer loading value > 0,50 and Average Variance Extracted (AVE) value > 0,50 thus can be concluded that indicator of GRDP Real Estate Activities and variable of GRDP CBD meet the convergence validity test.

Further, the result of discriminant test upon GRDP CBD can be seen in Table 2 as follow:

Table 2. Discriminant Test Result (Cross Loading Value) upon GRDP CBD

Indicator \ Variable	Airport Performance (X1)	Regional Gov Budget Performance (Y1)	GRDP Non CBD Performance (Y2)	GRDP CBD Performance (Y3)
Aircraft Arrival (x1.1)	0.995	0.127	0.411	0.412
Aircraft Departure (x1.2)	0.994	0.131	0.416	0.418
Passenger Departure (x1.3)	0.989	0.124	0.398	0.400
Passenger Arrival (x1.4)	0.989	0.117	0.392	0.393
Baggage Unloaded (x1.5)	0.988	0.115	0.396	0.401
Baggage Loaded (x1.6)	0.990	0.108	0.380	0.381
Cargo Unloaded (x1.7)	0.685	0.028	0.225	0.237
Cargo Loaded (x1.8)	0.828	0.037	0.214	0.219
GRDP Construction (Y1)	0.332	0.612	0.857	0.927
GRDP Transportation and warehousing (Y2)	0.329	0.666	0.921	0.958
GRDP Real Estate/Real Estate Activities (Y3)	0.425	0.881	0.922	0.896
GRDP Information and Communication (Y4)	0.376	0.561	0.893	0.938

The result above indicates that the biggest cross loading value is on the formed construct. Thus, it can be concluded that research indicators meet the discriminant validity.
Meanwhile, the result of discriminant test with AVE root value can be seen on Table 3 as follow:

Table 3. Discriminant Test Result (AVE root value)

Variable	Airport	GRDP CBD
Airport	1	
GRDP CBD	0.392408	1

Table 4 shows Discriminant Test Result (AVE root value) where each AVE root for each construct is bigger than the correlation among constructs within the model thus it is valid in the discriminant validity test. The result of discriminant validity test indicates that both cross loading and AVE root meet rule of thumb thus it is valid.

Further, the result of R Square test and T Value test can be seen in Table 4 and Table 5 as follow.

Table 4. Result Test of R Square

Dependent Variable	R-Square Value	
GRDP CBD Performance (Y)	0.948	Substantial

Based on the output result with the help of SmartPLS 2.0 software, it shows that R-Square value is 0,948 for GRDP CBD performance (Y).

Table 5. T-Value Test Result

Relationship between Variables	Original Sample (O)	Sample Mean (M)	Standard Deviation (STDEV)	Standard Error (STERR)	T Statistics (|O/STERR|)
Airport Performance (X) -> GRDP CBD Performance (Y)	-0.039	-0.050	0.016	0.016	2.408

Based on the result of T-Value test, T-Value value for Airport Performance variable is 2,408. The result compared to T-Table is 1,976 (the significant level is 0,05), that means T-Value > T-Table thus Airport Performance variable predicted to have significant influence on GRDP CBD Performance.

3.3. Aero City Development Model in Indonesia

Comprehensive analysis upon various aspects required in generating a model of Aero City development. Airport Development that consists of Instituto Superior Tecnico, Universidade do Porto Faculdade de Engenharia, Universidade de Beira Interior Portugal, Universiteit Anwerpen and TPR Department of Transport and Regional Economics generated airport business development using Osterwalder approach. The basic questions of Osterwalder approach are who is our consumer, how do we make money, types of fee should be spent, and how we generate value that can be utilized by other resources [15]. This Osterwalder approach then formulated in the Canvas Business Model. When this Canvas Business Model used to develop airport business in Indonesia, then it can be elaborated as follow:

Table 6. Aero City Development in Indonesia using Canvas Business Model

<table>
<tr>
<td rowspan="2">Key Partners
1. Passengers
2. Airline
3. Aircraft manufacture
4. Cargo
5. Fuel enterprises
6. Government
7. International regulations
8. Catering
9. Industries/ service provider related flight and airport services: safety, security, satisfaction, etc.
10. ATC, SAR, etc.</td>
<td></td>
<td>Key Activities
1. Flight Services.
2. Aircraft Maintenance
3. Airport Maintenance
4. Management Services
5. Cargo services
6. Safety and security Services
7. Commercial property: hotel, retail, apartment, office</td>
<td rowspan="2">Key Proposition
1. Interconnection, Indonesian and many countries coverage.
2. Accessibility
3. Reliability
4. On time
5. Tour packages/ destination
6. Additional services: pick up from airport to hotel, from hotel to airport, retail, public transportation, etc.
7. Visa on arrival
8. Taxes policy</td>
<td>Customer Relationship
1. Excellent service
2. Handling customer complaints</td>
<td rowspan="2">Customer Segments
a. Purpose:
1. Business
2. Leisure
3. Study

b. Budget:
1. Premium Class
2. Low-cost</td>
</tr>
<tr>
<td></td>
<td>Key Resources
1. Infrastructure
2. Human resources
3. Financial Resources
4. Management
5. Supplier
6. Software Management</td>
<td>Channel
1. Mediated Communication: internet, apps, phone.
2. Face to face
3. Commercial advertising</td>
</tr>
</table>

	Cost Structure 1. Infrastructure 2. Services 3. Human resources 4. Supplier 5. Software	**Revenue Streams** 1. Airlines services 2. Commercial properties 3. Subsidies 4. Travel Agents 5. Additional Services

Based on the components in this canvas business model, then Aero City in Indonesia can be developed through partnership system between the government and private sector that including commercial property construction, development, and implementation that include hotel, apartment, retail, offices, ect.

There is increasing opportunity of airport infrastructure and business centre at airport area investment, which shown on the realization of both domestic and international investment in 2016, especially construction sector reached Rp 14,2 trillions, restaurant and hotel sector reached Rp 2,4 trillions, transportation, storage and communication reached Rp 27,5 trillions, also real estate and service enterprise reached Rp 11,5 trillions [16].

The development of business centre at the airport area by government-private collaboration eventually will support tourism competitiveness by infrastructure quality enhancement, among others public facility, information technology, tourism promotion, the conducive business climate, connectivity and accessibility integrity. This collaborative work of government-private sector in developing business centre will also escalate the economic construction sector, the increasing trend in the latest 5 years; it will be also followed by increasing employment, supported by subsector of civil building development for public includes road, bridge, train, irrigation, and port with 60% portion, followed by subsector buildings include house, office, shopping centre, health, education, accommodation, and recreation place with 28% portion. The high involvement of private sector especially housing building construction and commercial, there is still up to 22% households has not own a house by themselves yet and in needs of 3 million houses of house supply.

4. Conclusion

Business centre development in the airport area will be related with the airport management and the commercial property (business centre) growth at the airport. Aero City in Indonesia can be developed by partnership system between government and private sector that consists of construction, development, and implementation of commercial property such as hotel, apartment, retail, office, etc. Airport is expected to play the role in enhancing the economic level of the region, especially the local people around the airport, by developing Aero City.

5. Acknowledgment

This paper is a part of a PhD Dissertation titled "Business Centre of Property Commercial Development Model of Airport Area in Indonesia" with Prof. Ir. Surjono Surjokusumo, M.Sc.F, Ph.D., Ir. Dadang M. Ma'soem, MSCE, Ph.D. and Dr. Ir. Johny Johan, M.Eng, MM. as the counselors.

6. References

[1] Kementerian Perhubungan, 2013. *Peraturan menteri perhubungan PM 69 tahun 2013 tentang tatanan kebandarudaraan nasional,* Jakarta, Jakarta: Republik Indoensia.

[2] Airport Council International, 2017, "Airport industry connectivity report 2017," ACI Europe, Paris, France.

[3] OAG Punctuality League, 2016. The 2016 OAG Punctuality League: On-time performance results for airlines and airports, 2016 OAG Aviation Worldwide Limited, Tokyo.

[4] J. Kasandra, 2013, Aerotropolis : Business Mobility and Urban Competitiveness in the 21 st

Century, *In Cultures of Mobility,* pp. 1-22.

[5] CHANGI airport group, 2016, " 2016 annual report: changi airport group. 60 Airport Boulevard #046-019-01 Changi Airport Terminal 2 Singapore 819643," CHANGI airport group. 60 Airport Boulevard #046-019-01 Changi Airport TeChangi Airport Group (Singapore) Pte, Singapore.

[6] Malaysia Airport, 2017, An Immersive airport experience: anual report 2016, Malaysia Airport, Selangor, Malaysia,.

[7] A. Nasution, D. Harisdani and P. Napitupulu, 2017. The implementation of aerotropolis concept on new town planning and design in Mebidangro, Sumatera Utara, *IOP Conf. Series: Materials Science and Engineering,* **180,** *012293.*

[8] J. Creswell and V. P. Clark, 2011, Designing and conducting mixed methods, Thousand Oaks, CA: Sage.

[9] Kementerian Perhubungan, 2017. Bandar udara Indonesia. http://hubud.dephub.go.id/?id/bandara/index.

[10] Menteri Perhubungan Republik Indonesia, 2015. *Pedoman penyusunan perjanjian tingkat layanan (service level agreement) dalam pemberian layanan kepada pengguna jasa bandar udara,* Jakarta, Jakarta: Republik Indonesia.

[11] Bank Indonesia, 2017, Perkembangan Properti Komersial, Triwulan II, 2017. Bank Indonesia, Jakarta.

[12] Bank Indonesia, 2017. Kajian ekonomi dan keuangan regional: laporan nusantara. Bank Indonesia, Jakarta.

[13] Badan Pusat Statistik Subdirektorat Konsolidasi Neraca Produksi Regional, 2016. Tinjauan regional berdasarkan PDRB Kabupaten/Kota 2011-2015: Buku 1,2,3,4,5. Badan Pusat Statistik, Jakarta.

[14] H. Sofyani, 2016. Modul praktik Partial Least Square (PLS): untuk penelitian Akuntansi, Yogyakarta: Universitas Muhammadiyah Yogyakarta.

[15] AIRDEV, 2012. The business model of an airport STRUYF Els, in *AIRDEV Conference 2012*, Lisbon (Portugal).

[16] Badan Pusat Statistik, 2017. Statistical yearbook of Indonesia 2016. Badan Pusat Statistik, Jakarta.

[17] S. J. A. &. J. Kasarda, 2013. The Airport City Phenomenon: Evidence from Large US Airports, *Urban Studies,* **50**, 6, pp. 1239-1259.

IOP Conf. Series: Journal of Physics: Conf. Series **954** (2018) 012025 doi:10.1088/1742-6596/954/1/012025

Analysis of Water Well Quality Drilling Around Waste Disposal Site in Makassar City Indonesia

R. Maru, I. I. Baharuddin, N. Badwi, S. Nyompa, and Sudarso

Department of Geography, Faculty of Mathematics and Natural Sciences, Universitas Negeri Makassar, Indonesia

rosminimaru@unm.ac.id

Abstract. Clean water is one of human need which is very important in carrying out its life. Therefore, this article analyzes the quality of the well water dug around the landfill. The method used is a well water well sample taken from 4 wells around a landfill taken by a purposive sampling at a different distance. The parameters measured are physical, chemical, and biological properties. The results of the analysis were then compared with the standard of drinking water quality criteria allowed under The Regulation of Health Minister of Indonesia No. 416 year 1990 on the Terms and Supervision of Water Quality of the Minister of Health of the Republic of Indonesia. The result of the research shows that there are two wells whose water quality does not meet the physical requirement i.e Location of Points II and III, based on the construction of wells also does not meet the requirements of the wells in general. While at the well Locations Point I and IV the quality of water physically, chemically and biologically as well as well construction qualify. From the result of this research, the researcher give suggestion of the need to improve the physical condition of dug wells, it is necessary to do the extension to the well water user community for drinking water about the physical condition of the dug well, the need to monitor and supervise the quality of drinking water, and should involve the community to independently meet the needs absolute i.e clean water to drink.

Keywords: water quality, well water, dug well, coli bacteria, Makassar city, clean water.

1. Introduction

Water is a major need for life processes on earth. There would be no life if there is no water on this earth. Clean water is highly coveted by human beings both for daily living purposes, for industrial purposes, for urban sanitation, and for agricultural purposes and so on [1].

The need for clean water is arranged in one of the government programs, namely the provision of clean water. The issue of water supply is one of the priorities in improving the health status of the community. Given the existence of water is vital needed by living things. Life on earth can only take place in the presence of water [2].

Many residents are forced to take advantage of water that is not good quality. Of course this will result in less good for public health in the short term, poor quality can lead to Vomitting, Diarrhea, Cholera, Typhoid, or Dysentery. This can occur in an environment of poor sanitation. If ground water and surface water are contaminated by dirt, germs are automatically spread to water sources used for house-hold use. In the long run, less quality water can lead to bone loss, dental corrosion, anemia, and kidney damage. This occurs because of the presence of heavy metal metals that are much toxic (toxic) and deposition in the kidneys [3].

Clean water coverage program in Makassar city In 2004 was 82,55%, while drinking water coverage in 2004 was 80,85% for physical, clean water quality coverage was 87,02% was increase compared to 2003 that was 83,43 % based on Regulation of the Ministry of Health of the Republic of Indonesia [4].

The Garbage Final Disposal Site (TPA) located at RT 04 RW 11 Parang Tambung Village Tamalate Subdistrict, Makassar City, utilizes 900 square meters (m^2) of land, and utilization by the community as a garbage dump for 20 years. To be evenly distributed so that it does not accumulate at one point, the local community usually burns the garbage, the people living around the landfill (TPA) of garbage still use dug well water as the source of clean water of the family in daily life (TPA RT 04 RW 11). The existence of dug wells (SGL) around the landfill both from the water quality and construction aspects of dug must meet the health requirements.

2. Basic Theory

Dug wells are one of the most common wells used to collect ground water for small communities and individual houses for clean water and drinking water with a depth of 7-10 meters from the ground. Dug wells provide water derived from a layer of soil relatively close to the surface of the soil, therefore easily exposed to contamination through seepage. Generally, seepage comes from human waste disposal (latrine/latrine), animal waste, well waste itself, either because of its floors or sewerage that is not water-resistant [26].

2.1. Water Quality

2.1.1. Water Quality Standards. Water Quality is a quality characteristic that is required for the specific utilization of water sources. With the water quality standard, people can measure the quality of the various kinds of water. Each type of water can be measured the concentration of elements contained in the standard of quality, thus can be known quality terms, in other words quality standards can be used as a benchmark. The standard of clean water quality can be defined as the provisions based on Regulation of the Ministry of Health of the Republic of Indonesia [5] which is usually set forth in the form of statement or number indicating the requirements that must be fulfilled so that water does not cause health problems , diseases, technical disturbances, as well as disturbances in aesthetic terms.

This regulation is made with the intention that drinking water that meets health requirements has an important role in the maintenance, protection and heightening of public health. With this regulation has been obtained legal basis and technical foundation in terms of water quality control. Similarly, the water used as daily water needs, the water should be colorless, tasteless, odorless, clear, and has a temperature in accordance with the standards set so as to create a sense of comfort.

2.1.2. Physical requirements. Regulation of the Minister of Health [5] on drinking water quality requirements states that water that is consumed and used in everyday life is water that has good quality as a source of drinking water and raw water (clean water), among others must meet the requirements physically , odorless, tasteless, not cloudy, and colorless. The physical properties of water can be affected by various factors including the following:

a) Temperature

Water temperatures will affect people's acceptance of the water and may also affect chemical reactions in the treatment especially if the temperature is very high. The desired temperature is 20ºC-30°C surrounding air temperature which can provide a sense of freshness, but the local climate or type of source of water will affect the water temperature. In addition, the temperature of the water directly affects the toxicity of many pollutant chemicals, the growth of microorganisms, and viruses. Temperature or water temperature is measured using a water thermometer.

b) Odor and taste

Odor and taste usually occur simultaneously and are usually caused by decaying organic materials, certain types of microscopic organisms, and chemical compounds such as phenol. The ingredients that cause these odors and flavors come from various sources. The intensity of odor and taste can be increased when there is chlorination. Since the measurement of odor and taste is dependent on individual reactions the reported results are not absolute. For drinking water standards and clean water is expected water does not smell and does not taste.

The color in the water is divided into two, apparent colors are the colors caused by the turbidity particles (soil, sand, etc.), fine particles of iron, manganese, microorganism particles, industrial colors, etc. The second is true color is a color derived from the decomposition of natural organic substances, namely humus, lignin, tannins and other organic acid.

Technical color removal can be done in various ways. Among them: coagulation, flocculation, sedimentation, filtration, oxidation, reduction, bioremoval, applied electro, etc. The degree of water dye can be known through laboratory examination by photometric method. For the color of clean water standard is 25 TCU and maximum standard for drinking water is 15 TCU.

c) Turbidity

Water is said to be turbid when the water contains so many suspended material particles that it gives a muddy and dirty color. The materials that cause this turbidity include clay, mud, organic materials dispersed from small particles that are suspended. Turbidity in water is one thing that should be considered in the provision of water to the public, given that the turbidity will diminish the aesthetic aspect, make it difficult in screening efforts, and will reduce the effectiveness of the disinfection effort.

The level of turbidity of water can be known through laboratory examination with Turbidimeter method. For a maximum permissible water turbidity standard of 25 NTU and 5 NTU for drinking water standards.

2.1.3. Chemical Requirements. Good clean water is water that is not contaminated with excessive chemical substances that are harmful to health such as iron (Fe), degree of acidity (pH), chloride and other chemicals. The content of chemicals in clean water used daily should not exceed the maximum levels allowed for drinking water quality standards and clean water.

a) Iron (Fe)

Iron is one of the more important elements in surface water and groundwater. Iron-containing waters are highly undesirable for domestic use, as they may cause rust marks on clothing, porcelain and other tools and cause unpleasant taste in drinking water at concentrations above about 0.31 mg / l. The aquatic chemical properties of iron are redox properties, complex formation, metabolism by microorganisms, and exchange of iron between phases and solid phases containing iron carbonate, hydroxide and sulfide.

b) Degree of acidity (pH)

The degree of acidity (pH) is a term used to express the intensity of the acid or base state of a solution. It is also a way of expressing the concentration of H + ions. In water supply, pH is a factor affecting the processing activity to be performed [6].

As a measure of the acidity and alkalinity nature of water is expressed by pH value, which is defined as the logarithm of the return of hydrogen ion concentration in moles per liter. The pure water at 24 is weighed with respect to the OH ions - each having a content of 10-7 moles per liter. Thus the pH of pure water is 7 [7].

c) Chloride (C1)

Chloride levels generally increase with increasing cadminerality. High chloride levels, followed by high levels of calcium and magnesium, can improve the corrosivity of water. This resulted in the occurrence of metal appliances. Chloride levels> 250 mg / l can give a salty taste in water because the value is a chloride limit for water supply, which is 250 mg / l.

IOP Conf. Series: Journal of Physics: Conf. Series **954** (2018) 012025 doi:10.1088/1742-6596/954/1/012025

d) Biological Requirements

In bacteriological parameters used indicator bacteria or sanitary indicator pollution indicators. Bacterial indicator sanitation is a bacteria that can be used as an indication of faecal pollution from humans and from animals, because the organism is an organism contained in the gastrointestinal tract of humans and animals. Water contaminated by human feces cannot be used for drinking, washing food or cooking because it is considered to contain pathogenic microorganisms that are harmful to health, especially pathogen-causing gastrointestinal infections.

3. Method and Data

Sample in this research is a part of all digging wells which have distance less than 50 meter from garbage dump RT 04 RW 11 Parang Tambung village Tamalate Subdistrict, Makassar city, that is as much 4 dug wells. The sampling method is done by purposive sampling with the following criteria: a) The owner of the dug well that still uses the well water as the clean water daily (bathing and washing), b) the location of the well from other pollutant sources, and c) The well owner is willing to have the well to be sampled.

Water samples are taken on dug wells that have been determined location or place. The water sample will then be under a laboratory to test the physics, chemical and biological parameters. Parameters analyzed in the laboratory are color, turbidity, pH, iron, chloride and total coliform. For temperature parameters, taste and odor are directly determined at each location of the dug well water sampling.

3.1. Physical Parameters

- Temperature (temperature) is measured in the field using a water thermometer.
- Flavors and odors are determined directly in the field using taste and olfactory senses.
- Color is measured in laboratory by using test method / technique SNI 06-6989.24-2005 [22]
- Turbidity is measured in collaboration by using SNI 06-6989.25-2005 test method

3.2. Chemical Parameters

- pH was measured in laboratory using test method/technique SNI 06-6989.11-2004 [23]
- Iron (Fe) is measured in laboratory using test method/technique SNI 6989.4-2009 [24]
- Chloride (CI) was measured in laboratory using test method/technique SNI 06-6989.19-2004 [25]

3.3. Biological Parameters

- Total coliform was measured in laboratory using standard total coliform fermentation technique.

Further data is analyzed in the laboratory to determine the nature of physics, chemistry, and biology. Furthermore, in comparison with the standards of water quality criteria in accordance with Regulation of The Minister of Health of the Republic of Indonesia [5] for water quality parameters. And for the construction of dug wells compared to the requirements of wells in general according to Suripin [7].

4. Result

4.1. Construction of Dug well

Field observation results can be explained about the construction of dug wells around the landfill of RT 04 RW 11 Parang Tambung Village Tamalate Sub District, Makassar City as follows. The drawing construction of wells dug around waste disposal site can be seen in the table 1.

Table 1. Drawing Construction of Wells Dug around Waste Disposal Site RT 04 RW 11 Parang Tambung Village, Tamalate Sub District, Makassar City, 2017.

No	Requirements of Construction of Dug Well	Analysis Construction Dug Well Result			
		Point I	Point II	Point III	Point IV
1	Have Waterproof Wall 3 Meters Go Down (Yes /No)	Yes	Yes	Yes	Yes
2	Have Lips Well (Yes/No)	Yes	No	Yes	Yes
3	Distance of Well From Source of Pollution More Than 11 Meters (Yes /No)	Yes	No	No	Yes
	Explanation	Q	NE	NE	NE

(Data Processing Result, 2017)
note: Q (Qualify); NE (Not Eligible)

Based on Table 1. it is known that all dug wells have a waterproof wall 3 meters down, and one digging well does not have well lips and two wells spaced from pollution less than 11 meters. Based on the observation of dug wells examined, overall after the analysis of the allocation of two wells that do not meet the health requirements that have been set the wells at the location of points II and III because it has no well lips and well distances from pollution sources less than 11 meters.

Table 2. Results of Analysis of Physical Parameter

No	Physic Parameters (Unit)	Water Quality Criteria (Regulation of The Minister of Health of the Republic of Indonesia	Sample Analysis Results			
			Point I	Point II	Point III	Point IV
1	Smell	No smell	No smell	No smell	Smell of soil	No smell
2	Flavors	no taste	no taste	Feels brackish	no taste	no taste
3	Temperature (° C)	30°C	30°C	31°C	29°C	29°C
4	Color (TCU)	50	2,5	2,5	2,5	2,5
5	Turbidity (NTU)	25	6,14	0,04	3,32	1,98
	Explanation		Q	NE	NE	NE

(Data Processing Result, 2017)
note: Q (Qualify); NE (Not Eligible)

4.2. Physical parameters
Physical parameters in this study are odor, taste, temperature, color and turbidity. The results of the analysis can be seen in the Table 2. Based on Table 2 it is known that from the four wells taken by water samples, there are two wells whose water quality physically does not meet the water quality requirements as determined by Minister of Health of the Republic of Indonesia [5].

4.2.1. Temperature
Based on the measurements at the four sample sites, the sample well water temperature ranges from 29 ° C - 31 ° C. The results of the analysis can be seen in the Figure 1.

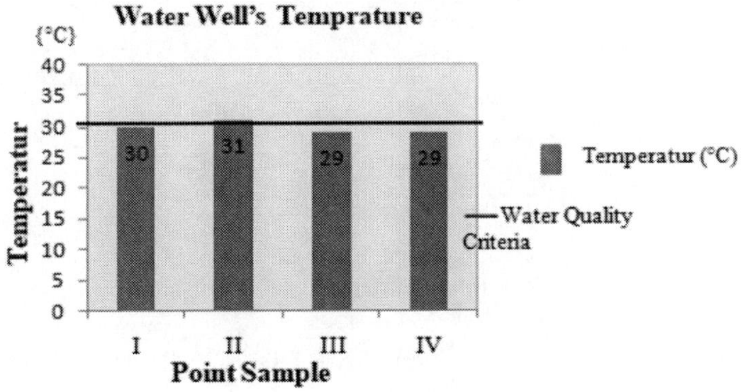

Figure 1. Histogram of Water Well's Temprature RT 04 RW 11 Parang Tambung Village, Tamalate Sub District, Makassar City, 2017

Based on the histogram or Figure 1 above, the observation of the temperature or temperature of the well water in the field samples I at 30 ° C, the second location of the well water temperature is 31 ° C and is the highest water temperature whereas at locations III and IV the water temperature is the most low is 29 ° C.

4.2.2. Color
Based on the results of the analysis in the laboratory showed that the color parameters for well water samples all locations meet the criteria of water quality criteria. The results of the analysis can be seen in the Figure 2.

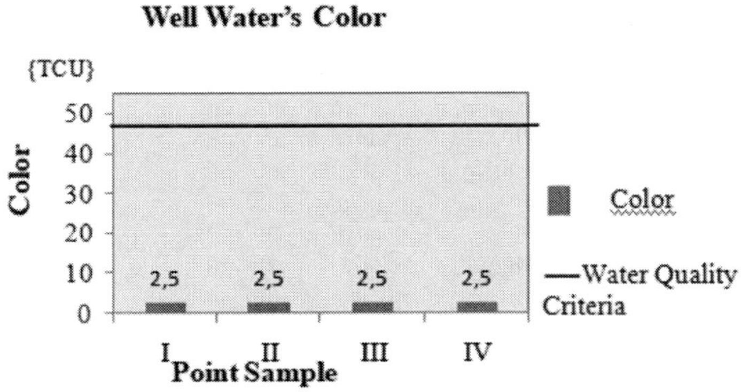

Figure 2. Histogram of Well Water's Color Parang Tambung Village, Tamalate Subdistrict, Makassar City, 2017. The results obtained in the color of well water samples at all locations are 2.5 Pt-Co color Unit.

4.2.3. Turbidity

Based on the results of the sample analysis in the laboratory, the turbidity parameters of all sampling points all meet clean water quality standards of 25 NTU. The results of the analysis can be seen in the Figure 3.

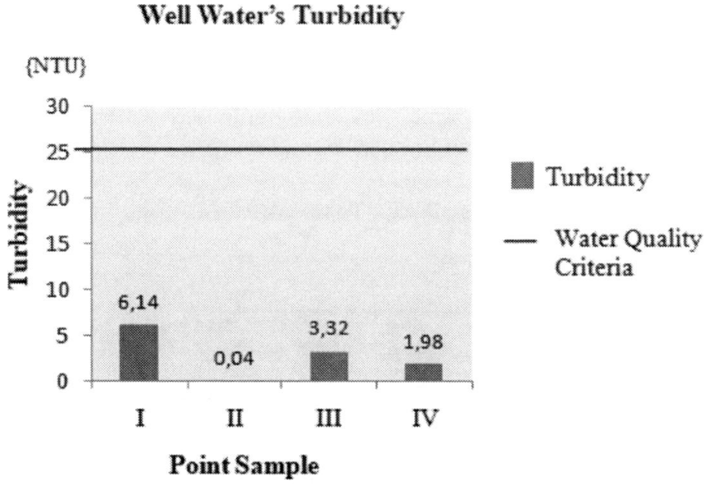

Figure 3. Histogram of Well Water's Turbidity RT 04 RW 11 Parang Tambung village, Tamalate subdistrict Makassar City, 2017.

Based on Figure 3 above shows that the highest turbidity value is located at location I that is 6.14 NTU, but it does not pass the water quality feasibility standard. While the lowest turbidity value is at location II that is 0.04 NTU.

4.3. Chemical Parameters

Chemical parameters here are pH, iron (Fe), chloride (CI), where if one of these variables is not eligible, it can be said that the water does not meet the water quality requirements determined in accordance with Regulation of the Minister of Health of the Republic of Indonesia [5]. The analysis of sample can be seen in the
Table 3.

Table 3. Analysis of Sampel Chemical Parameters Result

No	Chemistry Parameters (Unit)	Maximum of Water Quality Criteria Regulation of the Minister of Health of the Republic of Indonesia	Analysis Sample Result			
			Point I	Point II	Point III	Point IV
1	pH	6,5-9,0	6,678	6,747	6,757	6,690
2	Iron/Fe (mg/L)	1,0	0,1039	0,1340	0,0838	0,1178
3	Chlorida/CI (mg/L)	600	27,2256	34,7410	45,0924	36,4426
	Explanation		Q	Q	Q	Q

(Data Processing Result, 2017)
note: Q (Qualify); NE (Not Eligible)

Based on Table 3 it is known that from the four wells taken by water samples, all meet the criteria of clean water quality standards established in accordance with Regulation of the Ministry of Health [5].

4.3.1. pH

pH is a term used to express the intensity of the acid or base state of a solution. The results of the analysis can be seen in the Figure 4.

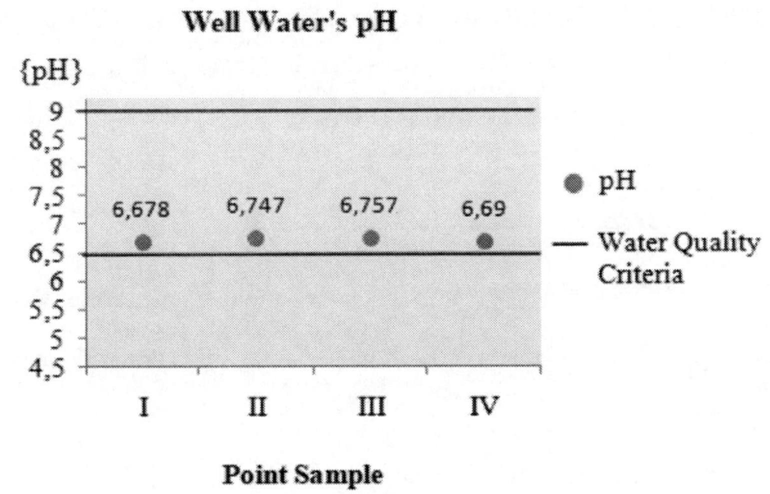

Figure 4. Histogram of Well Water's pH RT 04 RW 11 Parang Tambung village, Tamalate subdistrict Makassar City, 2017.

Based on the histogram above the highest parameter value of pH is location III that is 6.757 whereas the lowest pH value is location I that is 6,678.

4.3.2. Iron (Fe)

The permitted level of iron content (Fe) for clean water quality is 1.0 mg/L. The results of the analysis can be seen in the Figure 5.

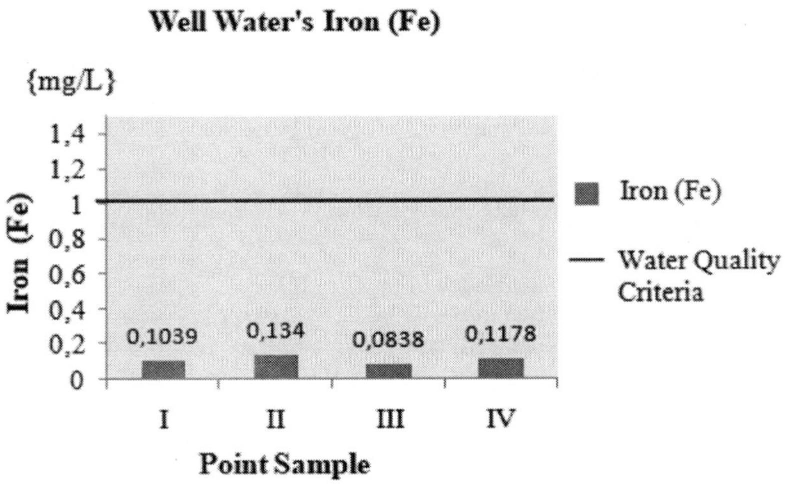

Figure 5. Histogram of Well Water's Iron (Fe) RT 04 RW 11 Parang Tambung village, Tamalate subdistrict Makassar City, 2017.

When viewed from the iron content of the highest value that is at the location II is 0.1340 mg / L while the lowest iron content value is the third location of 0.0838 mg / L. According to Cole, in [9], iron is only found in waters that are in anaerobic condition (anoxic). Iron compounds are generally soluble and quite abundant in the soil. Groundwater typically has a relatively large amount of carbon dioxide, presumably with low dissolved oxygen or even anaerobic atmosphere.

4.3.3. Chloride (Cl)
The content of chloride (Cl) allowed for clean water quality is 600 mg / L. The results of the analysis can be seen in the Figure 6.

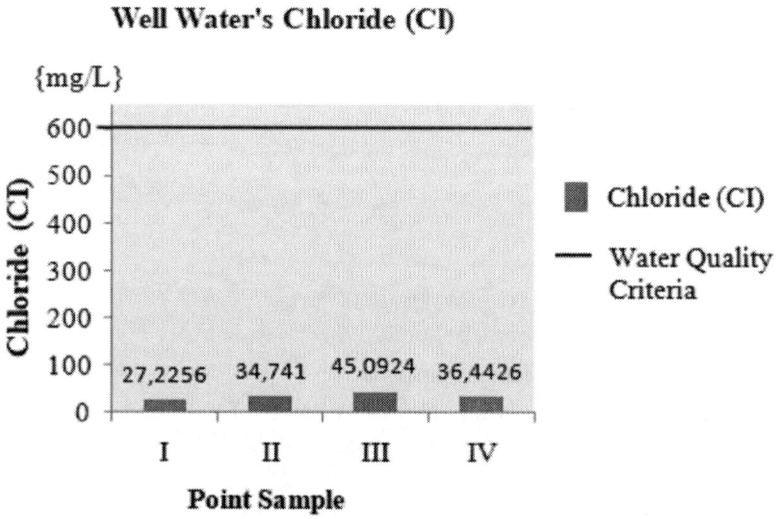

Figure 6. Histogram of Well Water's Chlorida (Cl) RT 04 RW 11 Parang Tambung Village, Tamalate Subdistrict, Makassar City.

When seen from Figure 6. The highest content of chloride that is at location III that is 45,0924 mg / L while the value of chloride content is lowest is location I that is 27,2256 mg / L.

4.4. Biology Parameter
The biological parameters are total coliform (E-coli). The results of the sample analysis can be seen in Table 4.

Table 4. Analysis Biological Parameters Result

No	Biological Parameters (Satuan)	Water Quality Criteria (Regulation of the Minister of Health of the Republic of Indonesia)	Analysis Sample Result			
			Point I	Point II	Point III	Point IV
1	Biology/Total Coliform (Total/100 ml)	50	48	< 1,8	47	24
	Explanation		Q	Q	Q	Q

(Data Processing Result, 2017)
note: Q (Qualify); NE (Not Eligible)

Based on Table 4. above it is known that from the four wells taken in water samples, all meet the criteria of water quality standards established in accordance with Minister of Health of the Republic of Indonesia [5]. Comparison of the total number of coliform can be seen in Figure 7.

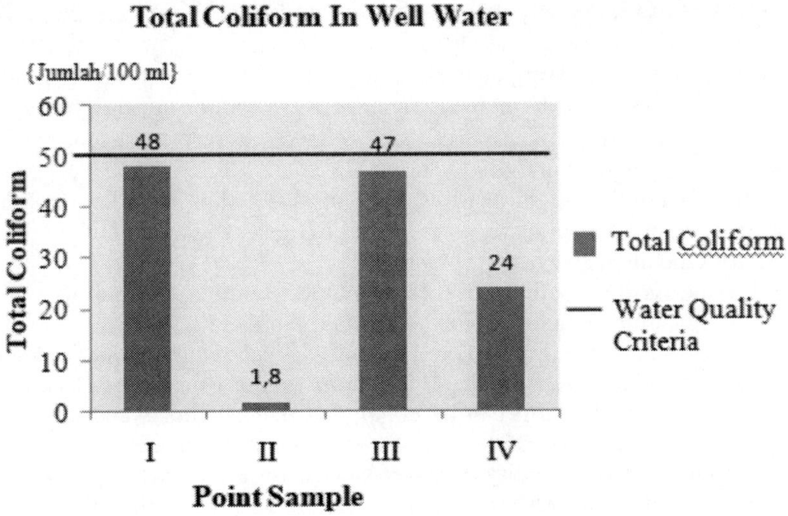

Figure 7. Histogram of Total Coliform In Well Water RT 04 RW 11 Parang Tambung Village, Tamalate Subdistrict, Makassar City.

Based on Figure 7 above it is known that the highest total coliform value is at location I that is 48 colony / 100 ml and the lowest at location II that is <1,8 colony / 100 ml.

5. Discussion

From the results of the research, it is known that dug wells are located around the garbage dump RT 04 RW 11 Parang Tambung Village, Tamalate Subdistrict, Makassar City, from four dug wells that have been checked the water quality physically there are two digging wells that the physical quality of the water does not meet the standard of clean water quality criteria in accordance with the decision of Regulation of the Minister of Health of the Republic of Indonesia [5], the location of Point II and location of Point III, where the physical condition of the well water at the Point II location, the taste in the well water feels slightly brackish and the temperature water exceeds the water quality standards set. While at the location of Point III the physical state of the well water smells the soil.

The odor of dug well water at the location of Point III, which smells of soil, may be caused by water derived from various sources such as waste material or caused by the process of organic compounds by bacteria and carcasses of animals suspended in well water so that water wells at the location of Point III smelled of soil.

According to Effendi [9], water is good and safe for consumption is water that has characteristic odorless when kissed from far or near. Groundwater is caused by water derived from various sources such as waste material or caused by the process of organic compounds by bacteria and animal carcasses.

The flavor of the well water in point II, which is slightly brackish, whereas at the location of Point I, III, and IV the taste of the water is not felt or in accordance with the water quality requirements in general. According to Hadipurwo and Danayanto [10], groundwater flavor is determined by the presence of salt or substances contained in the water, either suspended or dissolved. The slightly brackish water flavor at the site of Point II, the possibility of any salt or substances contained in the water, either suspended or dissolved.

Based on the temperature measurements of the four sampling sites, one well (Location of Point II) whose water temperature exceeds the threshold with the water temperature of 31° C. While the desired temperature in clean water is 20 ° C - 30 ° C. The high temperature is due to the local climate and the type of water source that will affect the temperature [7]. The state of well water temperature depends on the local climate including the temperature and rainfall conditions [11][12]. This is not much different from the results of the study by Maru, et. al. the average daytime temperature in Makassar is 31.29 °C [13],[15]. Even some other big cities have experienced similar things, as the study in Jakarta by Maru and Ahmad, that the average temperature during the day is 33.32°C [15]. Among the causes of rising air temperatures are reduced forest areas or green open spaces [16], and increasing anthropogenic activity [17].

From the water test of the four digging wells as clean water that has been done by the Center of Production of Makassar Plantation Industry is 2.5 TCU (Eligible) in accordance with the decision of Minister of Health of the Republic of Indonesia [5] is 50 TCU.

The color of the water is due to the presence of natural tannins and humic acid. This substance when exposed to chlorine can form toxic choloroform compounds [18].

From the fourth well water turbidity test, the turbidity level ranged from 0.04 - 6.14 NTU. The standard of turbidity parameter based on the Regulation of the Minister of Health of the Republic of Indonesia [5] is maximum 25 NTU, so it can be concluded that all four wells taken by water samples are all eligible according to Regulation of the Minister of Health of the Republic of Indonesia [5].

According to Hadipurwo [10], turbidity of water caused by the presence of not soluble substances contained. An example is the presence of clay particles, silt, as well as organic substances or microorganisms.

From the result of the research, it is known that the dug wells around the landfill of RT 04 RW 11 Parang Tambung Village Tamalate Subdistrict of Makassar City, from the four digging wells that have been checked the water quality chemically include pH, Iron (Fe), and Chloride (CI), the four wells all meet the water quality criteria criteria in accordance with the Regulation of the Minister of Health of the Republic of Indonesia [5].

From pH test on dug well water as clean water which has been done by Indonesian Center for Plantation Product of Makassar City is 6,690 - 6,757 (Normal) and feasible according to standard quality value from Decree of Minister of Health of the Republic of Indonesia [5] is 6, 5 - 9.0. The pH content in water does not exceed the maximum quality standard. Water should be neutral, not acidic or alkaline, to prevent the occurrence of heavy metal dissolution and corrosion of water distribution networks. Water is a very good solvent, so it is helped by a non-neutral pH to dissolve the various chemical elements it passes through [17].

From Iron (Fe) test on well water dug as clean water that has been done by Center of Production Industry of Makassar City is 0,0838 mg / L - 0,1340 mg / L (Eligible) as per standard value of Regulation of the Minister of Health of the Republic of Indonesia [5] is 1.0 mg / L. Iron is soluble at low pH and can cause yellowish water, cause stains on clothing and breeding place of creonothrinx bacteria, therefore iron content should not exceed 1.0 mg / L, because it can accelerate the growth of bacteria and can cause taste and smell [18].

From Chloride (CI) test on well water as clean water that has been done by Center of Production Industry of Makassar City is 27, 2256 mg / L - 45,0924 mg / L (Eligible) according to standard quality value of Minister of Health of the Republic of Indonesia [5] is 600 mg / L. What if the content of chloride (CI) in well water exceeds the quality standard in large amounts of chloride will cause saltiness, corrosion in the water supply system pipeline [17].

From the examination of Total Coliform in digging well water as clean water which has been done by Central of Makassar Plantation Product Industry is <1,8 - 48 per 100 ml of sample and (Eligible) according to value of clean water quality standard from Regulation of the Minister of Health of the Republic of Indonesia [5] is 50 Mg / L. According to Fardiaz [20], [21] the high amount of Coliform Total can occur due to the high contamination of pathogenic bacteria from human or animal digestive tract and other pathogenic agents.

6. Conclusion

Based on the research objectives and discussion, it can be concluded and suggestions as follows: Physical quality of well water around the landfill of RT 04 RW 11 Parang Tambung Village, Tamalate Sub-District, Makassar City, at location I and IV meet the criteria of clean water quality in accordance with Regulation of the Minister of Health of the Republic of Indonesia [5], based on the construction of dug wells meet the requirements. While on location II and III the physical quality of the water does not meet the criteria of clean water quality in accordance with Minister of Health of the Republic of Indonesia [5] based on the construction of dug wells do not meet the requirements. Description of the chemical quality of well water dug around the landfill of RT 04 RW 11 Parang Tambung Village, Tamalate Subdistrict, Makassar City, from 4 wells taken by water samples all meet the criteria of clean water quality in accordance with Regulation of the Minister of Health of the Republic of Indonesia [5], based on the construction of the dug wells at sites II and III although not eligible but the water quality does not exceed the threshold for the chemical quality of the water. The description of bacteriological quality (Total Coliform) of well water dug around garbage dump RT 04 RW 11 Parang Tambung Village, Tamalate Subdistrict, Makassar City, from 4 wells taken by water samples all meet the criteria of clean water quality in accordance with Regulation of the Minister of Health of the Republic of Indonesia [5], based on the construction of dug wells at sites II and III although not eligible but the water quality does not exceed the threshold of the biological quality of its water.

Acknowledgments

We would like to express our gratitude to the Head of Universitas Negeri Makassar and the Makassar City Government for their permission and support in completing this research.

References

[1] Wardhana, 2004. *Impact of Environmental Pollution.* Yogyakarta:Andi Offset.
[2] Ratna, S., 2006. Study of Dug Well Water Quality in Paria Village, Duampanua Subdistrict, Pinrang Regency, Unpublished Thesis, Faculty of Public Health Unhas.
[3] Kusnaedi, 2002. *Managing Water For Drinking Water*. Jakarta: Rineka Cipta.
[4] Ministry of Health of the Republic of Indonesia. 2004. Decree of the Minister of Health No. 128 / MENKES / SK / II / 2004 on the Basic Policy of the Center for Public Health. Jakarta
[5] Regulation of Health Minister of Indonesia No. 416 year 1990, *Terms and Supervision of drinking water quality.*
[6] Sutrisno, Totok C, 2006. *Water Supply Technology*. Rineka Cipta, Jakarta.
[7] Suripin, 2002. *Preservation of Soil and Water Resources*. Yogyakarta: Andi Off Set.
[8] Harter, T. 2003. Water Well Design and Construction. appropriate quality, drillers and groundwater consultants rely on their prior knowlPUBLICATION 8086 FWQP REFERENCE SHEET 11.3. by the Regents of the University of California, Division of Agriculture and Natural.
[9] Effendi, H, 2003. Water Quality Study for Resource Management and Water Environment. Yogyakarta: Kanisius.
[10] Hadipurwo, S and Danaryanto, 2006. *Conservation as a Ground Rescue Efforts in Indonesia.* Presented at: National Seminar of World Water Day 2006. Directorate of Geothermal Operations and Ground Water Management Directorate General Mineral Coal andGeothermal Department of Energy and Mineral Resources.
[11] Sanusi, W., Jemain, A. A., & Zin, W. Z. W. (2014). Fuzzy Clustering for Regionalization of Drought Proneness in Peninsular Malaysia. *Sains Malaysiana*, *43*(11), 1791-1800.
[12] Sanusi, W., Jemain, A. A., Zin, W. Z. W., & Zahari, M. (2015). The drought characteristics using the first-order homogeneous Markov chain of monthly rainfall data in peninsular Malaysia. *Water Resources Management*, *29*(5), 1523-1539
[13] Maru, R., Baharuddin, I.I., Umar, R., Rasyid., R., Uca, Sanusi, W., and Bayudin. 2015. Analysis of The Heat Island Phenomenon in Makassar, South Sulawesi, Indonesia. *American Journal of Applied Sciences*. 12 (9): 616.626. ISSN online 1554-3641. DOI: 10.3844/ajassp.2015.

[14] Maru, R. and Ahmad, S. 2015. The relationship between land use changes and the urban heat island phenomenon in Jakarta, Indonesia. *Journal of Advanced Science Letters*. Vol. 21, No. 2, pp. 150–152(3). ISSN 1936-6612 (Print)
[15] Maru, R. and Ahmad, S. 2015. The Relationship between Temperature Patterns and Urban Morfometri in the Jakarta City, Indonesia. *Asian Journal of Atmospheric Environment*. Vol. 9-2, pp. 128-136, June 2015. ISSN (Online) 2287-1160. ISSN (Print) 1976-6912. DOI: http://dx.doi.org/10.5572/ajae.2015.9.2.128.
[16] Maru, R., Abidin, M.R., Arfan, A., Nyompa, S., Sideng, U., and Hasja, S. 2016. Mapping of Protected Forests and Cultivated Area in North Luwu South Sulawesi, Indonesia. *Asian Journal of Applied Science*. Vol 9, issue 4, pp 189-195, 2016. ISSN1996-3343.
[17] Maru, R., & Ahmad, S. (2014). Daytime temperature trend analysis in the city of Jakarta, Indonesia. *World Applied Sciences Journal*, *32*(9), 1808-1813.
[18] Slamet, S, J, 1994. *Environmental Health*, Gadjah Mada University Press.
[19] Sutapa, I.D.A, 2000. Correlation Test of Effect of Tapioca Waste on Well Water Quality. Journal of Development Studies, Society & Environment, Vol. 2, Vo1; 47-65.
[20] Fardiaz, S. (1992). *Polusi air dan udara*. Kanisius.
[21] Fardiaz, Srikandi. 1992. *Food Microbiology I*. Jakarta: PT. Gramedia Pustaka Utama
[22] National Standarization Agency, SNI 06-6989.24-2005: *water and wastes: color test procedure by visual comparative*, National standarization Agency, Jakarta, 2005.
[23] National Standarization Agency, SNI 06-6989.11-2004: *Ph test procedure,* National standarization Agency, Jakarta, 2009.
[24] National Standarization Agency, SNI 06-6989.19-2009: *Iron (Fe) test procedure,* National standarization Agency, Jakarta, 2009.
[25] National Standarization Agency, SNI 06-6989.19-2004: *chloride test procedure by argentometry methode* , National standarization Agency, Jakarta, 2004.
[26] Adelekan B A. 2010. *International Journal of Water Resources and Environmental Engineering,* 2010, 2 (6): 137-147

Lecturers' Understanding on Indexing Databases of SINTA, DOAJ, Google Scholar, SCOPUS, and Web of Science: A Study of Indonesians

Ansari Saleh Ahmar[1,235], Nuning Kurniasih[2], Dasapta Erwin Irawan[3], Dian Utami Sutiksno[4], Darmawan Napitupulu[5], Harapin Hafid[6], Muhammad Ikhsan Setiawan[7], Janner Simarmata[8], Agung Wibowo[9], Ahmad Sururi[10], Akbar Iskandar[11], Andre Paulus Saleky[12], Citra Kurniawan[13], Danner Sagala[14], Dewiana Novitasari[15], Dian Eka Indriani[16], Early Ridho Kismawadi[17], Ende[18], Fabian Souisa[19], Febri Liantoni[20], Fitridawati Soehardi[21], Hardi[21], Haries Pribady[22], Hasan Baharun[23], Heny Vensye Saiya[24], Indrya Mulyaningsih[25], Irfana Diah Faryuni[26], Irwan Sugiarto[27], Joko Sampurno[26], Lusi Dwi Putri[21], Muh. Barid Nizarudin Wajdi[28], Muhammad Syahrul Kahar[29], Muthia Anggraini[21], Novia Nur Rosyida[30], Otto Fajarianto[31], Ratnadewi[32], Riski Amalia Madi[6], Ruzita Sumiati[33], Saeful Bahri[34], Tomi Listiawan[35], Uun Dwi Al Muddatstsir[17], Vivi Aulia[36], Wahyu Widyantoro[37], Wan Anita[38], Wellem A. Teniwut[19], A.A Gde Satia Utama[39], Aan Aswari[40], Abd. Ghofur[41], Abdul Atsar[42], Abdul Haris Nasution[43], Abdul Rahman Suleman[44], Abdurrozzaq Hasibuan[43], Achmad Daengs GS[45], Achmad Yusuf[46], Adi Santoso[47], Agung Nugroho Catur Saputro[48], Agus Darmuki[49], Agustinus Suradi[50], Ahmad Al Yakin[51], Ainur Rifqi[23], Aji Raditya[52], Albertoes Pramoekti Narendra[53], Amaliatulwalidain[54], Ambo Upe[6], Andi Riyanto[55], Andik Wijayanto[56], Andri Yanto[2], Angra Meta Ruswana[57], Anita Dewi Moelyaningrum[58], Anna Yuliana[3], Anto Ariyanto[21], Ardiansah[21], Ari Purwadi[59], Aria Hendrawan[60], Aria Wahyuni[61], Arjulayana[52], Arlis Dewi[62], As'adi[63], Asep Najmurrokhman[64], Asih Niati[60], Asrianti[6], Bahidin Laode M[65], Bambang Eko Hari Cahyono[66], Bayu Adhi Prakosa[67], Beta Ubaya Nindya[68], Budianto Hamuddin[21], Cahya Fajar Budi Hartanto[69], Castaka Agus Sugianto[70], Chairuddin[16], Cut Sriyanti[71], D. Purnomo[72], Dahlena Sari Marbun[43], Dahlia Andayani[73], Daniel Susilo[74], Darmawan[75], Desi Ulpa Anggraini[76], Dhanar Intan Surya Saputra[77], Dian Febrida Sari[78], Dian Rianita[21], Diana Arief[6], Didi Susanto[79], Didin Hadi Saputra[73], Didit Darmawan[80], Dini Afriansyah[81], Djohar Maknun[25], Djufri Rays Pattilouw[82], Dwi Cahyono[83], Dwi Ermayanti[84], E. Wityasminingsih[85], Eka Hartati[86], Eka Maya Sari Siswi Ciptaningsih[87], Ekbal Santoso[88], Eko Sumartono[89], Elfrida Ratnawati[90], Elida Gultom[91], Emah Khuzaemah[25], Emilia Dewiwati Pelipa[92], Endah Masrunik[93], Endang Fourianalistyawati[94], Eni Defitriani[95], Erfan Yudianto[96], Eric Kunto Aribowo[50], Erpidawati[97], Erwinsyah Satria[98], Ety Youhanita[41], Euis Erlin[57], Evi Gravitiani[48], Evi Sukmayeti[99], Fadlina[100], Fathimah[30], Fatimah[51], Fatimah Nur Arifah[101], Fauzi Farchan[102], Febrianty[103], Fedianty Augustinah[83], Feri Huda[88], Fidrayani[104], Firman[105], Folkes E. Laumal[106], Gumgum Gumelar[107], Gunawan Ali[108], Hadi Arnowo[109], Haida Dafitri[110], Halifia Hendri[111], Hani Tuasikal[112], Hardianto Djanggih[113], Harry Dhika[114], Heri Prabowo[115], Herlina Ahmad[51], Himmatul

IOP Conf. Series: Journal of Physics: Conf. Series **954** (2018) 012026 doi:10.1088/1742-6596/954/1/012026

Ulya[116], Ifit Novita Sari[117], Ilham Khairi Siregar[118], Indra Gunawan Pratama[119], Indriyani[120], Irfan Sudahri Damanik[121], Irma Setyawati[122], Irwan Bempah[123], Irwan Setyowidodo[124], Isma Coryanata[89], Itsar Bolo Rangka[114], Jef Rudiantho Saragih[125], Joyce Christian Kumaat[126], Jubilezer Sihite[127], Junita Sipahelut[128], Kadeni[88], Kamid[129], Ketut Sukiyono[89], Khairul Imtihan[130], Khuriyatul Husna[21], Kiky Nurfitri Sari[131], Kurniawan[21], La Ili[6], La Ode Asmin[132], La Ode Husen[40], Lala Nailah Zamnah[57], Lina Fitriani[133], Listian Indriyani Achmad[134], Lutfiyah Dwi Setia[135], M Chairul Basrun Umanailo[136], M. Ali Ghufron[49], Maria Purnama Nduru[137], Marsahip[138], Martina Napratilora[139], Maulana Yusuf Aditya[16], Meri Andriani[140], Merla Madjid[141], Mira Aliza Rachmawati[142], Mita Rosaliza[143], Moh. Affaf[16], Mohammad Rizki Fadhil Pratama[144], Mufida Khairani[145], Mufria J Purba[146], Muh. Ardian Kurniawan[147], Muh. Shofi[148], Muhamad Ratodi[149], Muhamad Yamin Noch[150], Muhammad Affan Ramadhana[151], Muhammad Alif[152], Muhammad Fajri[107], Muhammad Fuad[140], Muhammad Irfan Nasution[118], Muhammad Khoiruddin Harahap[153], Muhammad Luthfi H[154], Muhammad Sholahuddin[154], Murjainah[155], Mursyid[40], Mustafa[156], Musthafa Akbar[143], Mustofa Abi Hamid[157], Muthmainnah[51], Nancy Nopeline[127], Nani Hartati[158], Nanik Setyowati[89], Nastia[159], Nataliningsih[160], Ni Wayan Nursini[161], Nian Afrian Nuari[162], Nisaul Barokati Seliro Wangi[163], Nita Yessirita[164], Nopriadi[165], Nova Christian Mamuaya[126], Nova Susanti[129], Nugrahini Kusumawati[18], Nur Fauziah Siregar[166], Nurdiana Siregar[167], Nurhidayah[51], Nurmawati[45], Nurussama[168], Okma Yendri[169], Okti Purwaningsih[170], Patmawati[51], Prehatin Trirahayu Ningrum[58], Pudentiana Rr R.E[171], Purwo Subekti[172], Putu Sukma Kurniawan[173], Qashlim[51], R. Oktaviance S.[174], Rahmad Hidayat[175], Rahmat Hidayat[176], Rahmi Fauzia[152], Rahmi Lubis[177], Rama Kertamukti[178], Rama Oktavian[179], Ratna Dyah Suryaratri[107], Ratna Farwati[180], Ratna Said[159], Ratna Wati Simbolon[181], Reni Yunus[182], Retno Sari Dewi[183], Reza Rachmadtullah[107], Rezania Agramanisti Azdy[86], Rhini Fatmasari[184], Ria Marginingsih[185], Ricardo Freedom Nanuru[186], Rina Novia Yanti[21], Rina Septiani[187], Rinandita Wikansari[188], Ririen Wardiani[189], Rita Sari[190], Riyadh Arridha[191], Rizki Priya Pratama[192], Robbi Rahim[156], Rochmady[193], Rosanita Tritias Utami[35], Rosida Tiurma Manurung[32], Ruki Ambar Arum[194], Rustam[195], Saiful Bahri[196], Salniwati[6], Santirianingrum Soebandhi[197], Sarono Widodo[198], Sayyida[199], Selpida Handayani[40], Sitaresmi Wahyu Handani[77], Siti Juariah[200], Siti Nurina Hakim[154], Sitti Fithriani Saleh[201], Sri Wahyuni[21], Sri Winarni[129], St Handana Utari[202], Suhono[203], Sukardin[204], Sulfikar Sallu[195], Suliawati[43], Sumitro[205], Supriyadi[10], Supriyanto[206], Suriani[207], Susiana Susiana[208], Suwarsih[209], Syamsul Hidayat[18], Syifa Saputra[210], Tato Nuryanto[25], Taufiqurrahman[30], Teguh Trianton[211], Tien Zubaidah[212], Titing Magfirah[191], Tonni Limbong[213], Tri Listyorini[116], Trio Pahlawan[214], Tutut Suryaningsih[35], Upik Rahmi[98], Uus Ahmad Husaeni[215], Valentino Hary[119], Veronika Nugraheni Sri Lestari[83], Vieronica Varbi Sununianti[180], Vivi Novalia Sitijak[146], Wanti Arumwanti[216], Widodo[217], Yance Sonatha[33], Yansyah[218], Yenni Kristin Panjaitan[219], Yetti[21], Yohanes Dakhi[220], Yudithia Dian Putra[221], Yulia[222], Yuliyanto Budi Setiawan[60], Zainal Muktamar[89], Zulfikar[223], A. Farida Arsal[1], A. Irma Suryani Idris[1], Abd. Muis[1], Abdul Hadis[1], Abdul Mun'im[1], Abdurahman[224], Achmad Kusairi Samlawi[152], Ade Oktarino[225], Adiatma[1], Adnan[1], Adrian Faridhi[21], Agung Purnomo[226], Ahmad Bahtiar[227], Ahmad Fauzul Hakim Hasibuan[228], Ahmad Fudhail Majid[1], Ahmad Gunawan[158], Ahmad Hasyim[229], Ahmad Yani[1], Ahmad Zaki[1], Ainna Amalia FN[230], Alfian Saleh[21], Alimin[1], Alimuddin[1], Amandus Jong Tallo[231], Andi Akifah[232], Andi Aprasing[233], Andi Asmawati Azis[1], Andi Dian Angriani[234], Andi Nurani Mangkawani Arifin[235],

Andi Reni[236], Andista Candra Yusro[66], Andre Paulus Saleky[12], Andri Nugraha[237], Aneu Yulianeu[238], Anik Vega Vitianingsih[239], Anna Riana Suryanti Tambunan[8], Ankardiansyah Pandu Pradana[240], Ari Riswanto[241], Arif Rahman Putra[80], Arifah Novia Arifin[1], Arifin Ahmad[242], Army Auliah[1], Aryo Michael[243], Asdar Ahmad[1], Aswi[1], Aurellia Tatipata[82], Awi[1], Ayub Wijayati Sapta Pradana[244], Bambang Parmadi[89], Bayu Prabowo Sutjiatmo[188], Bernard[1], Bunga Dara Amin[1], Busro[245], Dahlan Abdullah[311], Daniel Silli Bataona[106], Dedi Muhtadi[246], Dedy Juliandri Panjaitan[247], Desi Ulpa Anggraini[248], Dessy Agustina Sari[42], Dewi Satria Ahmar[235,271], Diana Zuhroh[249], Dicky Nofriansyah[250], Dina Chamidah[59], Djadir[1], Dudung Ma'ruf Nuris[56], Dwi Ely Kurniawan[251], Dwi Nur Rachmah[152], Ebin Eyzer Danius[252], Edi Fitriana Afriza[246], Eka Fitriyana Hamsyah[229], Eka Selvi Handayani[253], Eka Zuliana[116], Eko Susanto[254], Eli Jamilah Mihardja[255], Ella Anastasya Sinambela[80], Elsa Rosyidah[256], Esther Kembauw[257], Eva Desembrianita[258], Eva Nur Rachmah[45], Evi Rinata[259], Evy Nurvitasari[260], Fajar Arwadi[1], Fajar Kusnadi Kusumah Putra[216], Fajria Fatmasari[188], Fandi Ahmad[229], Fathimah Az-zahra[261], Fika Megawati[259], Fitriana Destiawati[114], Fransiska Natalia Ralahallo[82], Frijona F. Lokollo[262], Gloria Christiana Manulangga[106], Halimah Husain[1], Halimatus Sakdiah[263], Hambali Thalin[40], Hamda[1], Hani Sirine[53], Hardin[1], Haris Munandar[229], Harisa Mardiana[264], Hary Hermawan[265], Hasina Fajrin R.[266], Hasri[1], Haviluddin[292], Heri Nurdiyanto[313], Hellna Tehubijuluw[82], Hendrati Dwi Mulyaningsih[267], Hendri Dony Hahury[82], Hengki Mangiring Parulian Simarmata[268], Herfa Maulina Dewi Soewardini[59], Heri Dwi Putranto[89], Herman[1], Hifdhotul Munawaroh[30], Hilda Karim[1], I Nyoman Sudiana[6], Hisyam Ihsan[1], Ibrahim[1], Ilham Minggi[1], Imas Srinana Wardani[269], Inderawati[6], Irma Suriani Idris[1], Irmawanty[270], Irwan[1], Irwana R. Badji[271], Ismail[1], Iwan Dini[1], Jacobus Cliff Diky Rijoly[82], Jajang Burhanudin[272], Jayanti Putri Purwaningrum[116], Jehan Safitri[152], Jogeneis Patty[82], Johanna Rimbing[126], Kaharuddin Arafah[1], Kamri Ahmad[40], Khadijah[229], Khoirul Ngibad[273], Kristiana Sri Utami[274], Kurniawan Harikesuma Ahmar[235], Kustanto[93], Laily Nurlina[211], Leon Andretti Abdillah[312], Lilik Istiqomah[275], Liliza Agustin[276], Lucky Nugroho[277], Lusy Tunik Muharlisiani[59], M. Agus Martawijaya[1], M. Mirza Abdillah Pratama[56], M. Taufik Qurohman[278], Maddatuang[1], Maria Maghdalena Diana Widiastuti[24], Marissa Putriana[95], Mas'ud Muhammadiyah[261], Masyitah Noviyanti[256], Maya Sari Wahyuni[1], Melda Gienardy[113], Memen Permata Azmi[279], Meri Azmi[33], Mila Hariani[80], Mirawati A. Tongko[113], Moch. Irfan[80], Mochammad Aldy Anwar[256], Mochammad Noviadi Nugroho[280], Mochammad Tanzil Multazam[259], Moh. Yamin[152], Mohammad Roesli[281], Mohammad Wijaya M.[1], Mugi Praseptiawan[270], Muh Yunus[229], Muh. Jaelani Al-Pansori[147], Muhafidhah Novie[282], Muhammad Agus[201], Muhammad Anwar[1], Muhammad Aqil Rusli[1], Muhammad Asfar[236], Muhammad Danial[1], Muhammad Darwis[1], Muhammad Jasri Djangi[1], Muhammad Junda[1], Muhammad Wiharto[1], Muhammad Yahya Matdoan[82], Muhammad Yusuf[283], Muhammad Yusuf[284], Mujiarto[314], Mukrimin[285], Mutahharah Hasyim[1], Nalal Muna[188], Nana Hendracipta[157], Nani Kurnia[1], Nany Noor Kurniyati[274], Natelda R. Timisela[82], Neka Erlyani[152], Netti Herawati[1], Nour Ardiansyah Hernadi[286], Nur Indah Sari[229], Nuraini[6], Nurhayati[1], Nurhayati B.[1], Nurmawati[45], Nurul Qamar[40], Okki Trinanda[224], Panji Kuncoro Hadi[66], Pariabti Palloan[1], Parma Hadi Rantelinggi[287], Pramita Laksitarahmi Isrianto[59], Prasetio Ariwibowo[114], Purwati Zisca Diana[288], Rahayu Mardikaningsih[80], Rahmadhani Fitri[289], Rainier Hendrik Sitaniapessy[257], Rais Hidayat[290], Ramadona Simbolon[43], Ramlawati[1], Ramli Umar[1], Ratnawaty Maming[1], Restu Rahmawati[105], Resy Nirawati[22], Rika Vira Zwagery[152], Rika

Yuliana[291], Rini Apriyani[292], Riski Muliyani[22], Rizal Bakri[293], Rizka Safriyani[294], Rohana[1], Rosdiana Ngitung[1], Rudi Abdullah[159], Rudi Syahputra[214], Rusli[1], Safrizal[140], Sahlan Sidjara[1], Riny Jefri[1], Said Fachry Assagaf[1], Samsul Arifin[80], Setiabudhi[45], Sherlock Halmes Lekipiouw[257], Sitti Hutari Mulyani[111], Sitti Saenab[1], Sonny Kristianto[59], Sowanto[295], Sri Elviani[43], Sri Maryanti[21], Sri Mukminati Nur[229], Sri Puspa Dewi[296], Sri Susilawati[113], Sri Wiwoho Mudjanarko[7], St. Humaerah Syarif[229], Stevanus Johan Gomies[4], Sudarmin[1], Sudding[1], Sugianto[297], Sugiarti[1], Suharto[298], Sukarna[1], Sulaiman Zhiddiq[1], Suman Sangadji[299], Suprapta[1], Surahman Nur[229], Susilo[300], Sutamrin[1], Syamsiah[1], Syamsurijal Basri[1], Taty Sulastry[1], Taufik Bidullah[113], Trian Pamungkas Alamsyah[157], Tristan Rokhmawan[301], Tuti Agustin[48], Ulinnuha Madyananda[48], Usmaedi[270], Usman[302], Uwes Fatoni[245], Vebtasvili[222], Vera Theresia Caroline Siahaya[4], Victor R. Pattipeilohy[4], Widhy Wahyani[303], Wahyudin Albra[311], Wiwiek Rindayati[304], Wiwik Handayani[305], Wiyanto[306], Yeni Ika Pratiwi[307], Yossie Yumiati[308], Yudi juniardi[157], Yuniningsih[305], Yuska Noviyanty[309], Yusuf Rahman Al Hakim[80], Yuuun Elizabeth Patras[290], Zakaria Husein Abdurrahman[310], Zulharbi Amatahir[113], Zulkifli Rais[229], Rahmat Hidayat[33], Juneman Abraham[315]

[1]Universitas Negeri Makassar, [2]Universitas Padjadjaran, [3]Institut Teknologi Bandung, [4]Politeknik Negeri Ambon, [5]Lembaga Ilmu Pengetahuan Indonesia, [6]Universitas Halu Oleo, [7]Narotama University, [8]Universitas Negeri Medan, [9]STMIK Nusa Mandiri Sukabumi, [10]Universitas Serang Raya, [11]STMIK AKBA Makassar, [12]Sekolah Tinggi Keguruan dan Ilmu Pendidikan Biak, [13]Sekolah Tinggi Teknik Malang, [14]Universitas Prof. Dr. Hazairin SH, [15]STIE Insan Pembangunan, [16]STKIP PGRI Bangkalan, [17]Institut Agama Islam Negeri Zawiyah Cotkala Langsa, [18]Universitas Bina Bangsa, [19]Politeknik Perikanan Negeri Tual, [20]Institut Teknologi Adhi Tama Surabaya, [21]Universitas Lancang Kuning, [22]STKIP Singkawang, [23]Universitas Nurul Jadid, [24]Universitas Musamus Merauke, [25]IAIN Syekh Nurjati Cirebon, [26]Tanjungpura University, [27]Sekolah Tinggi Hukum Bandung, [28]STAI Miftahul Ula Nganjuk, [29]Universitas Muhammadiyah Sorong, [30]Universitas Darussalam Gontor, [31]STMIK Bina Sarana Global, [32]Universitas Kristen Maranatha, [33]Politeknik Negeri Padang, [34]STMIK Nusa Mandiri Jakarta, [35]STKIP PGRI Tulungagung, [36]STKIP PGRI Banjarmasin, [37]Poltekkes Kemenkes Yogyakarta, [38]STIKES Tengku Maharatu Pekanbaru, [39]Universitas Airlangga, [40]Universitas Muslim Indonesia, [41]STKIP PGRI Lamongan, [42]Universitas Singaperbangsa Karawang, [43]Universitas Islam Sumatera Utara, [44]Universitas Graha Nusantara, [45]Universitas 45 Surabaya, [46]Universitas Yudharta Pasuruan, [47]Universitas Muhammadiyah Ponorogo, [48]Universitas Sebelas Maret, [49]IKIP PGRI Bojonegoro, [50]Universitas Widya Dharma, [51]Universitas Al Asyariah Mandar, [52]Universitas Muhammadiyah Tangerang, [53]Universitas Kristen Satya Wacana Salatiga, [54]Universitas Indo Global Mandiri, [55]AMIK BSI Sukabumi, [56]Universitas Negeri Malang, [57]Universitas Galuh, [58]Universitas Jember, [59]Universitas Wijaya Kusuma Surabaya, [60]Universitas Semarang, [61]STIKES Fort De Kock Bukittinggi, [62]STIE Muhammadiyah Jakarta, [63]Sekolah Tinggi Ilmu Ekonomi Gempol, [64]Universitas Jenderal Achmad Yani, [65]Universitas Muhammadiyah Luwuk, [66]Universitas PGRI Madiun, [67]Universitas Ibn Khaldun Bogor, [68]Universitas Kristen Satya Wacana, [69]Akademi Pelayaran Niaga Indonesia, [70]Politeknik TEDC Bandung, [71]Poltekkes Aceh, [72]Politeknik LP3I Jakarta, [73]Universitas Nahdlatul Wathan Mataram, [74]Universitas Dr Soetomo, [75]Universitas Budi Luhur, [76]STIE Rahmaniya Sekayu, [77]STMIK AMIKom Purwokerto, [78]STIKES Mercubaktijaya Padang, [79]Universitas Islam Kalimantan MAB Banjarmasin, [80]Universitas Mayjen Sungkono, [81]UIN Raden Fatah, [82]Universitas Pattimura, [83]Universitas Dr. Soetomo, [84]STIE PGRI

IOP Conf. Series: Journal of Physics: Conf. Series **954** (2018) 012026 doi:10.1088/1742-6596/954/1/012026

Dewantara Jombang, [85]Politeknik PIKSI Ganesha, [86]STMIK Palcomtech, [87]Universitas Pamulang, [88]STKIP PGRI Blitar, [89]Universitas Bengkulu, [90]Universitas Trisakti, [91]Sekolah Tinggi Ilmu Ekonomi Riau, [92]STKIP Persada Khatulistiwa Sintang, [93]Universitas Islam Balitar, [94]Universitas Yarsi, [95]Universitas Batanghari Jambi, [96]University of Jember, [97]Universitas Muhammadiyah Sumbar, [98]Universitas Pendidikan Indonesia, [99]Sekolah Tinggi Ilmu Administrasi Mataram, [100]AMIK Stiekom Sumatera Utara, [101]STMIK Bina Patria, [102]Universitas Wiralodra, [103]Politeknik Palcomtech Palembang, [104]UIN Jakarta, [105]Universitas 17 Agustus 1945 Jakarta, [106]Politeknik Negeri Kupang, [107]Universitas Negeri Jakarta, [108]Universitas Dharmas Indonesia, [109]Pusdiklat Kementerian Agraria dan Tata Ruang/Badan Pertanahan Nasional, [110]Sekolah Tinggi Teknik Harapan, [111]Universitas Putra Indonesia YPTK Padang, [112]Akademi Keperawatan Rumkit Tk. III Dr.J.A.Latumeten Ambon, [113]Universitas Tompotika Luwuk, [114]Universitas Indraprasta PGRI, [115] Balai Penelitian Tanaman Pemanis dan Serat Malang, [116]Universitas Muria Kudus, [117]Universitas Kanjuruhan Malang, [118]Universitas Muhammadiyah Sumatera Utara, [119]Universitas Nahdlatul Ulama Blitar, [120]Universitas Sany Bumi Ruwa Jurai, [121]STIKOM Tunas Bangsa, [122]Universitas Bhayangkara Jakarta Raya, [123]Universitas Negeri Gorontalo, [124]Universitas Nusantara Pgri Kediri, [125]Universitas Simalungun, [126]Universitas Negeri Manado, [127]Universitas HKBP Nommensen Medan, [128]STAKPN Ambon, [129]Universitas Jambi, [130]STMIK Lombok, [131]Sekolah Tinggi Ilmu Pertanian Rejang Lebong, [132]Institut Agama Islam Negeri Kendari, [133]STIKES Bina Generasi Polewali Mandar, [134]STAI Pelita Bangsa, [135]Politeknik Negeri Madiun, [136]Universitas Iqra Buru, [137]Universitas Flores, [138]Politeknik Medica Farma Husada Mataram, [139]STAI Auliya Urasyiddin Tembilahan, [140]Universitas Samudra, [141]Politeknik ATI Makassar, [142]Universitas Islam Indonesia, [143]Universitas Riau, [144]Universitas Muhammadiyah Palangkaraya, [145]Universitas Harapan Medan, [146]Universitas Methodist Indonesia, [147]Universitas Hamzanwadi, [148]Institut Ilmu Kesehatan Bhakti Wiyata Kediri, [149]Universitas Islam Negeri Sunan Ampel, [150]Universitas Yapis Papua, [151]Universitas Cokroaminoto Palopo, [152]Universitas Lambung Mangkurat, [153]Politeknik Ganesha Medan, [154]Universitas Muhammadiyah Surakarta, [155]Universitas PGRI Palembang, [156]Institut Teknologi Medan, [157]Universitas Sultan Ageng Tirtayasa, [158]STIE Pelita Bangsa, [159]Universitas Muhammadiyah Buton, [160]Universitas Winaya Mukti, [161]Universitas Dhyana Pura, [162]STIKES Karya Husada Kediri, [163]Universitas Islam Darul Ulum Lamongan, [164]Universitas Ekasakti, [165]Universitas Putera Batam, [166]Institut Agama Islam Negeri Padangsidimpuan, [167]Universitas Nahdlatul Ulama Sumatera Utara, [168]Politeknik Palcomtech, [169]Universitas Musi Rawas, [170]Universitas PGRI Yogyakarta, [171]Politeknik Kesehatan Kemenkes Jakarta I, [172]Universitas Pasir Pengaraian, [173]Universitas Pendidikan Ganesha, [174]STIKES Santa Elisabeth Medan, [175]Sekolah Tinggi Teknologi Mandala, [176]Sekolah Tinggi Ilmu Manajemen Sukma Medan, [177]Universitas Medan Area, [178]UIN Sunan Kalijaga, [179]Universitas Brawijaya, [180]Universitas Sriwijaya, [181]AMIK Medan Business Polytechnic, [182]Poltekkes Kemenkes Kendari, [183]Universitas Tulungagung, [184]Universitas Terbuka Jakarta, [185]Universitas Islam 45 Bekasi, [186]Universitas Halmahera, [187]Universitas Tama Jagakarsa, [188]Politeknik APP Jakarta, [189]STKIP PGRI Ponorogo, [190]IAIN Langsa, [191]Politeknik Negeri Fakfak, [192]Politeknik Kota Malang, [193]Sekolah Tinggi Ilmu Pertanian Wuna Raha, [194]Politeknik Informatika Nasional, [195]Universitas Sembilanbelas November Kolaka, [196]IKIP PGRI Pontianak, [197]Universitas Narotama, [198]Politeknik Negeri Semarang, [199]Universitas Wiraraja, [200]Akademi Analis Kesehatan Pekanbaru, [201]Universitas Muhammadiyah Makassar, [202]Universitas Madako, [203]Universitas Gadjah Mada, [204]Sekolah Tinggi Ilmu Kesehatan Mataram, [205]AMIK Labuhan Batu, [206]Bogor Agricultural University, [207]STIE Tri Dharma Nusantara

IOP Conf. Series: Journal of Physics: Conf. Series **954** (2018) 012026 doi:10.1088/1742-6596/954/1/012026

Makassar, [208]Universitas Maritim Raja Ali Haji, [209]Universitas PGRI Ronggolawe Tuban, [210]Universitas Al Muslim, [211]Universitas Muhammadiyah Purwokerto, [212]Poltekes Kemenkes Banjarmasin, [213]Universitas Katolik Santo Thomas Medan, [214]Politeknik Negeri Lhokseumawe, [215]Universitas Suryakancana, [216]Sekolah Tinggi Pariwisata Bandung, [217]Universitas Muhammadiyah Yogyakarta, [218]Universitas Muhammadiyah Banjarmasin, [219]Universitas Darma Agunga, [220]STIE Nias Selatan, [221]STIE IBMT Surabaya, [222]Universitas Bangka Belitung, [223]Universitas KH A Wahab Hasbullah, [224]Universitas Negeri Padang, [225]Universitas Adiwangsa Jambi, [226]BINUS Institute of Creative Technology, [227]UIN Syarif Hidayatullah, [228]Universitas Pembangunan Panca Budi Medan, [229]STKIP Pembangunan Indonesia Makassar, [230]STAI Miftahul Ula Nglawak Kertosono, [231]Universitas Agung Podomoro, [232]Universitas Tadulako Palu, [233]Universitas Sulawesi Barat, [234]Universitas Islam Negeri Alauddin Makassar, [235]AHMAR Institute, [236]Universitas Hasanuddin, [237]STIKES Karsa Husada Garut, [238]STMIK DCI, [239]Universitas Dr. Soetomo, [240]Universitas Borneo Tarakan, [241]STKIP PGRI Sukabumi, [242]Universitas Pasundan, [243]STMIK Kreatindo Manokwari, [244]Universitas Islam Malang, [245]UIN Sunan Gunung Djati Bandung, [246]Universitas Siliwangi, [247]Universitas Muslim Nusantara Al Washliyah, [248]STIE Rahmaniyah Sekayu, [249]Universtas 45 Surabaya, [250]STMIK Triguna Dharma, [251]Politeknik Negeri Batam, [252]Universitas Hein Namotemo, [253]Universitas Widya Gama Mahakam Samarinda, [254]Universitas Muhammadiyah Metro, [255]Universitas Bakrie, [256]Universitas Nahdlatul Ulama Sidoarjo, [257]Universitas Pattimura Ambon, [258]STIE Urip Sumoharjo Surabaya, [259]Universitas Muhammadiyah Sidoarjo, [260]Universitas Musamus, [261]Universitas Bosowa Makassar, [262]Univetsitas Pattimura, [263]STKIP M Sungai Penuh, [264]Universitas Buddhi Dharma, [265]STP ARS Internasional Bandung, [266]Balai Bahasa Sulawesi Selatan, [267]Telkom University, [268]Politeknik Bisnis Indonesia, [269]Universitas PGRI Adi Buana Surabaya, [270]STKIP Setiabudhi, [271]STKIP YAPTI Jeneponto, [272]Politeknik LP3I Bandung, [273]Universitas Ma'arif Hasyim Latif Sidoarjo, [274]Universitas Widya Mataram, [275]IAIN Surakarta, [276]UIN Sulthan Syarif Kasim Riau, [277]Universitas Mercu Buana Jakarta, [278]Politeknik Harapan Bersama, [279]Universitas Islam Negeri Sultan Syarif Kasim Riau, [280]UIN Syarif Hidayatullah Jakarta, [281]Univeritas Merdeka Surabaya, [282]Universitas NU Sidoarjo, [283]Institut Sains & Teknologi AKPRIND, [284]Universitas Asahan, [285]IAIN Sultan Amai, [286]Sekolah Yinggi Pariwisata Ambarrukmo, [287]Universitas Papua, [288]Universitas Sarjanawiyata Tamansiswa, [289]Universitas Pembangunan Panca Budi, [290]Universitas Pakuan, [291]UIN Ar-Raniry Banda Aceh, [292]Universitas Mulawarman, [293]STIEM Bongaya, [294]UIN Sunan Ampel Surabaya, [295]STKIP Bima, [296]Universitas Al Azhar Medan, [297]Universitas Islam Negeri Sumatera Utara, [298]Universitas Negeri Semarang, [299]Universitas Darussalam Ambon, [300]Universitas Muhammadiyah Prof.DR.Hamka, [301]STKIP PGRI PASURUAN, [302]Politeknik Bosowa, [303]Sekolah Tinggi Teknologi POMOSDA Nganjuk, [304]Institut Pertanian Bogor, [305]Universitas Pembangunan Nasonal Veteran Jawa Timur, [306]Sekolah Tinggi Teknologi Pelita Bangsa, [307]Universitas Merdeka Surabaya, [308]Universitas Dehasen Bengkulu, [309]Akademi Farmasi Al-Fatah Bengkulu, [310]Universitas Boyolali, [311]Universitas Malikussaleh Aceh, [312]Universitas Bina Darma, [313]STMIK Dharma Wacana, [314]Universitas Muhammadiyah Tasikmalaya, [315]Bina Nusantara University.

ansarisaleh@unm.ac.id

Abstract. The Ministry of Research, Technology and Higher Education of Indonesia has introduced several national and international indexers of scientific works. This policy becomes a guideline for lecturers and researchers in choosing the reputable publications. This study aimed to describe the understanding level of Indonesian lecturers related to indexing databases, i.e. SINTA, DOAJ, Scopus, Web of Science, and Google Scholar. This research used descriptive design and survey method. The populations in this study were Indonesian lecturers and researchers. The primary data were obtained from a questionnaire filled by 316 lecturers and researchers from 33 Provinces in Indonesia recruited with convenience sampling technique on October-November 2017. The data analysis was performed using frequency distribution tables, cross tabulation and descriptive analysis. The results of this study showed that the understanding of Indonesian lecturers and researchers regarding publications in indexing databases SINTA, DOAJ, Scopus, Web of Science and Google Scholar is that, on average, 66,5% have known about SINTA, DOAJ, Scopus, Web of Science and Google Scholar. However, based on empirical frequency 76% of them have never published with journals or proceedings indexed in Scopus.

1. Introduction

The tight competition in the globalization era has penetrated to all aspects of life, including at the aspect of science and technology development. In accordance with Indonesian Law No. 12 of 2013 on Higher Education, lecturers have the duty and the responsibility to develop science and technology through scientific research and dissemination. The Indonesian Government through the Ministry of Research, Technology and Higher Education *(Kemenristekdikti)* encourages Indonesian academics to compete with academics of other countries, especially in the scope of ASEAN. There are currently 277.564 lecturers in Indonesia. The number of lecturers and researchers in Indonesia is far more than other countries in ASEAN such as Singapore. However, the number does not correspond with publication in scientific journal. For example, the number of publications of Indonesian lecturers and researchers indexed in Scopus and/or Web of Science is still below Malaysia.

The low of publications number in reputable journals or proceedings has raised some important policies for the scientific publications development of lecturers and researchers in Indonesia such as the advices for publishing in journal or reputable proceedings, national journals accreditation, the consideration of impacts of journals, attention to number of citations, and H-Index of authors. Publications in the digital age have the consequences on the impact of a publication through digital media, so the impact of an article is no longer only seen from the number of citations, but can also be seen from how much of a scientific work is clicked, shared, saved or downloaded, discussed, adapted and reviewed in an online forum [1]. Figure 1 shows the development of Scopus-indexed Indonesian publications from year to year.

One of the references of Kemenristekdikti in determining the reputation of a journal or proceedings is the national and international accreditation and indexing bodies. For example, journals indexed in Scopus or Web of Science are categorized as high reputable Journals, journals indexed in DOAJ are categorized as medium reputable journals and journals indexed solely in Google Scholar are categorized as low reputation journals. This policy summary can be seen at the Regulation of the Minister of Research, Technology and Higher Education (Permenristekdikti) No. 20 of 2017 and the Guideline of Credit Score Assessment (PAK) of Indonesian lecturers.

Based on Permenristekdikti No.20 of 2017, to obtain the professional lecture allowances and the professors honorary allowances, a lecturer who has certain academic position, e.g. Associate Professor and Professor, are required to publish with reputable international journals as part of their duties and responsibilities of scientists, in this case at least 1 article per year. Furthermore, the Regulation of the Minister of State Apparatus Empowerment and Bureaucratic Reform Number 17 of 2013 about the Lecturer's Functional Position and the Credit Score obliges lecturers with the position of Instructor and Assistant Professor to conduct scientific publications. Based on Forlap PDDIKTI data recap, academic year 2016/2017, the number of lecturers who have academic position of Associate Professor is 29.287, where Professor is 4.949, as shown in Figure 2.

IOP Conf. Series: Journal of Physics: Conf. Series **954** (2018) 012026 doi:10.1088/1742-6596/954/1/012026

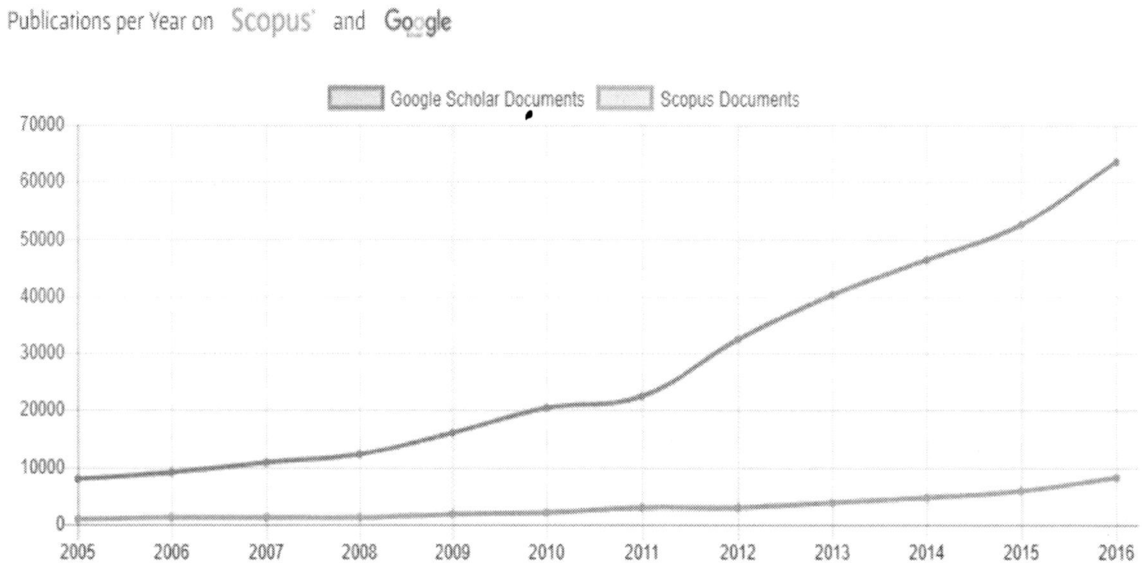

Figure 1. Indonesia-indexed publication Scopus and Google Scholar from 2005-2016 (Source: http://SINTA2.ristekdikti.go.id/)

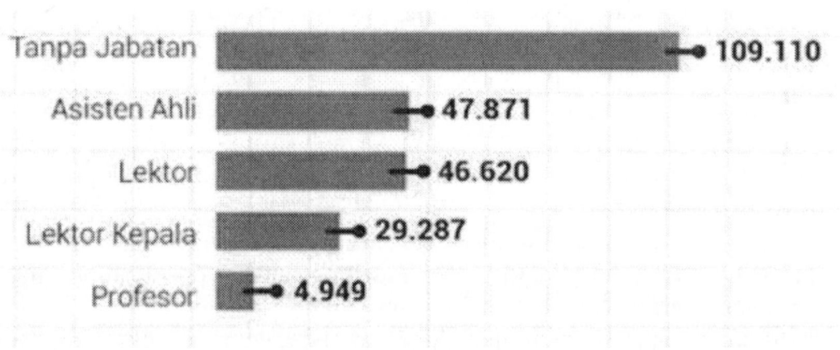

Figure 2. The distribution of lectures functional position
(Source: http://forlap.ristekdikti.go.id - Kemenristekdikti 2016/2017)
Notes: Tanpa Jabatan = Tutor/Lecturer without a position; Asisten Ahli = Lecturer; Lektor = Senior Lecturer/Assistant Professor; Lektor Kepala= Associate Professor; Profesor = Professor

The large numbers of lecturers who have position both the Associate Professor and Professor would be a great potential to increase the number of Indonesian publications at reputable international journals. Publication in international journals would be an achievement to increase the impact and lecturer's competitiveness in Indonesia. The capacity building and the lecturer competence need to be continuously encouraged. In addition, the journal publication is also a contribution of knowledge from a lecturer to be known by the public. Therefore, the results of research can be used in appropriate way in line with the original purpose of the research. By the existence of the publication bodies in Indonesia, inventions from abroad would be a complementary thing because the use of direct findings of Indonesian Scientists are so rich that could be found in the journals published in the country itself.

Moreover, the development and competence of lecturers should continue to be encouraged and also full supported by the universities where the lecturers are serving.

Amidst the publication incentive situation, lecturers are challenged to think strategically in producing scientific papers in order to get career paths ranging from faculty to professors, and in assessing their selves compared with the criteria included in the policies developed by the Government in the scientific field. This present study aimed at investigating the level of lecturers' understanding in Indonesia about indexing databases, i.e. SINTA, DOAJ, Scopus, Web of Science and Google Scholar. Lecturers need to increase their literacy on the knowledge of various types of indexing databases of scientific works.

2. Research Methods

This research used survey research method. The population in this study is lecturers and researchers of Indonesia. Primary data were obtained from a questionnaire filled by 316 lecturers and researchers from 33 Provinces in Indonesia taken with convenience sampling technique during October until November 2017. Data analysis was performed using frequency distribution tables, cross tabulation and descriptive analysis.

3. Result

In this section will explained about the level of understanding of Indonesian lecturers related to the publications in indexed by SINTA, DOAJ, Scopus, Web of Science, and Google Scholar.

3.1. Publication of Indonesian Lectures and Researchers

The lack of lectures' understanding toward indexed journal is caused by several things, e.g. lecturers are not familiar with publication metrics. For example, not all lecturers have known the citation counting technology. The function of indexed journal is not socialized massively. Only a few of lecturers have participated in the training on indexed journals. There is also a biased paradigm that the publication is only for filing work requirements, to get academic credit point, to meet the obligations of Lecturer Performance Indicator. When one finds his/her point "sufficient" in terms of ongoing credit point saving, the desire to continue the contribution of their publication to the globe is decreasing; local journal is then enough for them. This "comfort" feeling is what makes many lecturers lag in searching for the information and shocked by the development of indexed journals. It is not only about the lecturers' ability to understand and to follow the journal guideline but also the deficient writing culture. This is further exacerbated by the minimal research support facilities and low research funding at some of the new born universities. All of those factors influence make the research output is not of the best so that researchers are not confident to enter the article on indexed journals.

A good understanding of published and indexed journal publications will spark the lectures spirit to conduct the quality research and to improve publishing capabilities. Table 1 presents data on the publication experience of lecturer in a scientific journal. It shows that the experience of journal publication can quantitatively be said good although the quality of the papers could still be questioned and requires more examination. Almost all Indonesian lecturers surveyed in this research have published a scientific work in a scientific journal indicating that Indonesian lecturers know about scientific writing techniques the publication technique.

The interesting points from the data in Table 1 are that 4,23% of Tutor and 1,11% of Assistant Professor have never published their scientific work in a journal. Meanwhile, research activity is one of the elements assessed by the credit score assessors team in determining a functional position of Indonesian lecturers. This is still possible because the assessment is measuring not only publication in the journal but also in the form of books, monographs, journals, proceedings of seminars or conferences, scientific works saved in libraries, book translation or editing, and intellectual property rights. However, the credit point for the conference proceedings is generally lower than the publication in the journal. As many as 10% of Tutors have never published a scientific paper in a journal. It may be a natural thing because they are usually a new lecturer.

Table 1. Publication Experience in Journal

Functional Position	Ever published in a journal	
	No (%)	Yes (%)
Tutor	10,00	90,00
Lecturer	4,23	95,77
Assistant Professor	1,11	98,89
Associate Professor	0,00	100,00
Professor	0,00	100,00

3.2. Knowledge of Indonesian Lectures and Researchers on Indexing Databases

3.2.1. SINTA. Science and Technology Index (SINTA) is a portal that serves to measure the performance of science and technology development in Indonesia. Indicators assessed are the results of research that has been done by researchers, lecturers, and institutions. The advantages of SINTA compared to other indexing portals that is can automatically indexing the work that has been indexed in Google Scholar and Scopus. The Minister of Research, Technology and Higher Education, Mohamad Nasir said the portal was created because the both number of students and lecturers has not been comparable yet with the number of publications produced so effecting Indonesian researchers less well known by the global community.

Up to November 6, 2017, only 5.153 lecturers have registered and verified in SINTA (http://SINTA2.ristekdikti.go.id/). Some lecturers who have not registered with SINTA because they still did not get socialization yet about SINTA and some others still think that SINTA is only a product that is still in experimental stage and may change again both in terms of rules and systems at the future. Some other lecturers also consider SINTA not having their own original data and only depending on data held by Google Scholar and Scopus. Somehow, it makes the most of Indonesian lecturers did not know about another indexing database like DOAJ (see below). This also raises one big question that is how the fate of SINTA if then there are problems "arise" from Google Scholar and Scopus itself. If it is not anticipated, this will be a big problem, so the government needs to think about having their own database center. However, the government's effort to record the publication of lecturers in Indonesia through SINTA should be appreciated.

Through SINTA, lecturers can easily view the contribution of universities and research institutions as well as the lecturer's personal contribution toward indexed documents by Scopus and Google Scholar, as shown in Figure 3. Figure 3 shows that Institut Teknologi Bandung (ITB) is an educational institution that contributes most Scopus-indexed documents registered in the SINTA portal, while Institut Pertanian Bogor (IPB) is the largest educational institution that contributes most documents to Google Scholar registered in the SINTA portal. Figure 3 also shows the contribution of The Indonesian Institute of Sciences (LIPI) is ranked at 8th in the contribution of Scopus indexed documents and 7th position for the contribution of Google Scholar indexed documents.

Ranking process of colleges and authors in the SINTA could spur authors for publishing intensively on reputable journals and increasing the number of citation as well as achieving SINTA score of colleges and themselves. The knowledge of Indonesian lecturers and researchers on SINTA can be seen in Table 2. Table 2 shows that most lecturers with various levels of functional positions have known about SINTA (the new indexer launched by Kemenristekdikti) indicating that the socialization about SINTA is successful. One of the success factors of socialization is the existence of link between ownership accounts in SINTA with the lecturer existence as a researcher. The registration in the SINTA has raised. However, it is viewed by some lecturers only as an obligation.

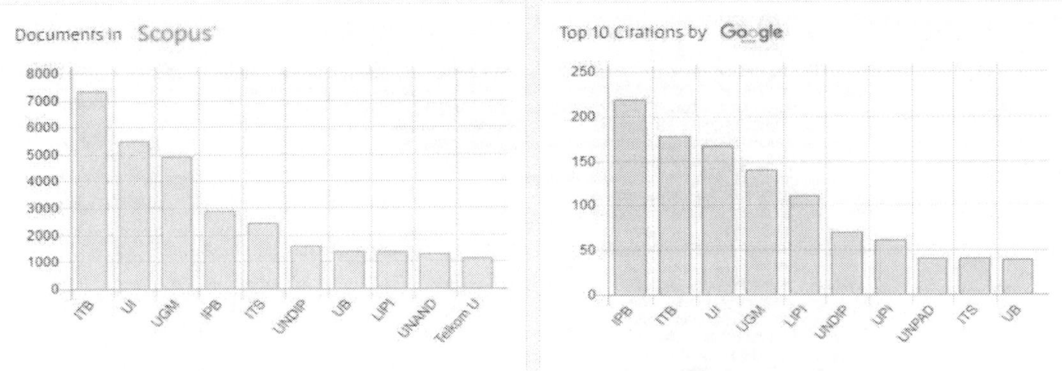

Figure 3. The Contribution of Universities and Research Institutions in Indonesia to Scopus and Google Scholar Indexed Documents
(Source: http://SINTA2.ristekdikti.go.id/)

Table 2. Knowledge on SINTA

Functional Position	**Knowledge on SINTA**	
	No (%)	Yes (%)
Tutor	10,00	90,00
Lecturer	12,68	87,32
Assistant Professor	11,11	88,89
Associate Professor	3,33	96,67
Professor	0,00	100,00

3.2.2. DOAJ. DOAJ (Directory of Open Access Journal) is a community-curated online directory that indexes and provides access to high quality, open access, peer-reviewed journals. DOAJ is independent. All funding is via donations, 50% of which comes from sponsors and 50% from members and publisher members. All DOAJ services are free of charge including being indexed in DOAJ. All data is freely available. DOAJ is launched in 2013 at Lund University, Sweden. At the time of launch there were about 300 journals open access that joined in it. Over time, the open access journals joined DOAJ range from 9.000 journals covering all areas of technology, science, medicine and social science studies. In order for a journal to be indexed in DOAJ then the main requirement is that the journal must be Open Access. Open Access is a condition where the journal can be read and accessed by everyone without any cost [2][3]. DOAJ measures the accessibility of a journal without a quality ranking system. However, DOAJ also provides a standard of *DOAJ Seal* to demonstrate the best practice of an Open Access (OA) journal where in Indonesia only 1 journal obtains DOAJ Seals similar to those occurring in the country Brazil.

Figure 4 shows the number of OA journals indexed by DOAJ from different countries (Top 50). It can be seen from the graphic that the countries order based on the number of journals indexed DOAJ (Top 5) are Brazil, United Kingdom, United States, Egypt, and Indonesia [4]. Table 3 shows that lecturers' knowledge on DOAJ is less when compared to other indexers.

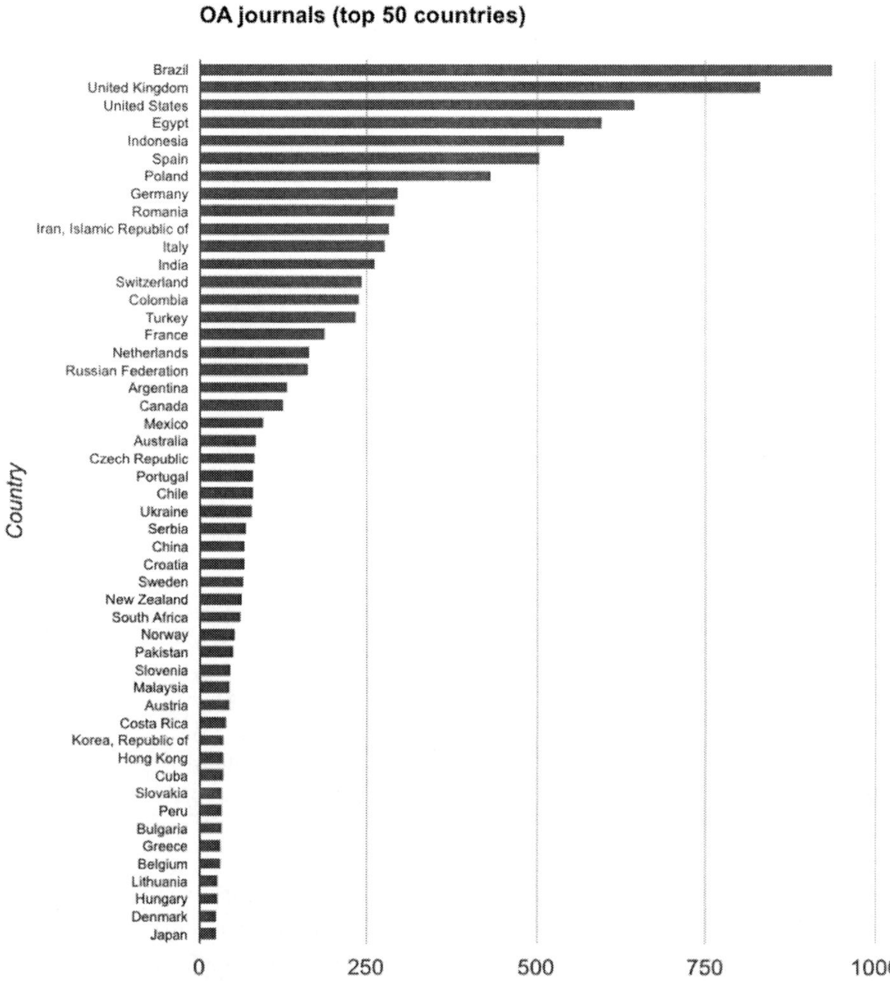

Figure 4 Number of Journals Indexed in DOAJ

Table 3. Knowledge on DOAJ

Functional Position	Knowledge on DOAJ	
	No (%)	Yes (%)
Tutor	26,00	74,00
Lecturer	21,13	78,87
Assistant Professor	21,35	78,65
Associate Professor	30,00	70,00
Professor	60,00	40,00

3.2.3. Google Scholar. Google Scholar is a service provided by Google for its users in searching scientific documents in the form of text, Google Scholar presents them in the various publication formats, including journal papers, conference papers, books, patents, and other publications. Google Scholar became one of the largest indexers of online articles in the world. Almost all journals in Indonesia are indexed in Google Scholar. The usefulness of indexation in Google Scholar can help lecturers or researchers to disseminate the research results to be cited by other researchers. Google Scholar will help one to identify the most relevant research of all academic research.

Table 4. Knowledge on Google Scholar

Functional Position	**Knowledge on Google Scholar** No (%)	Yes (%)
Tutor	16,00	84,00
Lecturer	7,75	92,25
Assistant Professor	6,74	93,26
Associate Professor	0,00	100,00
Professor	0,00	100,00

Table 4 shows that almost 90-95% of the total lecturers have knowledge about Google Scholar. This indicates that lecturers in Indonesia have been familiar with the least indexing database that functionate as a forum to publish the lecturer's scientific work in the field of research and community service. Google Scholar became the basis of lectures publication performance which recorded on SINTA. Therefore, for the 5% of the lecturers in the table who are not knowledgeable about Google Scholar, they will be encouraged to increase their literacy and insights related to Google Scholar's details. This is because the Google Scholar ownership publication profile is the main requirement to obtain the lecturer performance record through SINTA. Furthermore from 92,09% who know about publications in Google Scholar, based on empirical frequency about 80% have publications indexed in Google Scholar.

3.2.4. Scopus. Scopus is the largest abstract database and citation of a scientific publication such as scientific journals, books, and conference proceedings. Scopus contains publications in the form of science, engineering, social sciences, medicine, arts and humanities. At worldwide, Scopus is used by over 3,000 academic, government and corporate institutions and is the primary data source that supporting the Intelligence Research portfolio (Scopus.com). Based on the discussion results in the ResearchGate forum [5], there was a fierce debate about the pros and cons regarding to the question of "Is SCOPUS (Elsevier) a suitable indexing database to evaluate a researcher's career and academic productivity?". Some argue that "*Scopus is the best indexing platform among the available databases. Compared to the Web of Science, Scopus covers various journals and conferences. You can find several articles, including non-Elsevier, as soon as they are online as 'in print'. Of course, they have strict indexing rules to maintain quality, which is quite valid. I am not sure about the cost of indexing though.*" And on the other hand, the cons side with this statement says that (1) the process of its data update is not fast (conference may be waiting until six months before appearing in Scopus), and Elsevier's product may be updated more faster than non-users of Elsevier (e.g. ACM and IEEE); (2) includes only a portion of the scientific literature (e.g. not all IEEE conferences), and (3) Scopus is a commercial database.

Table 5 shows that the number of lecturers who did not know about Scopus is highest among 4 indexing platforms (SINTA, Google Scholar, DOAJ, Scopus) compared in this study at each functional position. Of 59,81% of lecturers/researchers who know about Scopus and based on empirical frequency there are 76% who have never done publication in Scopus. It is possible that lecturers / researchers only know about Scopus without ever doing a publication indexed in Scopus. The number of lecturers who did not know about Scopus is closely related to the access to related journals because the journal is paid. The number of Indonesian scientific articles published in Scopus indexed journals has increased significantly and it was the highest at the ASEAN scientific articles (76%) since 2011 to 2016 (Table 6).

Table 5. Knowledge on Scopus/Web of Science

Functional Position	Knowledge on Scopus	
	No (%)	Yes (%)
Tutor	42,00	58,00
Lecturer	47,18	52,82
Assistant Professor	32,58	67,42
Associate Professor	30,00	70,00
Professor	20,00	80,00

Table 6. Comparison of the number of scientific articles, citations, and H index of ASEAN countries before the publication obligation policy of scientific articles in international journals for Indonesian doctoral students (2010) and afterwards (2016)

Country	Documents		Citations		H-index	
	2010	2016	2010	2016	2010	2016
Malaysia	15740	28546	128548	19024	224	224
Singapore	15590	19992	321718	32504	454	454
Thailand	10049	14176	111214	11331	269	269
Indonesia	2704	11470	25644	4604	175	175
Viet Nam	2163	5563	28756	4970	167	167
Philippines	1342	2642	21430	2598	189	189
Cambodia	193	480	4598	525	86	57
Laos	136	377	2592	552	67	86
Brunei Darussalam	124	272	971	485	57	58
Myanmar	116	256	1318	338	58	67

(Source: http://www.scimagojr.com/countryrank.php)

3.2.5. Web of Science. Thomson Reuters is an information company formed through the purchase of Reuters by Thomson Corporation on April 17, 2008. Thomson Reuters shares are listed on the Toronto Stock Exchange (TSX: TRI) and the New York Stock Exchange (NYSE: TRI). Thomson Reuters is headquartered in Midtown Manhattan, New York City, USA. The Wood-bridge Company, a holding company for the Thomson family in Canada holds 53% of the group's shares, which operates in 100 countries and has more than 55.000 employees. Thomson Reuters became Canada's leading corporate brand on the list of Inter-brand Best Canadian Brands 2010. Web of Science is a Thomson Reuters service product that provides access for academics and scientists to access and to cite the research results more easily.

The knowledge of lecturers and researchers related to the overall publications indexed by Web of Science's, is still poorly understood about it. As many as 56,01% of lecturers and researchers were still less know about Web of Science, according to the research result.

3.3. Publication Issues

Scopus is not clean from predatory journals. Mart, as cited in Savina and Sterligov [6], revealed that the government and academia will be bothered with the fake papers that are claimed to appear in international journals. The development of fake or predatory journals is significantly improved even the predatory journals also present prominently in the 2,79% of journals indexed in Scopus database by 2015 [6]. Such information is often under-socialized by the government to the scientific community or academics till not a few lecturers who fall into the trap of predatory journals, of course with promise of indexation.

The ability of lecturers in generating appropriate articles with the indexed journal standards is still minimum and the lack of information on the procedure of the indexed journal become problems experienced by some lecturers / researchers studied. It is true especially for lecturers who are interested only in his or her environment and do not want to explore the world outside. Many lecturers conduct research but only when reporting it into Department of Research and Community Service, they are not brave or lazy to publish their research results. Even though, more of their research can be used directly for a businessmen and stakeholders. In addition, the language barrier is also one of the obstacles of lecturers to be able to publish on reputable international journals indexed in Scopus. Even if the contents of research is worthy to entry at international journals but because the language constraint, the researchers prefer a local journal that is not necessarily indexed and reputable.

The real phenomenon shows that lecturers are too preoccupied with teaching and paradigm of money source which is only obtained from teaching. One of the right strategies is to build the knowledge management and through it the culture of collecting, managing, and sharing publicity knowledge. The simple tactical step of the knowledge management is to make a schedule of workshops about the publications followed by publication accompaniment.

Furthermore, the accepted research report in reputable international journals that have a high-quality level might be contributed by adequate funding. Unfortunately, not all researchers in college can access the funding of the study. The research policy still requires competition so more lectures do not qualify than those who qualify. This becomes an imbalance that must be improved, where all lecturers should be able to access the source of research funds. The unevenness of educational facilities and infrastructure in universities is a problem that is still very strong. The most classic example is the laboratory facilities and infrastructures among universities are still showing a gap, as well as reliable laboratory resources. It must be admitted that there is a gap of infrastructure between universities in the region (outside Java) with universities on the Java Island.

The enforcement of rules related to reputable international publication for Associate Professor and Professor functional lecturers is sufficient to have a disastrous impact. It is because the sudden implementation of the policy is accompanied by the "punishment" of terminating the professional lecturers (including professor) allowances. Based on discussions with the researcher/lecturer who participate in this research, there are aspirations that lecturer publication policy to not be linked with the threat of the dismissal of the certification allowance and the honorarium of the professor. The scholarly publications achievement for lecturers who penetrate reputable international journals is the time to be motivated by the provision of additional reward in the form of research funding without selection, so it can motivate researchers and other prospective researchers. Performance of researchers, lecturers and institutions is not only measured by the number of published papers and citations but also measured by the number of unique paper readings.

4. Discussions

Is Knowledge / Understanding Sufficient? The Position of This Paper

The data contained in the four tables (Knowledge about SINTA, DOAJ, Google Scholar, and Scopus) indicated that consecutively the indexing databases known by Indonesian lecturers are 1) SINTA, 2) Google Scholar, 3) DOAJ, and 4) Scopus. SINTA is most widely known when this research was conducted. This is not surprising because of the endorsement by the Ministry of Researcher, Technology and Higher Education of The Republic of Indonesia policy which requires the lecturers to have SINTA account because of the practical implication to the promotion of academic position as well as the acceptance consideration in the future proposal of research grant in the Ministry.

SCOPUS is the least understood indexing database among the four. In this case, Scopus is still perceived as an "alien" by the majority of the lecturers, despite the fact that Scopus contains many useful functions such as publication search serving various boolean operations (so that the retrieval becomes very precise), citation tracking, and author identification. In addition to paid access, this lack of understanding may also be contributed by "the experience of contact" with Scopus. This experience in getting published with Scopus journal/proceeding is apparently able to explain *the gap* between the four tables (Table 2, Table 3, Table 4, Table 5) and Table 6. It has been known that the number of Indonesian scientific articles indexed in Scopus has increased sharply. The increasing number has been the highest in the ASEAN scientific articles. The numbers even exceeded Thailand, and according to the Ministry's strategic plan would surpass Malaysia.

Based on the empirical data collected through this present study, Scopus is the least understood database. However, the quantity of scientific publications of Indonesians is increasing rapidly in Scopus. How to bridge these two seemingly contradictive realities?

In this Discussion, the author presents a theoretical perspective of the field of psychology, namely the importance of *past experience*. The authors argue that the disparity between the two realities is highly likely to be due to the experience of "in touch or together with Scopus" is unevenly distributed, both among institutions/universities/research institutes and among lecturers/researchers. This unequal distribution of experience can contribute to the formation of so-called "elitism". The term "elitism" is used to illustrate the possibility that a particular group (both individuals and institutions) contributes to Scopus database with their publication significantly in higher frequency than any other rest group. These "elite groups" are instrumental in accelerating the increasing number of Indonesian publications indexed by Scopus.

Theoretically, these groups are very likely to be the group that has had a debut in the publication of Scopus indexed articles. While the other group, either for one reason or another, has not even debuted, either because of financial, psychological or other structural and/or situational restrictions or barriers.

Based on social psychology studies, one of the most potent predictors found to affect future behavior is *past behavior*, either through intention [7] or directly without intention [8]. Investigation of these variables is continuously carried out to this day within the framework of the Theory of Planned Behavior (TPB) [9][10]. Ajzen found that the explained variance grew from 41% to 54% after adding past / prior behavior into the TPB predictive model of behavior, without mediation of intention; in fact, its predictive power goes beyond the contribution of intentions, attitudes, and other variables in the TPB framework [9]. A number of alternative explanations, such as the high frequency of past behavior, constitute a habituation / routine process that leads to similar subsequent behavioral automation, without meaningful cognitive effort [11].

In that kind context, this paper is written and published. In the dialogues in the Telegram group, the authors of this paper agreed to do a crowd-writing to gain a first experience of obtaining Scopus ID. Related to the mentioned theory, this research contributes to the occurrence of "past behavior" in order to psychologically facilitate lecturers to come back into contact with Scopus in the future. In the psychological term, it is called "efficacy" [12]. In the process of compiling this paper, for example, lecturers are either (consciously) or indirectly (unconsciously) gaining shared learning experience in preparing the Introduction of research. The section containing the state of the art of this paper is the most contributed by lecturers from various scientific backgrounds and institutions.

Therefore, partially the preparation of this paper is actually a good example of a kind of *action research*. Research is conducted not only to explain phenomena but to make *meaningful social changes* for both the community and the actors. The writing of this paper would address the inequity of Scopus experience. Thus, knowledge / understanding on indexing databases, according to the results of this present study, might be not sufficient to support the Government's vision to develop the number of publications. It would be unfair if the experience and incentives of publications can only continuously be enjoyed by certain individuals and groups without a meaningful or systematic effort to take care lots of other inexperienced---vis a vis unknowledgable---lecturers/researchers. We could not rely solely on the Government programs and initiatives but as intellectuals need to jointly empower ourselves. *Shared or coordinated practices* become the unique key words, psychological breakthrough and contribution of this paper to improve the habits and enjoyment of proceeding to give birth to publishable articles indexed in Scopus. That is the position of this crowd-writing paper.

5. Conclusion

On average, 66,5% of Indonesian lecturers/researchers have known about SINTA, DOAJ, Google Scholar, Scopus, Web of Science. However, based on empirical frequency 76% of them have never published with journals or proceedings indexed in Scopus.

Acknowledgment

This article is a development of an article that have been accepted in Joint Workshop of KO2PI 2017 & ICMSTEA 2016 and then undergone 100% change with the latest data in the 2017 involving more contributors. Thanks to the committee of 2nd ICMSTEA 2016 and the committee of KO2PI Workshop 2017 to provide opportunities for publication with IOP Publisher.

References

[1] Kurniasih, N 2016 The implementation of Altmetrics in library as an alternative measurement method for scholarly article impact in social web ecosystem *Proceedings of International Conference on Science Mapping and the Development of Science* 63-73

[2] Björk, B C and Solomon D 2012 Open access versus subscription journals: a comparison of scientific impact *BMC Medicine* **10** 73

[3] Chua, S, Qureshi, A M, Krishnan, V *et al.* 2017. The impact factor of an open access journal does not contribute to an article's citations [version 1; referees: 2 approved] *F1000Research* **6** 208

[4] Irawan, D E, Multazam, M T, and Mulyaningsih, I 2017 Status jurnal berbahasa Indonesia di DOAJ https://www.authorea.com/users/87737/articles/160896-status-jurnal-berbahasa-indonesia-di-doaj

[5] Ghezzi, A 2012 Is SCOPUS (Elsevier) a suitable indexing database to evaluate a researcher's career and academic productivity? *ResearchGate* https://www.researchgate.net/post/Is_SCOPUS_Elsevier_a_suitable_indexing_database_to_evaluate_a_researchers_career_and_academic_productivity.

[6] Sterligov, I and Savina, T 2016 Riding with the metric tide: Predatory journals in SCOPUS *Higher Education in Russia and Beyond* **1**(7) 9-12 https://herb.hse.ru/data/2016/03/02/1125175286/3.pdf

[7] Xu, Y, Li, Y, and Zhang, F 2013 Pedestrians' intention to jaywalk: Automatic or planned? A study based on a dual-process model in China. *Accident Analysis & Prevention* **50** 811-819

[8] Franklin, K 2013 The best predictor of future behavior is … past behavior: Does the popular maxim hold water? *PsychologyToday* https://www.psychologytoday.com/blog/witness/201301/the-best-predictor-future-behavior-is-past-behavior

[9] Ajzen, I 2002 Residual effects of past on later behavior: habituation and reasoned action perspectives. *Personality and Social Psychology Review* **6**(2) 107-122

[10] Gadiraju, T 2016 Investigating the determinants of recycling behavior in youth by using Theory of Planned Behavior *Unpublished Master's Thesis* (Florida, USA: College of Arts and Sciences, University of South Florida)

[11] Ouellette, J A and Wood, W 1998 Habit and intention in everyday life: The multiple processes by which past behavior predicts future behavior *Psychological Bulletin* **124**(1) 54

[12] Abraham, J and Pane, M M 2014 The mediating role of social competition identity management strategy in the predictive relationship between susceptibility to social influence, internet privacy concern, and online political efficacy. In: Linawati, Mahendra, M S, Neuhold, E J, Tjoa, A M, and You, I. (eds) *Information and Communication Technology*. ICT-EurAsia *Lecture Notes in Computer Science*, **8407** 492-499 (Berlin, Heidelberg: Springer)

Method Improving Reading Comprehension In Primary Education Program Students

Rohana

Department of Primary School Teacher Education, Universitas Negeri Makassar, Indonesia

rhnsyamsuddin@gmail.com

Abstract. This study aims to determine the influence of reading comprehension skills of English for PGSD students through the application of SQ3R learning method. The type of this research is Pre-Experimental research because it is not yet a real experiment, there are external variables that influence the formation of a dependent variable, this is because there is no control variable and the sample is not chosen randomly. The research design is used is one-group pretest-post-test design involving one group that is an experimental group. In this design, the observation is done twice before and after the experiment. Observations made before the experiment (O1) are called pretests and the post-experimental observation (O2) is called post-test. The difference between O1 and O2 ie O2 - O1 is the effect of the treatment. The results showed that there was an improvement in reading comprehension skills of PGSD students in Class M.4.3 using SQ3R method, and better SQ3R enabling SQ3R to improve English comprehension skills.

Keywords: reading comprehension skills, application of SQ3R method

1. Introduction

English is the International language used to communicate in various fields and aspects of the life of the international community, the communication process also includes some activities such as reading letters, reading reports, writing and talking and listening. This activity is certainly in English. English reading is a routine activity, however, this activity needs to be careful, careful, to know the information submitted by news writers. Cognitive and learning style is one of the important things in the process of thinking process [1]-[6].

Reading has many benefits, by reading humans acquiring a lot of knowledge, developing speaking skills fluently in speaking, developing creative reasoning, enhancing comprehension of problems, improving the ability to comprehend conceptual concepts of learning or reading, window of the world means obtaining various information from various sources and various directions. Linguistic reading is a process of retrieval and decoding process, in contrast to speaking skills and writing skills that involve encoding an aspect of decoding is to connect the written word, with the meaning of oral language (oral language meaning) which includes the conversion of writing/prints into meaningful sounds [7].

In language skills there are four main skills, namely speaking skills, reading, writing and listening, these four skills are very useful in communicating. The preferred reading skill on the skill of understanding the contents of the text, discourse. Ideas and information submitted by the author. By reading can improve intelligence and broaden the horizon of reading skills is very supportive of improving the quality of human resources and the quality of education.

In the process of reading a person is trying to understand the contents of the author's message contained in the reading (Hodgson in [8]) read is a process done by the reader to obtain a message conveyed by writers who scroll word media or writing the language. Tarigan found that reading there are two ways of reading aloud and silent. says that reading is a very complex and complicated process, because the reading process involves various factors, internal and external, internal factors include interest, motivation, intelligence, talent, purpose, and other, while external factors are the environment, the means, the level of legibility, the habit and the culture of reading., reading is bridge for who and wherever that want to progress and success both at school and an environment of work [8].

The various obstacles faced when there is a desire to cultivate reading the unavailability of interesting reading material so that the desire to read less, unmotivated reading desire resulted in low learning achievement in reading skills, lack of creative development and innovation. This is a challenge faced when learning English because English is one of the subjects that is considered difficult for PGSD students, English has things to be mastered by learners such as grammar, vocabulary, tenses, intonation, pronunciation, reading, writing, meaning. Proposition, conjunction etc. When given reading comprehension learning is considered very difficult, because learners must use dictionaries to translate the meaning of discourse, reading, and finding the main idea, analyze the content of discourse, and explain the contents of the discourse. This work is very boring for learners, only 12% are able to implement in enough categories, while 60% in low category and 28% are very low. This is a challenge that should be immediately addressed to anticipate the low achievement of learners in reading comprehension skills and low achievement of English. So the need to improve the reading comprehension skills of English through the application of appropriate methods such as the SQ3R method that is assumed to increase interest in reading learners is expected through the SQ3R method of reading comprehension ability of English learners.

Strategically approaching textbook assignments can help a reader not only move through his reading more efficiently but it can also make him a more effective reader and learner. One strategy that has been proven to help readers comprehend and think about texts is SQ3R. The SQ3R method is a 5-step process for working through texts that allows the reader to understand and process information at the same time. The name, SQ3R, is an acronym for the five steps: Survey, Question, Read, Recite, and Review [9]. The method of reading SQ3R introduced by French P. Robinson in 1948, Effective study and in Indonesia introduced by Tampubolon in 1986 this method includes five steps: survey, question, read, recite or recall and review. Survey readings to ask you questions that the answer we hope to find in these readings will be easier to understand the reading, and then by trying to express in our own words the main points, we will master and remember it longer, or read the survey (survey reading) is to survey the reads to be studied by examining or examining the indexes, list of words, examine the title, examine the scheme, the outline of the book in question.

Questions ask as many questions about the contents of the reading, by changing the subtitle and subtitle and subtitle into a question. Use the words who, what, when, where, or why. Read by the preparation process before reading, then the process of reading the entire contents can be done at high speed. Recite, this step is done to test the understanding of what has been read. This process is done by recounting the thoughts discussed in the book in your own language style. Furthermore, the review and review process is done after the process of reading finish so that what is read not only entered in the short-term memory but into the long term.

The SQ3R method is very useful in applying to English comprehension readers to measure improving English reading comprehension skills of learners in PGSD. Therefore, the authors conducted a study with the title of an influence of the application of the SQ3R method of reading comprehension skills of English for PGSD students. In order to know the effect of using SQ3R on reading comprehension on learning English

SQ2R involves 5 steps: (1) Surveying the material in the table of contents or by reviewing the titles and subtitles in advance, (2) Questioning, which means asking questions, in advance, concerning the key topics that the student is preparing to read about, (3) Reading, which involves reading and taking notes that answer the questions.(4) Reciting, where students can record any additional facts that help them make sense of the material, help them draw links to information they already know, or help them to remember what they read, (5) Review, where the student is asked to summarize the answers to each of the questions in one paragraph of three to 5 sentences [10].

It has been proven that the success of the SQ3R learning method has been obtained through several studies that have been conducted in America. In one experiment, several parts of the learning class measured their reading ability (reading level and accuracy of understanding) on tests related to Canadian history. They were then given training in the use of the SQ3R Method for several days, after which they performed a comparable reading test. Prior to training at SQ3R, the average reading rate for the class was only at the 34th percentile, but after training, at the 53rd percentile. In another experiment, two quizzes with similar difficulties have been prepared. For the first quiz, the students learn their own way, but for the second quiz, they are shown how to predict quiz questions with the SQ3R method. These five steps of the SQ3R Method, if applied and practiced, should result in an increase in reading comprehension, an improved ability to identify important points and better retention of the material. You should also discover one other worthwhile outcome: happily, test questions will seem familiar, because the headings you turned into questions are usually the points the instructor will emphasize on exams [11].

The most efficient students read with purpose and set goals. The SQ3R Method. Is designed to help you read faster and retain more information than ordinary reading methods.SQ3R stands for the steps in reading: survey, question, read, recite, review. It might seem like it takes more time to use the SQ3R method, but you'll find that you remember more and have to reread less often [12].

SQ3R is an acronym for the five steps: Survey, Question, Read, Recite, and Review. Here's how SQ3R works [13].

1) Survey: While it would seem that a reader should begin reading at the top of the first page of the text and then move sequentially through each page of it until the done, according to SQ3R, starting off by surveying the text features of the text is more beneficial because it keys the reader's brain into the main ideas of the text. Text features are the visual supports a writer uses to add meaning to her text. Examples of common text features are bold print, italics, colored text, headings and subheadings, charts, graphics, illustrations, and glossaries or indexes. In SQ3R, the reader surveys or scans the text features prior to reading to preview its information and to focus his reading.
2) Question: After surveying the text features, the reader has a pretty good idea of where the text might be heading. Now, he needs to write questions to further focus his reading. In SQ3R, the reader writes down questions he thinks might be answered in the full text based on his survey of the text features.
3) Read: Finally, the reading begins! Now, the reader reads the text from start to finish, but instead of reading it cold without any real direction, he has his questions to guide him through. As he reads, he thinks about the questions he wrote down in the previous step and notices their answers.
4) Recite: The recipe step takes place concurrently with the Read step. As the reader is reading through the text looking for answers to his questions, he pauses when he comes to information he believes may answer each question. At this point, he recites the information to himself and then writes it down next to the question. While this step may seem unusual to some students, it is very important because the recitation helps cement the information into their long-term memories.
5) Review: The final step in SQ3R is the review. Once the reader has moved through the entire text, reciting and writing down answers to his questions, he takes a moment to re-read or review

IOP Conf. Series: Journal of Physics: Conf. Series **954** (2018) 012027 doi:10.1088/1742-6596/954/1/012027

the text. As he reviews the text, he looks for answers to questions he was unable to respond to in the first reading and recites the answers to those he did during the first reading.

The benefit of SQ3R it helps students to discover the important facts and ideas that are contained in a textbook, and master and retain that information so that they are prepared for an examination.

2. Methods

The research approach used is quantitative approach because the researcher wants to know the influence of the use of learning video media. The type of research used is Pre-Experimental research because it is not yet a real experiment. There are still external variables that contribute to the formation of dependent variables, this happens because the absence of control variables and samples are not selected randomly. The research design used in this research is one-group pretest-posttest design involving one group that is as experiment group. The experimental group applied direct learning model using SQ3R learning method. The design of this study did not involve any control group. In this design, the experimental group was given pretest then applied the treatment by using SQ3R method after it was held posttest. In this design, the observation is done twice before and after the experiment. Observations made before the experiment (O1) are called pretests and the post-experimental observation (O2) is called posttest. The difference between O1 and O2 i.e. O2 - O1 is assumed to be the effect of the treatment.

2.1. The stage of SQ3R
The stage of SQ3R method for thorough study:

1) Step 1: Survey
Skim through the book and read topical/sub-topical headings and sentences. Read summaries at the end of chapters and books. Try to anticipate what the author is going to say. Write these notes on paper, then look it over to get an overall idea.
2) Step 2: Questions
Turn paragraph headings into questions (e.g. "Basic Concepts of Reading" to "What are the Basic Concepts of Reading?"). Write these questions out.
3) Step 3: Read
Read with alertness to answer the questions you came up with. Write notes, in your own words, under each question.
4) Step 4: Recall
Without looking at your books or notes, mentally visualize, in your own words, the high points of the material immediately upon completing the reading
5) Step 5: Review
Look at your questions, answers, notes, and book to see how well you did recall. Finish up with a mental picture of the whole Adapted from F.P. Robinson. Effective Study. New York: Harper and Bros. 1948. Chapter II

2.2. Location, Population, and Sample
The place of this study in PGSD Makassar FIP UNM Tidung Campus, time is done at time in this study is the even semester of the academic year 2014/2015 which is in may which lasted for 2 weeks. The population in this study are all PGSD students force year 2014/2015 consisting of 6 classes or a number of 240 students and the sample selection in this research is done directly by choosing one class (without random) that is class M4,3 with amount of 40 participant. The amount of table as follows sample research. Data collect using the test of learning result reads comprehension of English. The sample selection in this research is done directly by choosing one class (without random) that is class M 4,3 with amount of 40 Participant The amount of table as follows.

Table 1 Number of students Class of M.4.3

Class	Sex	Number
M.4.3	Male	12
	Female	32
Total		44

2.3. Data Analysis Technique

2.3.1. Descriptive Statistics. The descriptive statistical analysis aims to describe the learning achievement of writing skills obtained by participants from the experimental group. The learning outcomes are then compared by classifying the following learning outcomes (the following sheets):

Table 2 Guidelines for Categorizing Student Results

Interval Nilai	Category
86 – 100	Very Good
71-85	Good
56-70	Medium
41-55	Less
≤ 40	Very Less

2.3.2. Inferential statistics. Inferential statistical analysis is used to test the research hypothesis by using the t-test. Prior to hypothesis testing, prerequisite analysis was performed, ie normality test and homogeneity test where all data was processed in Statistical Package for Social Science (SPSS) version 20.0.

- Normality test
  Normality test is used to determine whether the samples examined are normally distributed or not. Testing of data normality of learning the result of reading skill reading using One-Sample Kolmogrov-Smirnov Test on SPSS version 20.0 system. The learning outcomes of the population will be normally distributed when the sig (2-tailed) $\geq \alpha$ with a real level of $\alpha = 0.05$.
- Homogeneity Test
  Testing the homogeneity of learning data aims to know the data in this study have the same variance (homogeneous) or not. Testing homogeneity using learning skill reading comprehension using Test of Homogeneity of Variances with Statistical Package for Social Science (SPSS) version 20.0. The test criteria used are sig (2-tailed) $\geq \alpha$ with real level $\alpha = 0.05$ data is homogeneous.
- Hypothesis testing
  Hypothesis testing of this study was conducted to determine whether there is a significant influence between the uses of the SQ3R method of reading comprehension skills PGSD students 2014/2015 academic year. To make it easier to see how the influence of independent variables on dependent variables, in this study used t-test combined using Statistical Package for Social Science (SPSS) version 20.0, by comparing the t-count with t-table ($\alpha = 5\%$).

3. Result and Discussion

The results obtained from the results of evaluation tests conducted early before treatment or treatment in the experimental class and the results of evaluation tests after treatment or treatment in the learning process using. Evaluation test in the form of multiple choice as many as 11 numbers, 1 essay-shaped number, and there is a Likert scale attitude scale questionnaire, as well as other instruments, namely Learning Plan and Student Worksheet validated by the expert validator. In this Pre-Experimental study, researchers conducted a study on class M 4.3 as an experimental class with a total of 44 students consisting of 12 male and 32 female. The research design used in this study is one-group pretest-posttest design.

Table 3 Recapitulation of Learning Outcomes Learning Skills Reading Pretest and Posttest Understanding by Using SQ3R Method

Descriptive Statistics	Cognitive	
	Pretest	Posttest
Sum of Sample (N)	44,00	44,00
Minimum	28,00	72,00
Maximum	83,00	100,00
Mean	62,68	88,34
Range	55,00	28,00
Standard Deviation	12,853	6,888
Median	66,00	90,40
Mode	68,00	93,97

Table 3 shows the value of learning outcomes at the lowest and highest values of pretest cognitive domains of 28.00 and 83.00 after being given repetitive treatment in the cognitive domain of 72.00 and 100.00, means the lowest and highest values in the experimental class on posttest learning outcomes after being given repeated treatments were higher than pretest learning outcomes before treatment.

The median value at pretest of the cognitive domain is 66,00 while the median value in posttest of the cognitive domain is 90,40 mode value when pretest of the cognitive domain is 68,00 while mode value at posttest of the cognitive domain is 93,97 Mean median value and experimental class posttest mode higher than the median and pretest mode. The standard deviation value (standard deviation) in the experimental class when the cognitive domain pretest is 12,853, the standard deviation value in the cognitive domain posttest is 6.888. The posttest default deviation value indicates that the value of variation size is lower than the standard deviation of the pretest. The standard deviation that the value of variation size close to zero then the data uniformity more perfect and if the size of variation away from zero means more uniform data owned.

In table 4, the categories of learning outcomes during the pretest of cognitive domains show the number of students who have a good category score of 12 students with a percentage of 27.27%, quite as many as 19 students with a percentage of 43.1%, less as many as 9 learners with percentage 20.45% and very less as many as 4 students with a percentage of 9.1%. Unlike the posttest category of cognitive domain learning results showed the number that has a very good category score of 32 participants with the percentage of 72.73% and good as many as 12 students with a percentage of 27.27%. Based on the categorization of learning result of reading comprehension of pretest English in cognitive domain experiment class, it was found that descriptive statistic analysis only show or show value in descriptive statistic analysis only show or show value on pretest and posttest were given only one experiment class that is class M. 4.3 given treatment using the SQ3R Learning method and not to test the hypothesis. Descriptive statistics only present the statistics calculated on the sample, but if descriptive statistics are used to test the hypothesis (the provisional assumption must still be tested) then it has entered the inferential statistical region. This means that descriptive statistics attempt to describe and analyze the given group without making or drawing conclusions about a larger population or group. Inferential statistics relate to decision making. Inferential statistics are based on descriptive statistics.

IOP Conf. Series: Journal of Physics: Conf. Series **954** (2018) 012027 doi:10.1088/1742-6596/954/1/012027

Table 4 Frequency Distribution and Percentage Category Learning Outcomes Learning Skills Reading Understanding by Using SQ3R Learning Method

Category	Cognitive			
	Pretest		Posttest	
	Number of Students	%	Number of Students	%
Very Good (86-100)	-	0%	32	72,73%
Good (71-85)	12	27,27%	12	27,27%
Medium (56-70)	19	43,1%	-	0%
Less (41-55)	9	20,45%	-	0%
Very Less (≤ 40)	4	9,1%	-	0%
Total	44	100%	44	100%

Based on hypothesis test results with inferential statistics indicate the influence of the learning methods SQ3R on learning outcomes reading comprehension PGSD students class M 4.3. Decision-making on hypothesis testing is done in two ways: comparing t-table with t-count and comparing probability value. From the statistical result using manual calculation for t-test two dependent samples (paired sample t-test) combined with SPSS 20.0 program obtained t-table value for N (44) = 2.01669 while t-count of cognitive domain learning -16,99, t-count (16.99) & t-table (2.01669), so H_0 is rejected and H_1 accepted, noting that the price of t-count is the absolute price, so it is not positive (+) or negative (-). Whereas by comparing the probability value, it is obtained the significance value of learning result of reading skill of understanding the English language cognitive mean Ho is rejected Ha accepted. So there is a significant influence between the uses of the SQ3R Method on the learning outcomes of comprehension reading comprehension.

Based on the results of research that has been done that the use of learning methods SQ3R can affect the results of learning comprehension reading skills can be seen from the comparison of pretest and posttest learning outcomes. Test of learning outcomes obtained by learners at the time of pretest cognitive domains at most are in the category enough with the percentage of 43.1% and good with the percentage 27.27When the posttest cognitive domain is at most in the category very good with the percentage of 73.73% and good with a percentage of 27.27%. From this result, it can be concluded that the use of learning use SQ3R Method can influence learning achievement reading comprehension result of class M.4,3 PGSD students. considering 70% and after then only able to remember 10% of what the students heard, but if a student learn by using the sense of hearing and sight, then after 3 hours he was able to remember 85% and after learn he was still able to remember 65% of what he hears and he sees. Based on this, it is clear that the use of SQ3R learning method is very helpful for learners in learning. How not in addition to helping to learn in the realm of cognitive (knowledge), SQ3R review method is in fact very effective to improve learning outcomes reading English comprehension.

4. Conclusion

Based on the results of research, data analysis, and discussion, it can be concluded that: The result of reading skill study Understanding on pretest in cognitive domain is in enough category while learning result of reading comprehension skill at end of learning (posttest) in cognitive domain is in very good category. The increase in the average of learning achievement in English reading comprehension at the end of the learning (posttest) shows that there is a significant influence on the use of learning method of learning SQ3R to the learning achievement of reading comprehension of the students of class M.4.3 PGSD Year 2014/2015.

References

[1] Mulbar, U., Rahman, A., & Ahmar, A. S. 2017. Analysis of the ability in mathematical problem-solving based on SOLO taxonomy and cognitive style. *World Trans. Eng. Technol. Educ.*, **15**, 1, pp. 68-73. doi:10.26858/wtetev15i1y2017p6873

[2] Rahman, A., & Ahmar, A. S. 2016. Exploration of Mathematics Problem Solving Process Based on The Thinking Level of Students in Junior High Schoo. *Int. J. Environ. Sci. Educ.*, **11**, 14, pp. 7278-7285.

[3] Rahman, A., & Ahmar, A. S. 2017. Relationship between learning styles and learning achievement in mathematics based on genders. *World Trans. Eng. Technol. Educ.,* **15**, 1, 74-77. doi:10.26858/wtetev15i1y2017p7477

[4] Rahman, A., Ahmar, A. S., & Rusli. 2016. The Influence of Cooperative Learning Models on Learning Outcomes Based on Students' Learning Styles. *World Trans. Eng. Technol. Educ.*, **14**, 3.

[5] Ahmar, A. S., & Rahman, A. 2017. Development of teaching material using an Android. *Glob. J. Eng. Educ.*, **19**, 1, pp. 73-76. doi:10.26858/gjeev19i1y2017p7376

[6] Arsyad, N., Rahman, A., & Ahmar, A. S. 2017. Developing a self-learning model based on open-ended questions to increase the students' creativity in calculus. *Glob. J. Eng. Educ.*, **9**, 2, pp. 143-147.

[7] Anderson, R.C. and Kulhavy, R.W., 1972. Learning concepts from definitions. *American Educational Research Journal*, *9*(3), pp.385-390.

[8] Tarigan, H.G., 2008. Berbicara. *Bandung: Angkasa Bandung.*

[9] Joelle Brummitt-Yale. Reading Comprehension. http://www.k12reader.com/what-is-reading-comprehension/, Retrieved on July 11, 2011 at 13.30.

[10] Whetzel, J. M.. 2012. Idea Generating for Writers. USA: CreateSpace Independent Publishing Platform.

[11] Fraser, L. 2009. Making Your Mark: High-Performance College and Career Success. USA: LDF Publishing Inc

[12] Kuther, T. L. (2015). *The psychology major's handbook*. Cengage Learning.

[13] Fleming, G. 2016. SQ3R: A Reading Comprehension Strategy. https://www.thoughtco.com/sq3r-reading-comprehension-strategy-1857535

IOP Conf. Series: Journal of Physics: Conf. Series **954** (2018) 012028 doi:10.1088/1742-6596/954/1/012028

Optimization of Candidate Selection Using Naive Bayes: Case Study in Company X

JA Kadar*, D Agustono, D Napitupulu
Research Center for Quality System & Testing Technology, Indonesian Institute of Sciences, Tangerang Selatan, Banten, Indonesia

*jimmy.abdel.kadar@lipi.go.id

Abstract. This research was conducted as a decision-making system, and an alternative solution to complete the candidate assessment for a particular position. The human resources (HR) section on company X is responsible and initiative in selecting candidates in accordance with the assessment of their superiors. Selection by using the method of filling out the manager's assessment questionnaire on the candidate's subordinate. Three (3) managers have been determined to assess the 11 candidates for subordinates. By using questionnaire of quality classification of human resources and formula naive bayes it will get result which finally grouped using criteria scale as final grouping. The HR department has also determined that what is received is that which meets criteria 5. The result is three (3) candidates who can be proposed as candidates for certain positions in company X, and have met all required calculations. Furthermore the candidate will be given to management as an alternative input data in the selection of candidates.

1. Introduction

The role of decision-making by management to select division heads within an organization is common, either tightly sealed, direct appointments or using a method of calculation. The development of a company forced management to organize and prepare competent personnel in their field to take part and work further. The human resources department (BSDM) of a company is the part that regulates and distributes the capabilities of the workforce in accordance with the field and its competence. The HR department is also responsible for selecting candidates from internal companies in ensuring the careers of their employees [4].

According to Argue [1] the problematic functions in employee planning are recruitment and selection. This is the most complex function in the BSDM section of an organization, and it is important to concentrate on this function, as this is time-consuming, not least costly and most often a complicated process [1] [19]. BSDM in selecting the personal required to hold a certain position can use a scientific approach by examining the needs of what is needed and the solution can be done to meet the answer to the problems in selecting employees to be placed in certain positions with specifications that have been determined.

Assigning and searching for candidates required for a particular position can be made with decision-making methods (PK) or more commonly known as decision support (DS), but at this point the PK to be used is not as complicated as imagined. Step PK use in the selection of candidates is based on the selection of candidates with scientific methods, transparent and can be more effective and efficient with adjusted at certain times, this provision is contained in the company's management policy X. Simple steps can be done is by selecting candidates from the manager that one slice with the position to be occupied. Implementation can be through several stages and / or ways, in this paper using questionnaire method by assessing candidates by superiors, the data have been collected and processed and classified according to rating scale rules. The rating scale classification is formed by using the classification defined by the HR department. Next use the calculation of naive bayes to classify candidates positioned according to their calculations. Use of naive bayes as part of statistics to calculate candidate chances for a particular position in a hypothesis. Thus understanding Bayes Optimal Classifier refers to the probability of a classification of the attribute criteria available and determines which classification is optimal [25].

The process of selecting and selecting candidates helps the organization in several ways. The whole purpose of the selection process and selection so that the organization can find a suitable person with a job with personal competence, thus the organization will be managed with the right people with the right skills [24]. Therefore, the aim of this study is to help management in providing alternative solutions or one reference information that can be a reference management in deciding the choice of candidates for certain positions especially in company X.

2. Literature Review

2.1 Decision Making System (DSS)

Decision-making system (SPK) is part of computer science, now SPK is one part of science that can be a solution for management. Decision-making system as a support for the solving of structured and unstructured problem analysis [12] [21]. According to Hasan [10] basically the SPK is used in support of all stages of decision making; can start from the identification of problems, data and information formed in several relevant classifications, determine the process approach, to the SPK evaluation. Syamsi [17] argues that decision-making should take into account the elements: (1) the purpose of decision-making, (2) identification of alternative decision-making solutions, (3) taking into account external factors, (4) the means used in evaluating or measuring decision-making.

Based on the above knowledge then the decision can be based on the measurement or assessment of management of its employees. The organizational structure, culture, and personal inclinations of managers influence the decision-making process [23]. This assessment is an alternative to the many measurement methods for employees in a company. This alternative solution reaches all the information to be assessed. Information based from the human resources section will get detailed information and still raw. This information needs to be processed and developed in accordance with the needs to be used in producing the desired output in accordance with the overall plan.

2.2 Naive Bayes

The naive bayes classification is one of the simplest opportunity classifications of the bayes theorem, which predicts future probabilities with previous experiences. In a simple sense, it can be assumed that a classification of naive bayes is assumed that the presence (or absence) of a particular feature of a room is unrelated to the presence (or absence) of other features. In Bayes terms, X is considered a "proof." As always, it is explained by the measurements performed on a set of attributes n. Let C be some hypothesis, such as X tuple data included in class F. For classification problems, the goal is to determine P (C | X), the probability that hypothesis C is applicable, given the "proof" or observed tuple data X. P (C | X) is the posterior probability of C conditioned on X. Posterior probability is estimated as follows [15]:

$$P(C \mid X) = P(X \mid C) \cdot P(C) / P(X) \qquad (1)$$

Where:
P (X) is constant for all classes
P (C) is a frequency. relative sample of class C
Searchable P (C | X) is maximum, as well as P (X | C) • P (C) is also a maximum value

The common flexibility of the probability model, the naive bayes classifier can be efficiently directed, for example a supervised learning. As another example of a simple application, the estimation parameter for the naive bayes model uses the maximum likelihood method, in other words, one can work with the naive bayes model without believing in bayesian probability or using bayesian methods. In the naive bayes classifier, another advantage feature is in the classification stage only requiring a small amount of training data to estimate the parameters (means and variances of the variables) required for classification. Since the independent variable is assumed, only the variance of the variables for each class needs to be determined and not the entire covariance of the matrix.

2.3 *State of the art*

Organizational dynamics in line with employee career dynamics, the selection of candidates to be placed in a certain position requires careful planning. Company X, represented by the HR department, has determined in the selection of candidates using a decision-making system as input for management. Based on candidate behaviour data and interviews as one of the basic assessments [7], a questionnaire was distributed to managers.

Questionnaire design to measure employee behaviour / competence and manage it as accumulated data to predict related achievement [14]. Several factors influence decision making. These factors, including the past experience of the candidate [11], cognitive differences [18], age and individual differences [5], believe in personal relevance [2], and the escalation of commitment, influence the choice of others. Understanding the factors that affect the decision-making process is important to understand what decisions are made. That is, the factors that influence the process can affect the outcome [6]. The HR section has a basis for determining the criteria in selecting candidates according to the needs of the position. The criteria can be seen in Table 4. The determination of the criteria has been done with the previous review, this study is internal and has been recognized by management. Based on these criteria, candidate assessment and placement can be done according to plan.

3. Research Methodology

Methodology used in this study is quantitative method [16] while the instrument used was a questionnaire compiled based on the needs of the firm as the concept measurement framework in this study. Quantitative research begins with a problem statement and involves the formation of hypotheses, literature review, and quantitative data analysis [22]. Primary data collection is done to managers who will assess candidates for specific positions. The number of respondents obtained from data collection is 3 managers to assess 11 candidates. The rating scale in the questionnaire was based on Muchinsky's theory [13] by assessing the candidate's perceptions of 5 choices of answers: "1" Unsatisfactory Performance, "2" Improvement Desired, "3" Meets Expectation, "4" Exceeds Expectation, "5" Outstanding Performance. Based on Muchinsky's theory [13], it is translated for the kusioner making as in Table 1.

Table 1. Questionnaire of Human Resource Quality Classification

a.	**Quality of Work**
5	The quality of work always exceeds the standard that has been set.
4	The quality of work sometimes exceeds the established standard.
3	The quality of the work always meet the standards that have been set.
2	The quality of the work sometimes does not meet the established standards.
1	The quality of the work is always less than the standard set.

3	Have enough passion to learn for self-improvement and improvement of work ability, but need to request support from superiors to apply it in their duties.
2	Less enthusiasm for learning for self-improvement and improvement in work ability, and or need to be controlled / reminded always to really apply it in its duties.
1	Less have the spirit to learn for self-development and improvement of work

b. Quantity of Work

5	The quality of the work in the form of quantities, or the volume or frequency or speed of completion of work always exceeds the standard set.
4	The quality of work in the form of quantities, or the volume or frequency or speed of completion of work sometimes exceeds the established standard.
3	Quality of work in the form of quantities, or volume or frequency or speed of completion of work always meet the standards that have been set.
2	The quality of the work in the form of quantities, or the volume or frequency or speed of completion of work is sometimes insufficient to meet established standards.
1	The quality of work in the form of quantities, or the volume or frequency or speed of completion of work is always less than the standard set.

c. Knowledge and Skills on Work

5	Very good in mastering and understanding the knowledge and skills of the field of duties and responsibilities.
4	Mastering and understanding the knowledge and skills of the field of responsibilities.
3	Simply master and understand the knowledge and skills of the field of responsibilities.
2	Lack of mastering and understanding the knowledge and skills of the areas of his duties.
1	Very lacking in mastering and understanding the knowledge and skills of his field of duties.

d. Initiative

5	Has the ability to be very fast, precise, correct in acting to carry out and complete his duties without waiting for orders and instructions.
4	Has the ability to be very fast, precise, correct in acting to carry out and complete the tasks, and sometimes still waiting for instructions with a relatively global instruction.
3	Has the ability to be very fast, precise, correct in acting to carry out and complete the tasks, and sometimes wait for a relatively detailed command.
2	Slow and always wait-for-command capability and with detailed instructions.
1	Slow and always wait-for-command capabilities and with very detailed instructions.

e. Cooperation

5	Be active in helping and supporting each

	ability, although already reminded.

h. Attendance Level

5	During the performance appraisal period is never too late, leave early, work lost, leave the work permit.
4	During the 1-2 late performance appraisal period, leave early and or 1 day lost to work.
3	During the period of achievement assessment 3-5 late, home early and or 2-3 days lost to work.
2	During the period of the 6th to 10th weekly achievement, leave early and or 4-5 days lost to work.
1	During the performance appraisal period more than 10 times late, leave early and or more than 5 days lost to work.

i. Planning & Organizing

5	Excellent in setting task priorities and placing and managing existing resources, so that the realization of their work is very effective and efficient.
4	Have a good ability in setting task priorities and placing and managing existing resources, so that the realization of work is always effective and efficient.
3	Have a pretty good ability in setting task priorities and placing and managing existing resources, so the realization of the work is good enough.
2	Less able to set priorities and less effective in managing existing resources, so that the planned work is rarely realized properly
1	Not able to assign task priorities and manage existing resources, so that the planned work is not realized.

j. Control

5	It is good to monitor and control all resources under their responsibility in the rules / regulations / procedures in carrying out their duties so that the resources are very optimal in doing actuality.
4	Can do well in monitoring and controlling all the resources under his responsibility remain in the rules / regulations / procedures in carrying out its duties so that the resources are optimal enough.
3	It is good enough to monitor and control all the resources under its responsibility in the rules / regulations / procedures in performing its duties, but sometimes it needs guidance so that sometimes it is less than optimal.
2	It is good enough to monitor and control all resources under its responsibility in the rules / regulations / procedures in performing its duties, but must always be

IOP Conf. Series: Journal of Physics: Conf. Series **954** (2018) 012028 doi:10.1088/1742-6596/954/1/012028

	co-worker and / or giving very seriously towards the achievement of company goals.
4	Active in helping and supporting co-workers only in the section / section / group only and or give serious attention to the achievement of the goal group / section / section only.
3	Need to be reminded / reprimanded, to assist and support co-workers and or need to pay serious attention to the achievement of group / company goals.
2	Less enthusiastic to help and support colleagues despite being reminded / reprimanded and or less attention to the achievement of group goals / company.
1	No matter to colleagues and / or always become obstacles in achieving the goal group / company.

f. Reliability

5	Always willing without reasons if given the tasks and responsibilities outside of the main duties and responsibilities and can solve it as expected.
4	Always willing without reasons if given the task and responsibility outside of the main duties and responsibilities, but in completing the results are relatively in accordance with what is expected.
3	Always willing for certain reasons if given the task and responsibility outside of the main duties and responsibilities, and in resolving the results are less in line with what is expected.
2	Always willing but for some reason given assignments and responsibilities outside of the main duties and responsibilities, and or when executing the results rarely as expected.
1	Always unwilling to be assigned duties and responsibilities outside of the main tasks and responsibilities, and or carrying out the tasks in resolving them rarely in accordance with what is expected.

g. Have Learning Ability and Willingness

5	Have a great passion for learning for self-improvement in order to improve work ability and earnestly apply what has been learned in his duties.
4	Have a good spirit to learn for self-development and improvement of work ability and seriously want to apply it in its duties.

	reminded for completion so that it is less than optimal in performing its activities.
1	Simply monitoring and controlling all resources under his / her responsibility remain in the rules / conditions / procedures in performing his duties, but must always be reminded to make the settlement so that less than optimal.

k. Decision-Making

5	In making a decision on a problem very quickly and precisely as well as the results can be justified.
4	In making decisions on a problem is always fast and accurate as well as the results can be justified.
3	Quite capable in making decisions quickly, precisely although several times need support and direction and the results can be justified.
2	Sometimes it takes great support and direction to dare to make decisions quickly, appropriately and / or decisions are sometimes less accountable.
1	Always slow in making decisions, even if they have been given their dukurang and their directives or decisions are not accountable.

l. Development of Staff

5	Be very attentive in the improvement and development of subordinates and or have a planned program for staff.
4	Attentive in the improvement and development of subordinates, but lacks a planned program for staff.
3	Just enough attention in the improvement and development of subordinates, but does not have a planned program for his staff.
2	Lack of attention in the improvement and development of subordinates, and does not have a planned program for his staff.
1	No attention in the improvement and development of subordinates, and does not have a planned program for his staff.

IOP Conf. Series: Journal of Physics: Conf. Series **954** (2018) 012028 doi:10.1088/1742-6596/954/1/012028

In Table 1, it can be seen that the indicators of Work Quality, Quantity, Knowledge and Work Skills, Initiative, Cooperation, Reliability, Ability and Learning Will, Attendance, Planning and Organizing, Controlling, Decision Making, and Development of Subordinates, each have variables which can directly measure the intended employee. This measurement is intended for the selection of candidates for certain positions, becoming one of the inputs for the leadership in determining the personal to be selected.

The HR department has an important and responsible role in this assessment activity. The need for a lightweight interview with candidates with the basics: values, beliefs, feelings, and motives [20]. Next Steps distributed the research questionnaire. Managers are asked to rate their subordinates as well as subordinates from other sections, the results of the manager's appraisal are returned to BSDM. The next step is to make the category of choice from the classification of the quality of human resources into the criteria scale. Scale criteria can be seen in Table 2. The HR section has determined that for each position requires certain criteria, and in the position to be occupied by the candidate has been established yag incorporated in the criterion scale 5, in other words calculations in this study determine the candidate that matches criterion 5. The other side also sees opportunities by prioritizing the personality approach of the candidate [8]. This determination on the scale of this kirteria (table 2) has become a reference in the selection of candidates and established by company X. The next step is to calculate the naive bayes formula (1) and the calculation will be shown in Table 3.

Table 2. The Division of Criteria Based on the Assessment Aspect

Criteria of Scale	Description
Criteria 1	Have good learning ability, high morale and quality of work satisfactory
Criteria 2	The results obtained are very effective with near perfect level of accuracy
Criteria 3	Self-organization and subordinates are optimal and responsible for all the work done
Criteria 4	Having a high level of creativity and initiative in working with the task of applying the knowledge mastered
Criteria 5	Have the courage to take decisions and dare to account for all the risks that occur

4. Result and Discussion

The questionnaire that has been given to the managers and has been filled in compliance with the terms of compliance, that this assessment is confidential during the assessment process for the assessed employees. Calculations using the naive bayes formula based on the assessment criteria can be seen in Table 3 as follows:

Table 3. Naive Bayes of Human Resource Quality Classification

Criteria		Human Resource Quality Classification											
		a	b	c	d	e	f	g	h	i	j	k	l
1	1/5	9/58	8/58	7/58	1/58	1/58	1/58	10/58	11/58	7/58	1/58	1/58	1/58
2	1/5	11/59	10/59	4/59	1/59	1/59	9/59	1/59	9/59	4/59	1/59	7/59	1/59
3	1/5	7/75	1/75	4/75	8/75	11/75	1/75	1/75	10/75	11/75	1/75	9/75	11/75
4	1/5	8/69	1/69	11/69	11/69	4/69	1/69	9/69	11/69	1/69	4/69	4/69	4/69
5	1/5	7/48	1/48	1/48	10/48	4/48	4/48	7/48	9/48	1/48	4/48	11/48	1/48

Description: a = Quality of Work; b = Quantity of Work; c = Knowledge and Skills on Work; d = Initiative; e = Cooperation; f = Reliability; g = Have Learning Ability and Willingness; h = Attendance Level; i = Planning and Organizing; j = Control; k = Decision-Making; l = Development of Staff

Table 3 is a series of calculation processes with naive bayes formulas, by combining the classification of human resource quality and criteria scales. The results of calculations and comparisons with Table 1 can be seen in Table 4. In table 4, the criteria column is the result of calculation with naive bayes, whereas in the candidate column is the personnel judged by the managers. This assessment is subjective from managers, assuming it has worked enough time to allow it to be assessed. In addition, the assessment is based on good performance, skills and attitudes, as the initial basis for personnel selection.

Table 4. Classification of Human Resources Quality Based on Assessment Aspect with Naive Bayes Calculation

Candidate	Criteria	Classification of Human Resource Quality
K-1	Criteria 1	Quality (4), quantity (3), knowledge and skills (3), ability and willingness of learning (5), attendance (5), planning and organizing (3)
K-2	Criteria 4	Quality (4), knowledge and skills (5), initiative (5), ability and willingness of learning (3), attendance (5), decision making (3), development of subordinates (3)
K-3	Criteria 2	Quality (5), quantity (5), reliability (4), attendance (4), planning and organizing (3), decision making (3)
K-4	Criteria 3	Quality (3), initiative (3), cooperation (5), attendance (5), planning and organizing (5), decision making (4), development of staff (5)
K-5	Criteria 5	Quantity (3), initiative (5), cooperation (3), ability and willingness to learn (3), attendance (4), control (3), decision making (5)
K-6	Criteria 4	Quality (3), knowledge and skills (5), initiative (5), cooperation (3), ability and willingness of learning (5), attendance (5), controlling (3)
K-7	Criteria 1	Quality (4), quantity (4), knowledge and skills (3), ability and willingness of learning (4), attendance (5), planning and organizing (3)
K-8	Criteria 3	Quality (3), knowledge and skills (3), initiative (4), cooperation (5), attendance (4), planning and organizing (5), decision making (4), development of staff (5)
K-9	Criteria 5	Quantity (3), initiative (4), reliability (3), ability and willingness of learning (3), attendance (4), decision making (5)
K-10	Criteria 2	Quality (5), quantity (4), knowledge and skill (3), reliability (4), attendance (4), decision making (3)
K-11	Criteria 5	Quantity (3), initiative (3), cooperation (3), ability and willingness to learn (5), attendance (4), decision making (5)

In table 4 can be seen in the candidate column is the person who is assessed and will be screened. Column Criteria is the result of calculation with naive bayes so that the results obtained with certain criteria. The classification column of the HR criteria is the result of the manager's assessment of each individual by the manager. In the classification of the quality of human resources, the results of the assessment of the managers in the form of questionnaires. Questionnaires actually contained 12 indicators, but managers did not entirely select and fill out the questionnaire, and it was still considered valid as it had represented more than 6 indicators. In this study, from 12 indicators used as measurement materials, specific decision-making indicators become important and occupy the top priority because related positions that will be held. Other indicators become supportive and remain important to note.

At this stage it has been found that there are 3 candidates who can be used as reference material for placement in the position to be occupied (figure 1). Candidate 5, candidate 9 and candidate 11. The three candidates have fulfilled the calculation after the assessment by the relevant managers, and all three are eligible to be further promoted to the management as a step to select this activity. Among all candidates in criterion 5, only Candidate 5 (K-5) has more indicators than the other candidates, and becomes an added value for the next calculation. Thus candidate 5 (K-5) can be recommended in management to be established as an alternative decision-making materials for management. Thus the task of the HR department in selecting candidates for a particular position has been enforced. Alternatively, candidates with criteria 5 (K-9, K-11) may be good assistants to new officials or replace other parts of the Assistant with their expertise and competencies.

Persons who have followed this stage of measurement as in Figure 1, who occupy criteria 4 will then be placed on the division that has been prepared as an alternative to other solutions. At this stage criterion 4 will be placed as assistant to the division chairman. This task becomes important because as a counterweight of the elected head of office. Personnel with criterion 3 scale will be placed in the area around the factory machine. Handles and is responsible for the machines and schedules of his subordinates. Likewise with personnel mapped on criteria 2 and 1 that assist personnel with criteria scale 3.

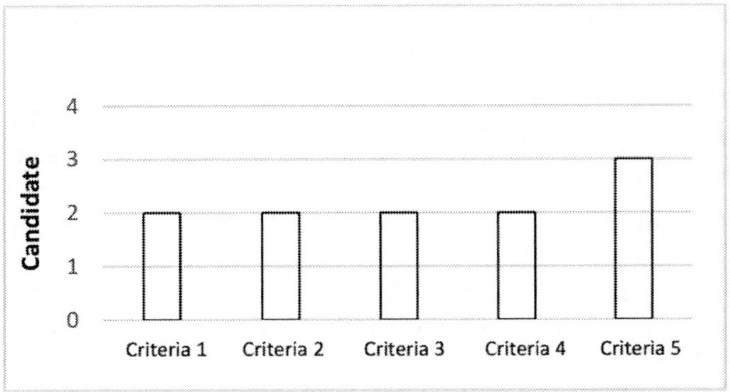

Figure 1. Classification of Human Resource Quality

5. Conclusion

The selection of candidates in company X for certain positions is done by assessing candidates to be submitted and judged eligible by the HR Section after going through several stages that have been determined. The initial stage is to look at the length of time working, performance, skills and good attitude, as the initial basis of selection. The next stage of assessment of managers through questionnaires distributed by the HR Section. With the distribution of questionnaires will be obtained assessment results for all candidates. Recapitulation of the assessment results are processed by naïve bayes formula and placed in accordance with the criteria set by the HR Section.

The provisions of the HR Section are only candidates who occupy criteria 5 that will be promoted to certain positions to be determined by management. Three candidates have been selected from criterion 5, and one candidate (K-5) has been selected as the result of calculation with this method, will be submitted to management to be selected and determined, or otherwise established by management. The selection of candidates as an alternative solution for management in determining the right candidates placed according to the field and competence and is part of decision making (PK). This research is quantitative with the assessments made by managers. In this research method candidates who belong to criterion 4 will be the assistant of the elected head of office. While candidates with criteria 3, 2, and 1 will be placed around the factory with the responsibility of holding machines and scheduling required.

There are 12 indicators to be assessed by the candidates, such as Work Quality, Quantity, Knowledge and Work Skills, Initiative, Cooperation, Reliability, Ability and Learning Will, Attendance, Planning and Organizing, Controlling, Decision Making and Development of Subordinates, -an indicator has a certain variable. After the following classification is obtained by calculation using the naive bayes formula, then the criteria for the candidate are assessed.

References

[1] Argue M 2015 The Importance of The Strategic Recruitment and Selection Process on Meeting an Organisations Objectives *Dissertation Master of Business Administration General Management Dublin Business School*
[2] Acevedo M and Krueger J 2004 Two Egocentric Sources of The Decision to Vote: The Voter's Illusion and the Belief in Personal Relevance *Political Psychology* 25(1) p 115-134
[3] Anosh M, Hamad N and Batool A 2014 Impact of Recruitment and Selection of HR Department Practices, *European Journal of Business and Management* 6(31)
[4] Chapman D S and Mayers D 2015 Recruitment Processes and Organizational Attraction, *Psychology Press, London And New York*
[5] De Bruin W B, Parker A M and Fischhoff B 2007 Individual Differences in Adult Decision-Making Competence *Journal of Personality and Social Psychology* 92(5) p 938-956.
[6] Dietrich C 2010 Decision Making: Factors That Influence Decision Making, Heuristic Used, and Decision Outcomes *Inquiriesjournal.Com* 2(2) p 1-3.
[7] Edenborough R 2005 Assessment Methods in Recruitment, Selection & Performance: A Manager's Guide to Psychometric Testing Interviews and Assessment Centres *Kogan Page Limited*
[8] García A L 2015 Work Analysis for Personnel Selection *Psychology Press London And New York*
[9] Gaudens T R 2013 Main Challenges Faced by Multinational Companies when Planning Humanresources in Venezuela Thesis Business Administration in Business Management Liverpool John Moores University
[10] Hasan I 2002 Pokok-Pokok Materi Pengambilan Keputusan *Ghalia Indonesia Jakarta*
[11] Jullisson E A, Karlsson N and Garling T 2005 Weighing the Past and the Future in Decision Making, *European Journal of Cognitive Psychology* 17(4) p 561-575
[12] Loudon K C and Loudon J P 2008 Sistem Informasi Manajemen: Mengelola Perusahaan Digital Edisi 10 Buku 2 *Penerbit Salemba Empat*
[13] Muchinsky M and Paul 1993 Psychology Applied to Work *California: Pacific Grove*
[14] O'meara B and Petzall S 2013 Handbook of Strategic Recruitment and Selection: A Systems Approach *Emerald Group Publishing*
[15] Sobhani F M and Madadi T 2015 Studying the Suitability of Different Data Mining Methods for Delay Analysis in Construction Projects *Journal of Applied Research on Industrial Engineering* 2(1) p 15-33
[16] Sugiyono 2008 Metode Penelitian Kuantitatif Kualitatif dan R&D *Penerbit Alfabeta Bandung*
[17] Syamsi I 2000 Pengambilan Keputusan Sistem Informasi *Jakarta: Bumi Aksara*
[18] Stanovich K E and West R F 2008 On the Relative Independence of Thinking Biases and Cognitive Ability *Journal of Personality and Social Psychology* 94(4) p 672-695
[19] Taylor T 2015 Managing People in Sport Organizations: A Stratgic Human Resource Management Perspektive *Routledge Business & Economics Second Edition*
[20] Taylor I 2007 A Practical Guide to Assessment Centres and Selection Methods: Measuring Competency For Recruitment And Development *Kogan Page*
[21] Turban E 2004 Decision Support Systems and Intelligent Systems *Prentice-Hall Inc 7th Edition*
[22] Williams C 2007 Research Methods *Journal of Business & Economic Research* 3 p 65-72.
[23] Yurtseven M K and Buchanan W W 2016 Decision Making and Systems Thinking: Educational Issues *American Journal of Engineering Education* 7(1) p 19-28.
[24] Zinyemba A Z 2014 The Challenges of Recruitment and Selection of Employees in Zimbabwean Companies *International Journal of Science and Research* 3(1) p 29-33
[25] Mitchell T M 1997 Machine Learning *McGraw-Hill New York*

Acknowledgments
We would like to thank the institution and especially Quality Technology Research Group for the support that has been given. Acknowledgments and appreciation to Company X which has allowed us to research and follow the process of measurement in the selection of candidates for certain positions, so that this research can run smoothly.

IOP Conf. Series: Journal of Physics: Conf. Series **954** (2018) 012029 doi:10.1088/1742-6596/954/1/012029

The Abstraction Ability in Constructing Relation Within Triangles by The Seventh Grade Students of Junior High School

Suwardi Annas[1*], Djadir[2], Sitti Mutmainna Hasma[2]

[1]Department of Statistics, Universitas Negeri Makassar, Indonesia
[2]Department of Mathematics, Universitas Negeri Makassar, Indonesia

*suwardi_annas@unm.ac.id

Abstract. Abstraction is an activity to organize a mathematical concept that has been previously owned into a new mathematical structure. Activites in abstraction are recognizing, organizing and constructing. Recognizing is a process of identifying a mathematical structure that had existed before. Organizing is a process of using structural knowledge to be assembled into a solution of a problem and constructing is a process of organizing the characteristics of the object into a new structure that does not exist. In abstraction process, the students use attributes to address the object, including routine attribute, nonroutine attributes, and meaningless attributes.This research applied descriptive qualitative research which aimed to describe the abstraction ability of students from high, moderate, and low groups to construct a relation within triangle. In collecting the data, this research used students' pre-ability math test, abstraction test, and guided interview. The sampling technique in this research was based on the students' scores in pre-ability math test, which were divided into three groups. Two students from each group were opted as the subjects of this research. Questions of the test are based on the indicators of steps in abstraction activity. Thus, based on the data gained in this research, researcher determined the tendency of attributes used in each abstraction activity.The result of this research revealed that students from high, moderate and low groups were prone to use routine attributes in recognizing triangles. In organizing the characteristics within triangles, high group tended to organize the triangle correctly, while the moderate and low groups tended to organize the triangle incorrectly. In constructing relation within triangles, students in high, moderate and low groups construct it incompletely.

1. Introduction

Mathematics is a science studying the abstract structure and its relationship pattern. Nurhasanah states that mathematics is a science focusing on abstract object. Mathematics is considered abstract because the objects and symbols do not exist in the real life [1]. Learning mathematics actually are to learn the concept, the structure of the concept, and to seek the relationship among concepts and their structures. Learning mathematics can learn using cognitive style [2][3], problem solving [4][5], problem posing. Learning mathematics can doing using online media [6][7]. The emphasis of good mathematics learning process is how to assist students to comprehend mathematical concepts well. It is in line with Cooney's statement (in [8]) who indicates that students' ability in learning mathematics is directly linked to their understanding about mathematical concepts and principles. Thus, all definitions, basic concepts, and even the most complicated concepts in mathematics are built on the abstract foundation.

It is different from students' condition at school, they are accustomed to think with evident objects. They are not accostumed to mathematics, which focusing on abstract objects.

There is gap between mathematics and students' thought at school. Marti in [9] describes that abstract mathematical object is a trouble that students face in learning mathematics. It is not only the students, teachers also get difficulties in teaching mathematics with its abstract disposition. Concepts in mathematics can be easily understood if they are in the concrete form. Therefore, teaching mathematics must be undertaken step by step. Learning mathematics should begin with concrete step, then direct into semi-concrete step. Eventually students can think and understand mathematics in its abstract features [10]. The process is termed as abstraction process.

Abstraction is a fundamental process in mathematics and mathematics education, the existance of abstraction in the learning process is a must, as it has an important role in forming mathematical concepts. Abstraction process needs much attention. Through the process, we can measure how far students understand a concept and how students construct the previous concept with the new one to find relationship.

Geometry occupies distinctive position in the mathematics learning at school with many concepts it has. Nurhasanah desribes, geometry is one of the mathematics' branches which discusses about objects like point, line, surface, dimension and their connections where all the objects are abstract [1]. Nurhasanah proposes that abstract object in geometry is frequently visualized and linked with real object empirically. Juwita adds, the visualization makes students get difficulties in learning geometry [8].

The concept of triangle is one of the materials in geometry in school mathematics in which many students get trouble in, especially in forming definition of triangle, sorting and drawing triangle based on its type. Yezita et al. also reveal that material in identifying the characteristics of triangle based on its side and angle is a difficult material for students to understand [11]. Thus, abstraction process is considerably needed in mathematics learning specifically the concept of triangle in geometry [12].

Based on the above elaboration, the problem statement in this research is "How is the abstraction ability of the high, moderate and low group of VII grade junior high school students in constructing the relation within triangles?". The objective of this research is to know the abstraction ability of the high, moderate, and low group of VII grade junior high school in constructing relation within triangles. The significance of the research is expected to give constibution to in investigating how the abstraction ability of the VII grade students junior high school in triangles material in mathematics learning and as a consideration for the upcoming similar research.

2. Content and Method

Soedjadi in his book "Kiat Pendidikan Matematika di Indonesia" concludes that an abstraction happens when we notice several objects and we "abolish" the unnecessary features and characteristics of the objects, and only the important characteristics taken [10]. Suharnan (in [8]) adds that abstraction is a process of separating the essential features or the same features of some objects and discarding the nonessential features of the objects to conclude at the end.

Cooney (in [13]) states that abstraction happens when someone realizes the similarities among the differences. Soedjadi explains the process of abstraction through the following scheme [10]:

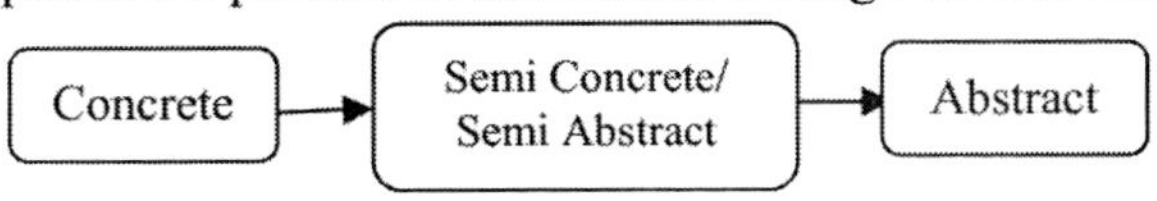

Figure 1.The Scheme of Abstraction process

In Geometry, the explanation of triangle concept can begun by using "cardboard triangle" then "wire triangle/leaf-rib triangle" can be used, then a more abstract image of triangle. Therefore, it is inferred that **abstraction** is the outcome and the process of organizing existed mathematical concept into new mathematical structure.

IOP Conf. Series: Journal of Physics: Conf. Series **954** (2018) 012029 doi:10.1088/1742-6596/954/1/012029

Budiarto explains more about the activities in the abstraction process as collecting, compiling, organizing, and developing the elements into novel elements so the activities in the abstraction process are recognizing, organizing, and constructing [14].

Recognizing in abstraction means identifying an existed mathematical structure in the similar or different activities. Recognition of a mathematical structure which has been learnt by the student before, happens when the student realizes that the structure he/she constructed of used before is appropriate with the mathematics situation given.

Organizing is an application uses structural knowledge which then organized to be the solution of the problem gained. Organizing is combining structural elements to achieve the appointed objective. Knowledge uses in organizing step is recognizing features of a structure and its definition.

Constructing is a necessary stage in abstraction. Constructing is organizing features of an object into new structures that its never owned. The activity of organizing characteristics of square changes into a network of relation within squares.

Budiarto adds that activity in the abstraction process which elaborated previously, is not linked chain [14]. Constructing does not only happen after recognizing and organizing in linear but recognizing and organizing can be undertaken simultaneously to do the structure of constructing.

Furthermore, Budiarto explains that in the abstraction process, students use attributes and characteristics of the objects [14]. Attributes used are divided into three, namely routine attribute, nonrutin attribute, and meaningless attribute.

1. Routine attribute is an attribute which commonly learnt at school in the initial process of compiling the definiton of a concept.
2. Nonroutine attribute is an attribute which not commonly learnt at school in the initial process of compiling definiton of concept.
3. Meaningless attribute is an attribute which cannot be used in the initial process of compiling definition of concept.

This research described students' abstraction as a process and outcome concerning how students recognize triangles, characteristics and definition of triangles, organize the characteristics of some triangles to construct the relationship of triangles. To know the abstraction process, this research used abstraction activities which cover recognizing, organizing the characteristics of triangles, and constructing the relationship of triangles. In the activities students possibly use routine attributes, nonroutine attributs and meaningless attributess.

Abstraction process in mathematics benefits as a process which spurs mathematics development so all things in the universe can be utilized prudently. Besides, abstraction also elaborates in depth the relationship among branches and parts in mathematics [15].

This research was a descriptive research using qualitative approach which aimed at describing or elaborating students' abstraction ability in constructing relation within triangles. This research was undertaken in SMP Negeri 2 Takalar with the research subjects were students of VII.1 class of 2015/2016 who had learnt the material of triangles. The procedure of collecting data was done by administering pre-ability test to the 30 subjects of the research individually. 8 students were categorized as high group, 8 students were in low group and 14 other students were in moderate group. Then, two subjects were opted from the three groups as the representatives. Students who became the representatives were given abstraction test and they were interviewed. The data analysis process began by analyzing the whole data provided from different sources, including the outcome of the written test, interview and field observation. The next step was the elaboration and the last stage was concluding the data of abstraction from the subject of the research.

3. Result and Discussion

3.1. Abstraction Ability of High Group

In recognizing activity of trianges, the first and second subjects were able to classify 6 (six) groups of triangle. Consideration used by the first and second subject was the same in classifying the triangle as both used routine attributes. Thus in the recognizing activity, subjects of high group were prone to use routine attributes.

In the organizing stage, students began by recognizing the charactertistics of each triangle and define it. In the recognizing activity, the first and second subjects used routine attributes for each group. In defining, the first and second subjects defined 6 (six) group of triangles completely and correctly. Thus, in defining triangle high group were prone to be precise. While in the organizing activity, the first and second subjects were able to organize 7 (seven) connections within triangles correctly. Therefore, in organizing the same characteristics within triangles, high group tended to organize well.

In the activity of constructing relation within triangles, the first and second subjects were prone to do it incompletely.

3.2. Abstraction Ability of Moderate Group

In recognizing triangles, the first and second subjects were able to categorize 6 (six) groups of triangle. Consideration used by the first and second subjects in categorizing triangles was the use of routine attributes. Thus, in recognizing activity, subjects of moderate group tended to use routine attributes.

In organizing the same characteristics of triangle, students began by recognizing the characteristics of each triangle and define it. In recognizing the characteristics of triangle, the first subjects used routine attributes for each triangle and the second subjects used routine attributes of each triangle group. In defining phase, the first and second subjects defined 6 (six) triangles completely and correctly. Thus, in defining triangle, moderate group tended to be accurate. While in organizing characteristics within triangle, the first subjects could only organize the same characteristics of 6 (six) triangles, while the second subjects could only organize the same characteristics of 5 (five) triangles. Thus, in organizing the same characteristics of triangle, moderate group tended to do organizing activity incorrectly.

In constructing relation within triangle, the first and second subjects tended to do it completely.

3.3. Abstraction Ability of The Low Group

In the recognizing activity, the first and second subjects were able to classify 6 (six) groups of triangle, but the activity was still imprecise. Consideration used by the first and second subjects in classifying triangles was routine attributes. Thus, in recognizing triangles, subjects of the low group tended to use routine attributes.

In organizing the same characteristics of triangle, students started by recognizing characteristics of each triangle and defined it. In the recognizing activity, the first and second subjects used routine and nonroutine attributes of each group of triangles. In defining, the first subjects defined 5 (five) groups of triangle correctly and the second subjects were only able to define 5 (five) triangles correctly. Thus, in defining triangles, low group tended to define it incompletely. While in organizing the characteristics of triangles, the first subjects only organized the same characteristics of 3 (three) triangles, the second subjects only organized the same characteristics of 1 (one) triangle. Thus, in organizing the same characteristics within triangles, low group tended to organize it incorrectly.

In constructing relation within triangles, the first and second subjects tended to do it incorrectly

The summary of abstraction activity among students' groups in constructing relation within triangles is shown in the table 1.

Based on the table 1, in recognizing activity, students from high group, moderate group and low group tended to use routine attributes. In organizing the same characteristics of triangles, students began by recognizing the characteristics of each triangle and defined it. High and moderate group were prone to recognize the characteristics of each triangle through routine attributes, while low group used routine and nonroutine attributes. In defining triangles, high and moderate group tended to define triangles precisely, while the low group tended to be inaccurate. Then, in organizing the characteristics of triangle, high group tended to organize it correctly while moderate and low groups tended to organize it incorrectly. In constructing relation within triangles, high, moderate and low groups tended to construct incompletely.

Table 1. Abstraction ability among groups

Subject Code	Recognizing Triangle	Sequencing the same characteristics of triangle			Constructing relation within triangle
		Recognizing the characteristics of triangle	Defining triangle	Sequencing the same characteristics	
High	Routine attributes	Routine attributes	Accurate	Correct	Incomplete
Moderate	Routine attributes	Routine attributes	Accurate	Incorrect	Incomplete
Low	Routine attributes	Routine and nonroutine attributes	incomplete	Incorrect	Incomplete

4. Conclusion

Based on the result of this research, it is concluded that in recognizing triangles process, students in high group tended to use routine attributes, in the process of organizing the characteristics of triangle, students of high group tended to organize correctly. In the process constructing relation within triangles, students in high group tended to construct incompletely. In the process of recognizing triangles, students in moderate group tended to use routine attributes, in the process of organizing the characteristics of triangle, students in moderate group tended to relate it incorrectly. In the process of constructing relation within triangle, students of i moderate group tended to construct incompletely. In the process of recognizing triangle, students in low group tended to use routine and nonroutine attributes, in organizing the characteristics of triangle, the low group tended to relate the triangle incorrectly. In constructing relation within triangle, students in low group tended to construct the triangle incompletely. For the future researchers who are interestedin studying the relevant studies, it is recommended to research on the idea of abstraction in reconstructing the definition of relation within triangles.

References

[1] Nurhasanah, F. 2010. Abstraksi Siswa SMP dalam Belajar Geometri Melalui Penerapan Model Van Hiele dan Geometers Sketchpad. Thesis has been published. Bandung: FKIP UPI Bandung.

[2] Mulbar, U., Rahman, A., & Ahmar, A. S. (2017). Analysis of the ability in mathematical problem-solving based on SOLO taxonomy and cognitive style. *World Transactions on Engineering and Technology Education, 15*(1), 68-73. doi:10.26858/wtetev15i1y2017p6873

[3] Rahman, A., & Ahmar, A. S. (2017). Relationship between learning styles and learning achievement in mathematics based on genders. *World Transactions on Engineering and Technology Education, 15*(1), 74-77. doi:10.26858/wtetev15i1y2017p7477

[4] Rahman, A., Ahmar, A.S, & Rusli. 2016. The influence of cooperative learning models on learning outcomes based on students' learning styles. *World Transactions on Engineering and Technology Education*, 14(3), 425-430.

[5] Rahman, A. & Ahmar, A.S. 2016. Exploration of Mathematics Problem Solving Process Based on The Thinking Level of Students in Junior High School. *International Journal of Environmental and Science Education, 11*(14), 7278-7285.

[6] Ahmar, A. S., & Rahman, A. (2017). Development of teaching material using an android. *Global Journal of Engineering Education, 19*(1), 72-76. doi:10.26858/gjeev19i1y2017p7376

[7] Arsyad, N., Rahman, A., & Ahmar, A. S. (2017). Developing a self-learning model based on open-ended questions to increase the students' creativity in calculus. *Global Journal of*

Engineering Education, *9*(2), 143-147.

[8] Juwita, E. R. 2012. Profil Abstraksi Siswa dalam Mengonstruksi Hubungan Antar Segitiga. Thesis has been published. Surabaya: Faculty of Tarbiyah, UIN Sunan Ampel Surabaya.

[9] Sundayana, R. 2015. Media dan Alat Peraga dalam Pembelajaran Matematika. Bandung: Alfabeta.

[10] Soedjadi, R.. 2000. Kiat Pendidikan Matematika di Indonesia.

[11] Yezita, E., 2012. Mengkonstruksi Pengetahuan Siswa pada Materi Segitiga dan Segiempat Menggunakan Bahan Ajar Interaktif Matematika Berbasis Konstruktivisme. *Jurnal Pendidikan Matematika*, *1*(1).

[12] Shulhany, A., Sukirwan, & Syamsuri. 2014. Abstraksi siswa SLTA pada Materi Dimensi Tiga dengan Bantuan Geogebra. Journal on Research and Mathematics Learning, 7(2), 31-42.

[13] Shadiq, F. 2014. Pembelajaran Matematika. Yogyakarta: Graha Ilmu.

[14] Teguh, B. M., 2005. Proses Abstraksi Siswa Smp Kelas 1 Dalam Mengkonstruksi Kubus Dari Rangkaian 6 Persegi. In *Seminar Nasional Penelitian, Pendidikan & Penerapan MIPA 2005*. Fakultas Matematika dan Ilmu Pengetahuan Alam UNY.

[15] Haryono, Didi. 2014. Filsafat Matematika. Bandung: Alfabeta.

Daily Suspended Sediment Discharge Prediction Using Multiple Linear Regression and Artificial Neural Network

Uca[1*], Ekhwan Toriman[2], Othman Jaafar[3], Rosmini Maru[1], Amal Arfan[1], Ansari Saleh Ahmar[4,5]

[1]Department of Geography, Faculty of Mathematics and Natural Sciences, Universitas Negeri Makassar, Indonesia
[2]Faculty Sciences Social and Humanities, Universiti Kebangsaan Malaysia, Malaysia
[3]Faculty of Engineering and Built Environment, Universiti Kebangsaan Malaysia, Malaysia
[4]Department of Statistics, Faculty of Mathematics and Natural Sciences, Universitas Negeri Makassar, Indonesia
[5]AHMAR Institute, Makassar, Indonesia

*ucasideng@unm.ac.id

Abstract. Prediction of suspended sediment discharge in a catchments area is very important because it can be used to evaluation the erosion hazard, management of its water resources, water quality, hydrology project management (dams, reservoirs, and irrigation) and to determine the extent of the damage that occurred in the catchments. Multiple Linear Regression analysis and artificial neural network can be used to predict the amount of daily suspended sediment discharge. Regression analysis using the least square method, whereas artificial neural networks using Radial Basis Function (RBF) and feedforward multilayer perceptron with three learning algorithms namely Levenberg-Marquardt (LM), Scaled Conjugate Descent (SCD) and Broyden-Fletcher-Goldfarb-Shanno Quasi-Newton (BFGS). The number neuron of hidden layer is three to sixteen, while in output layer only one neuron because only one output target. The mean absolute error (MAE), root mean square error (RMSE), coefficient of determination (R^2) and coefficient of efficiency (CE) of the multiple linear regression (MLRg) value Model 2 (6 input variable independent) has the lowest the value of MAE and RMSE (0.0000002 and 13.6039) and highest R^2 and CE (0.9971 and 0.9971). When compared between LM, SCG and RBF, the BFGS model structure 3-7-1 is the better and more accurate to prediction suspended sediment discharge in Jenderam catchment. The performance value in testing process, MAE and RMSE (13.5769 and 17.9011) is smallest, meanwhile R^2 and CE (0.9999 and 0.9998) is the highest if it compared with the another BFGS Quasi-Newton model (6-3-1, 9-10-1 and 12-12-1). Based on the performance statistics value, MLRg, LM, SCG, BFGS and RBF suitable and accurately for prediction by modeling the non-linear complex behavior of suspended sediment responses to rainfall, water depth and discharge. The comparison between artificial neural network (ANN) and MLRg, the MLRg Model 2 accurately for to prediction suspended sediment discharge (kg/day) in Jenderan catchment area.

Keywords: Suspended sediment discharge, multiple linear regression, artificial neural network, Jenderam catchment

1. Introduction

The every time the flowing of water river always of bring a variety of material. The quantity of material flowing by Water River it influenced by the season. Concentration of suspended sediment load high during the rainy season, while the dissolved load high of concentration in dry season, because the water in a river source from soil water that contain a high concentration of chemical elements [1]. The flowing of water river brings a variety of materials such as suspended sediment load, bed load and dissolved load [2][3]. This material sourced from by the erosion process which occurs in the subsurface and or surface of soil. Erosion that occurs on the surface soil caused by transformation rainfall water became to be surface runoff. Suspended sediment concentration from different time it varies depending on the season condition [4]. The situation is strongly influenced by climatic events, such as rainfall events [5]. Furthermore, the occurrence of rainfall is influenced by other climate factors such as temperature and air humidity [6]-[8]. Climate change in a region is influenced by the conversion of green areas into constructed areas, such as protected forests into production and settlement forest areas [8].

Predict the amount of daily suspended sediment discharge which occurs in the catchments area is very important as an indicator to assess the level of erosion hazard, management of the water resources, water quality, hydrology project management (dams, reservoirs, and irrigation) and to determine the level of damage in the catchment area [9].

The models predict the amount of sediment in a catchment area has a lot of developed such as regression methods, InfoWork Rs, stormwater & wastewater management model (XPSWMM), watershed modeling systems (WMS) and artificial neural network (ANN) [10]. Prediction of sediment suspension using ANN has been widely carry out in the catchments area with the exact result [11][12]. In a study conducted in the Jenderam catchment using multiple linear regression (MLRg) and artificial neural network for to predict of suspended sediment discharge. The multilayer perceptron feed forward training algorithm is used, with three different learning algorithms to predict the amount of suspended sediment discharge is: Levenberg-Marquardt (LM), Scaled Conjugate Gradient (SCD) and Broyden-Fletcher-Goldfarb-Shanno Quasi-Newton (BFGS). Although many researchers have previously been using ANN [11][12], but only use one to two training algorithms. Therefore in this research using three training algorithms . The purpose of this research is to compare the accuracy of the regression analysis and ANN models in the small catchments area .

2. Multiple Linear Regression

Multiple linear regression (MLRg) analysis is one of the most widely used of all statistics [11]. A regression model that involves more than one regressos variable is called a multiple linear regression model [14]. Multiple linear regression modelling has been widely used for modeling such as urban runoff pollutant load [15], wash load sediment concentrations [16], suspended-sediment discharge [17], prediction of swell potential of clayey soils [18]. The general form of regression model for k independent variables with have two or more regressor variables is given by [19]:

$$Y = \beta_o + \beta_1 X_1 + \beta_2 X_2 + ... + \beta_k X_x + E \qquad (1)$$

where $\beta_0, \beta_1, \beta_2, ..., \beta_k$ are the regression coefficients respectively, that need to be estimated, while E, Y and X_1, X_2, ..., X_k are the error, dependent and independent variable. The value of the independent variable (X) is always related with a value of the dependent variable (Y). The gooddnes performance of multiple linear regression model can be expressed by the value of the error (differences between observed and predicted values), coefficient correlation (r) and coefficient of determinations (R^2). In general determining the best estimate of the multiple regression equation using the least-squares method chooses as the best-fitting model the one that minimizes the sum of squares of the difference between the observed and predicted by the fitted model [14][19].

3. Artificial Neural Network

An artificial neural network is an information processing systems that has certain performance characteristics in common with biological neural network such as the human [20]. Artificial neural network has been developed as generalization of mathematical model of human cognition or neural biology, based on the assumptions that: Information processing occurs at many simple elements known as neurons, signals sent between neurons is through connection link, each connection link has an associated weight, which in a typical neural net, multiplies the signal transmitted, each neuron applies an activation function (usually non-linear) to its net input (sum of weighted input signals) to determine its output signal [21].

A artificial neural network is characterized by its pattern of connections between the neurons (called its architecture), the method of determining the weights on the connections (called its training, or learning, algorithm), and its activation function. The multilayer neural network architecture consists of input layer, hidden layer and output layer. Input layer consists of nodes or neurons that will receive the data. Hidden layer consists of nodes or neurons that receive input from the input layer. The output layer consists of neurons receiving data from the output of the hidden layer [22].

3.1. Multilayer perceptron feedforward

Multilayer perceptrons feedforward (MLP) is a systematic method for training multilayer neural networks. This method has a good mathematical basics, objectives and get the shape formula algorithm and the coefficient in the equation by minimizing the sum of squares error value through models developed by the net. The input layer, hidden layer and output layer with the interconnection weights value v and w between layers of neurons [23]. Backpropagation artificial neural network architecture is given in Figure 1.

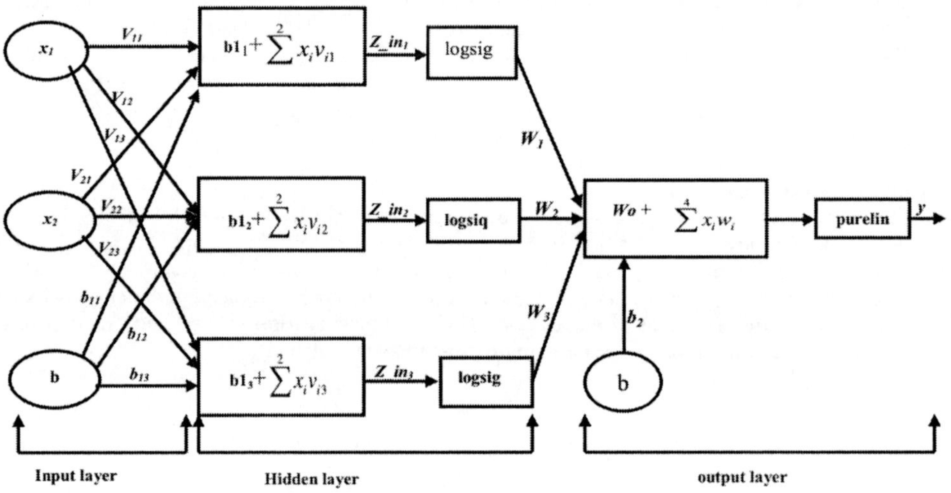

Figure 1. Schematic diagram of an MLP feed forward network

3.2. Selection of Training Algorithms

Some training in multilayer perceptrons feedforward algorithm in artificial neural networks that are often used: gradient descent, gradient descent with momentum, resilient backpropagation, powell-beale restarts, scaled conjugate gradient, BFGS quasi-newton, and levenberg-marquardt. Each training algorithms has different characteristics. In this research used training algorithms: gradient descent (GD), lavenberg-marquardt (LM), scaled conjugate descent (SCD) because not many researchers who compare the three algorithms for the training of predicting the amount of daily suspended sediment discharge in a small basin in the tropical area. Results of the analysis of the training algorithm will be used for compare the performance of each MLP training algorithm, RBF and MLRg.

3.2.1. Lavenberg-Marquardt. In mathematical levenberg-marquardt (LM) algorithm, also called as damped least-squares method (DLS), as part of troubleshooting to minimize numerical problems are generally non-linear functions. Algoritmh levenberg-marquardt (LM) is they gauss newton interpolation algorithm (GNA) and gradient descent (GD) methods. LM method more robust than the GNA found means in many cases can solve this problem if started very far from the minimum end. LM can also describe, as a Gauss Newton using the trust region approach. LM is very popularly used in curve fitting algorithm to solve the problem of generic curve fitting. But only found in local minimum is not at the global minimum. Levenberg-marquardt algorithm is designed by using the second derivative approach without having to compute the hessian matrix. LM training algorithm to update the weights using the following equation:

$$w_{t+1} = w_t - \Delta w_t \tag{2}$$

$$\Delta w_t = \left[J^T(w_t)J(w_t) + \mu I\right]^{-1} J^T(w_t)e(w_t) \tag{3}$$

where:

w_{t+1} = the update weight vector.
w_t = the weight vector before updating.
I = represents an identity matrix
J = the jacobian matrix that contains first derivatives from network error toward the weights value.
μ = the learning rate
$e(w_t)$ = the matrix error function evaluated at previous iteration.

When learning rate value 0, then this approach will be the same as Newton's method, while learning rate is large then this same approach with gradient descent. Newton's method is very fast and accurate for minimum error as the algorithm is expected to quickly change the learning rate to be equal to 0. For that, after a few iterations, the algorithm will reduce the learning rate. Increase the rate of learning will be done when it takes a step (temporarily) to reduce function performance.

3.2.2. BFGS Quasi Newton. BFGS Quasi-Newton algorithm is one of the conjugate gradient alternatives that can be used to get optimal value faster. Newton's method can work fast to get the optimum value completion, this method requires a large computer memory for each iteration calculating the second derivative. To overcome these obstacles are made improvements with quasi-newton method. In quasi-newton methods the hessian matrix does not need to be computed. The hessian is updated by analyzing successive gradient vectors instead. BFGS Quasi-Newton methods are a generalization of the secant method to find the root of the first derivative of multidimensional problems. In multi-dimensional the secant equation is under-determined, and quasi-newton methods differ in how they constrain the solution, typically by adding a simple low-rank update to the current estimate of the hessian. This method was developed by Broyden, Fletcher, Goldfarb, and Shanno so-called BFGS Quasi-Newton. BFGS. Quasi-Newton is the have the basic concepts:

$$w_{k+1} = w_k - A_k^{-1} \times g w_k \tag{4}$$

where:

w_{k+1} = is the update weight vector

w_k = is the weight vector before updating

A_k = is the second derivative hessian matrix

g = gradien

3.2.3. Scaled Conjugate Gradient. Scale conjugate gradient (SCG) is a supervised learning algorithm for feedforward neural networks, and is a number of the class of conjugate gradient methods. They are general purpose second order techniques that help minimize goal functions of several variables. Second order means that these methods make use of the second derivatives of the goal function, while first-order techniques like standard bacpropagation only use the first derivatives. A second order technique generally finds a better way to a local minimum than a first order technique, but at a higher computational cost. The development in the family of SC training algorithm for neural networks named with a SCG training algorithm [3]. All using the conjugate gradient algorithm will do the on lines search process continuously during the iteration process. For large amounts of data, then the iteration will also be large, so it takes a long time. Because the SCG will make improvements matter. In SCG training algorithm the weights are updated using the following set of equation [24]:

$$w_{m,t} = w_t + \gamma_t d_t \quad (5)$$

$$w_{t+1} = w_t + \alpha_t + d_t \quad (6)$$

$$s_t = E''(w_t)d_t \approx \frac{E'(w_{w,t}) - E'(w_t)}{\gamma_t} \quad (7)$$

$$\alpha_t = \frac{-d_t^T E'(w_t)}{d_t^T s_t} \quad (8)$$

where,

$w_{m,t}$ = is a temporal weight vector which lies between w_{t+1} and w_i.

d_t = is the conjugate direction vector of the temporal weight at t[th] iteration

γ_t = is the temporal weight updating step size called the short step size such that $0 < \gamma_t < 1$

w_{t+1} = is the next weight update vector.

w_t = is the vector of current weight.

α_t = is the actual weight updating step size called the long-step size

s_t = is the second order information

$E' \, dan \, E''$ = are the first and second derivative of error information with respect to respective weight vectors.

d_t^T = is the transpose d_t.

In SCG algorithm in any iteration, the temporal weights $w_{m,t}$, is calculated fist using the short- step size γ_t (equation 7). The temporal weight is then used to determine the long step-size size α_t (equations 6 and 8). The final weight update is computed using equation 8.

3.3. Radial Basis Function

Radial basis functions (RBF) network is training using a supervised training algorithm which can application for classification problems, function approximation, noisy interpolation and regularization [25][26]. The RBF applications in the field of hydrology has been widely used to non-linear rainfall-runoff model, suspended sediment load, stream flow forecasting to short-term and long-term and prediction of pore-water pressure [27]-[29]. Radial basis function the generally consists of three layers, namely input layer, hidden layer (radial basis layer) and output layer. The input layer consist nodes, which contain the input variable in the form dimensionality of the input vector, while the hidden layer consists computation unit, where each unit is mathematically described by radial basis function. The hidden to output layer part operates like a standard feed-forward multilayer perceptron network, with the sum of the weighted hidden unit activations giving the output unit activations. The link between nodes input and nodes hidden layer are direct with no weight value, while connected between hidden layer and output layer using the weight value. The output layer consist nodes is connected with the previous nodes in the hidden layer by linear weight, no restriction on the size on the output layer, but that typically the size of the output layer has been always smaller than that of the hidden layer [3][20].

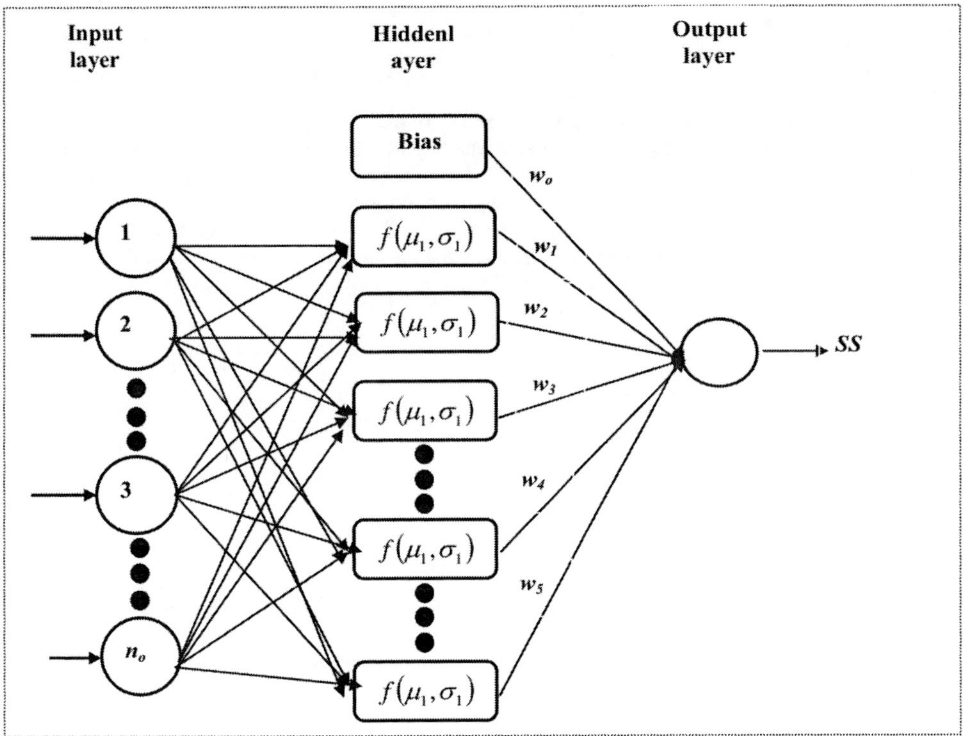

Figure 2. The structure of radial basis function neural network

In this study use the popular Guassian basis function as the transfer function at the hidden layer, most commonly used in supended sediment load [28], rainfall-runoff model [27] and streamflow forescasting [26].

$$\phi_j(x) = \exp\left(-\frac{1}{2\sigma_j^2} \| x_i - \mu_j \|^2\right) \tag{9}$$

where:
x_i : n_o = dimensional input vector, i = 1,2 ...P

μ_j = mean (center), j = 1,2 ...n

σ_j = standard deviation (spread), j = 1,2 ...n

ϕ_j = basis function value

n = number of hidden nodes.

The linear mapping between the hidden layer and output layer is given by:

$$SSQ = \sum_{k=1}^{n} \phi_j w_k + w_o, \quad i = 1,2..., P \tag{10}$$

where:
SSQ = output values, in this study suspended sediment discharge corresponding to X_i input vector.
w_k = connection weight;
w_o = bias term;

4. Selection of Activation Function/Transfer Function

Multilayer networks often use the log-sigmoid transfer function (logsig) in the hidden layer, while in output layer use linear transfer function (purelin). The function logsig generates outputs between 0 and 1 as the neuron's net input goes from negative to positive infinity [22]. Logistic sigmoid transfer function are more suitable for use on non-linear data trained with feedforward and more often used as the function between 0 and +1 and simple for derivatives. While activation function in the output layer transfer function using purelin, as desired a form of network outputs any number of real values, not on values between -1 and 1 or 0 and 1. The logistic sigmoid transfer function in hidden layer and linear transfer function in output layers are represented as [3][20][30].

$$f(x) = \frac{e^{x} - e^{-x}}{e^{x} - e^{-x}}, \text{ with derivative } f'(x) = [1 + f(x)][1 - f(x)] \tag{11}$$

$$f(x) = x, \text{ with derivative } f'(x) = 1$$

where, x is the input data.

5. Material and methods

5.1. Data source and study area

In this study water depth data is measured starting in January 1, 2011 until December 31, 2013. The changes of water depth are monitoring every 10 minutes using Omega-CPM instrument. The sampling of suspended sediment and flow discharge is taking during the dry period (no rain) and rainfall period. The number of suspended sediment sampling and flow of discharge is 1095 (2011-2013). The rainfall data is monitored using tree tipping-bucket rain gauge and every event drop rainfall data will the stored in Hobo Event Data Logger.

Jenderam catchment are is located in Selangor Malaysia . The catchments lies between at the 101^0 40 ' 00" to 101^0 50 ' 00" E longitude and 02^0 45'00 " to 02^0 50'00 " S latitude. Jenderam catchment is 20.51 km^2 in size. The texture characteristics of soils its a fine sandy loam, weak to moderate grade, the soils susceptible to erosion when rains occur mainly on slopes that disorders by human, such as land clearing and conversion of forest into residential and agricultural areas. Jenderam catchment has an elongated shape, with the elongation ratio of 0.44. The peak of hydrograph runoff is characterized by up and down a quickly. The number of stream order and drainage frequency is a 183 and 8.93 respectively. The study area consists of the Serdang - Kedah soil series , classified as: typic paleudult, fine loamy, kaolinitic, isohyperthermic (USDA Soil Taxonomy).

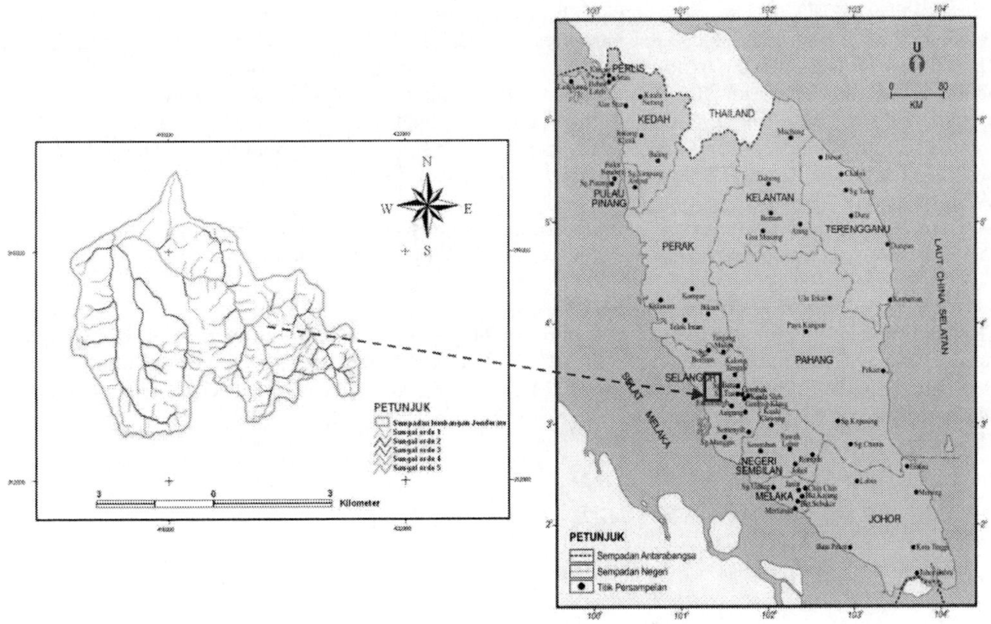

Figure 3. Location map of study area (catchment Jenderam) in Selangor Malaysia

5.2. Normalization of data

The normalization the raw data it is necessary to make neural network more efficient every step process on the network inputs and target [22]. Normalize data to ensure fast convergence and minimalization of global error during network training [31]. Most studies on the suspended sediment discharge prediction using ANN, have a used data normalization (scaling) relationship of the form [32]:

$$X_n = \left(\frac{X_i - X_{\min}}{X_{\max} - X_{\min}} \right) \tag{12}$$

X_n: is the normalized value
X_i: is the original value
X_{max}, X_{min} : are the maximum and minimum original value data respectively.

Generally, the normalization step is applied to both the input vectors and the target vectors in the data set. The input and target data were normalized before is using for training and testing by transforming the data to the range of 0 to 1. The network output can then be reverse transformed back into the units of the original target data when the network is put to use in the field. Usually the input data should be normalized to commensurate with the limit of the activation function used in the network [33].

5.3. Input data selection, input layer combination and statistical analyses

Input data selection and input layer combination play a vital role in the development of an appropriate neural network [3][32]. The modeling using artificial neural network, when the using multilayer networks the generally data is divided into three subsets namely training, validation and testing data. The first subset is the for the training set, which is used for computing the gradient and updating the networks weights and biases. The second subset is the validation set. The error on the validation set is monitored during the training process. The test set error is not used during training, but it is used to compare different models [22]. For this study only use two subset data, about 66.7% (January 1, 2011 to December 31, 2012) have been selected for training and 33.3 % (January 1, to Dismember 31, 2013) for testing the model.

The statistical parameters for the training and testing subsets, include maximum (X_{max}), minimum (X_{min}), mean (X_{mean}), standard deviation (SD), skewness (Skn) and coefficient of variation (CV).

5.4. Selection of model performance evaluation criteria

The performance of multiple linear regression, multilayer perceptron feedforward and radial basis function model evaluation criteria using the standard statistical measures, namely root mean square error (RMSE) [34][35], mean absolute error (MAE) [34][35], coefficient of determination (R^2) and coefficient of efficiency (CE) defined as:
Mean Absolute Error

$$MAE = \frac{1}{n}\sum_{i=1}^{n}\left(SSQ_{pi} - SSQ_{oi}\right) \tag{13}$$

Root Mean Square Error

$$RMSE = \sqrt{\frac{\sum_{i=1}^{n}\left(SSQ_{pi} - SSQ_{oi}\right)^2}{n}} \tag{14}$$

Coefficient of Determination

$$R^2 = \left[\frac{\sum_{i=1}^{n}\left(SSQ_{oi} - \bar{SSQ}_o\right)\left(SSQ_{pi} - \bar{SSQ}_p\right)}{\sqrt{\left(\sum_{i=1}^{n} SSQ_{oi} - \bar{SSQ}_o\right)^2}\sqrt{\left(\sum_{i=1}^{n} SSQ_{pi} - \bar{SSQ}_p\right)^2}}\right]^2 \tag{15}$$

Coefficient of Efficiency

$$CE = 1 - \frac{\sum_{i=1}^{n}\left(SSQ_{oi} - SSQ_{pi}\right)^2}{\sum_{i=1}^{n}\left(SSQ_{oi} - \bar{SSQ}_p\right)^2} \tag{16}$$

where SSQ_{oi} and SSQ_{pi} are the observed and predicted values respectively, $\bar{SSQ}_o$ and $\bar{SSQ}_p$ are the mean observed and predicted value respectively, and n is the number of observations. Ideally , the value of MSE, RMSE and MAE should be zero, while R^2 and CE should be one [3]. The error value can be zero and the value of the coefficient of determination can be one, if the observed and predicted value is the same.

Table 1. Model data input for MLRg, MLP and RBF

	Data Input	Total data	Model	Input layer variables	Target
Training	1. Rainfall (mm/day), 2. Water depth (m/day) 3. Discharge (m^3/day) 4. Suspended sediment discharge (kg/day)	January 1, 2011 – December 31, 2012	MLP RBF	1. P_t, H_t, Q_t 2. P_t, H_t, Q_t, P_{t-1}, H_{t-1}, Q_{t-1} 3. P_t, H_t, Q_t, P_{t-1}, H_{t-1}, Q_{t-1}, P_{t-2}, H_{t-2}, Q_{t-2} 4. P_t, H_t, Q_t, P_{t-1}, H_{t-1}, Q_{t-1}, P_{t-2}, H_{t-2}, Q_{t-2}, P_{t-3}, H_{t-3}, Q_{t-3}	SSQ
Testing	1. Rainfall (mm/day) 2. Water depth (m/day) 3. Discharge (m^3/day) 4. Suspended sediment discharge (kg/day)	Januari 1, 2013 – Desember 31, 2013	MLRg MLP RBF	1. P_t, H^t, Q_t 2. P_t, H_t, Q_t, P_{t-1}, H_{t-1}, Q_{t-1} 3. P_t, H_t, Q_t, P_{t-1}, H_{t-1}, Q_{t-1}, P_{t-2}, H_{t-2}, Q_{t-2} 4. P_t, H_t, Q_t, P_{t-1}, H_{t-1}, Q_{t-1}, P_{t-2}, H_{t-2}, Q_{t-2}, P_{t-3}, H_{t-3}, Q_{t-3}	SSQ

Where:
Qt , Pt, Ht : discharge, rainfall and average of the water depth at the current time respectively.
Qt-1, Qt-2 , Qt-3: discharge at one, two and three previous day.
Pt-1, Pt-2, Pt-3: rainfall at one, two and three previous day.
Ht-1, Ht-2, Ht-3:water depth at one, two and three previous day.

Table 2. Summary of statistics analysis on rainfall, discharge and suspended sediment discharge data.

Data set	Data Type	Min	Max	Mean	SD	Skewness	CV
Training	Rainfall (mm/day)	0.0000	93.9000	7.4321	14.5637	2.8528	1.9596
	Water depth (m/day)	0.0046	0.5840	0.1350	0.0600	2.6549	0.4446
	Discharge (m^3/day)	1,707.7532	1,2189,923.526	30,902.2424	70,005.8353	13.8342	0.4445
	Suspended sediment discharge (kg/day)	4.3762	3,550.3815	190.9396	276.8457	6.4555	1.4499
Testing	Rainfall (mm/day)	0.0000	141.1000	5.1956	12.7295	5.4183	0.5629
	Water depth (m/day)	0.0299	0.6530	0.0995	0.0560	3.2889	2.2581
	Discharge (m^3/day)	1703.7532	1,218,992.353	30,952.9815	69,990.7580	13.8412	2.2612
	Suspended sediment discharge (kg/day)	0.4723	4465.3437	87.8384	253.3504	14.5819	2.8843

6. Result and discussion

6.1. Multiple Linear Regressions

The analysis of the rainfall, water depth, discharge and daily suspended sediment discharge height the variability of the value the Jenderam catchment. The maximum, minimum, mean, standard deviation and coefficient of variation is show in Table 2. The general purpose of the multiple linear regressions (MLRg) is to learn more about the relationship between several independent or predictor variables and a dependent or criterion variable [18]. Multiple linear regression analysis was carried out to correlate the measured suspended sediment discharge to the variables independed, namely rainfall, water depth and discharge. The formula multiple linear regression models to predict the suspended sediment discharge (SSQ) are given below (Table 3). The performance analysis multiple linear regression using four models data independent with given in Table 4.

Table 3. Formula MLRg for prediction SSQ

Variable Independent	Variable dependent	Formula
P_t, H_t, Q_t	SSQ	$SSQ = 27.242 - 0.067P_t - 635.862H_t + 343.303Q_t$
P_t, H_t, Q_t, P_{t-1}, H_{t-1}, Q_{t-1}	SSQ	$SSQ = 33.074 - 0.033P_t - 550.436H_t + 343.394Q_t - 0.108P_{t-1} - 164.980H_{t-1} + 10.869Q_{t-1}$
P_t, H_t, Q_t, P_{t-1}, H_{t-1}, Q_{t-1}, P_{t-2}, H_{t-2}, Q_{t-2},	SSQ	$SSQ = 35.336 - 0.004P_t - 534.859H_t + 342.920Q_t - 0.092P_{t-1} - 133.771H_{t-1} + 9.279Q_{t-1} - 0.028P_{t-2} - 81.602H_{t-2} + 5.236Q_{t-2}$,
P_t, H_t, Q_t, P_{t-1}, H_{t-1}, Q_{t-1}, P_{t-2}, H_{t-2}, Q_{t-2}, P_{t-3}, H_{t-3}, Q_{t-3}	SSQ	$SSQ = 36.094 + 0.074P_t - 532.906H_t + 342.674Q_t - 0.056P_{t-1} - 119.269H_{t-1} + 9.043Q_{t-1} - 0.009P_{t-2} - 50.851H_{t-2} + 3.931Q_{t-2} - 0.121P_{t-3} - 51.285H_{t-3} + 0.676Q_{t-3}$

The performance of the statistics value for to four MLRg models with the input variable independent 3 (P_t, H_t, Q_t), 6 (P_t, H_t, Q_t, P_{t-1}, H_{t-1}, Q_{t-1}), 9 (P_t, H_t, Q_t, P_{t-1}, H_{t-1}, Q_{t-1}, P_{t-2}, H_{t-2}, Q_{t-2},) and 12 (P_t, H_t, Q_t, P_{t-1}, H_{t-1}, Q_{t-1}, P_{t-2}, H_{t-2}, Q_{t-2}, P_{t-3}, H_{t-3}, Q_{t-3}) were established (Table 4). The data set training and testing used for ANN, while for MLRg model only using testing data to develop the model. The result predicted suspended sediment discharge by MLRg compared with the predicted by ANN models.

Table 4. The Performance of statistical value Multilinear Regression

No	Input Variable	MAE	RMSE	R^2	CE
Model 1	P_t, H_t, Q_t	0.0000002	14.4113	0.9970	0.9967
Model 2	P_t, H_t, Q_t, P_{t-1}, H_{t-1}, Q_{t-1}	0.0000002	13.6039	0.9971	0.9971
Model 3	P_t, H_t, Q_t, P_{t-1}, H_{t-1}, Q_{t-1}, P_{t-2}, H_{t-2}, Q_{t-2},	-3.8356	72.8125	0.9774	0.9172
Model 4	P_t, H_t, Q_t, P_{t-1}, H_{t-1}, Q_{t-1}, P_{t-2}, H_{t-2}, Q_{t-2}, P_{t-3}, H_{t-3}, Q_{t-3}	-4.0274	76.4106	0.9747	0.9088

The MAE, RMSE, R^2 and CE of the MLRg model are given in the Table 4. The MAE, RMSE, R2 and CE of the MLRg value Model 2 (6 input variable independent) has the lowest the value of MAE and RMSE (0.0000002 and 13.6039) and highest R^2 and CE (0.9970 and 0.9971). The comparison between observed and predicted suspended sediment discharge by MLRg are plotted in Figure a1 – d2. The perfect line using MLRg model 2 with a 6 variable independent produced R^2 value 0.9971, with highest than the model 1 (R^2 0.9970), while model 3 (R^2 0.9774), and the model 4, R^2 and CE (0.9747 and 0.9088) only a slightly different. Even between MLRg model 1 and model 2 (3 and 6 variable independent) there was no significance difference in prediction, because the value MAE, RMSE, R^2 and CE difference only slightly (Table 4). Increasing the number of independent variables in the model MLRg, suspended sediment discharge prediction has a trend less accurate. Prediction obtained the overall MLRg model (four model) were completely free of negative prediction. The MAE of MLRg models 3 and 4 have a negative value, this means the total value of the prediction is lower than the total value of observation. While the MAE of MLRg models, has a positive value, this means the total value of prediction is higher than the total value of the observation. Based on result show that the MLRg model can be success applied to prediction daily suspended sediment discharge in Jenderam catchments with has size 20.51 km^2. This mean that suspended sediment concentration positively correlated with the daily rainfall, water depth and discharge the current time, likewise the one, two and three previous day.

IOP Conf. Series: Journal of Physics: Conf. Series **954** (2018) 012030 doi:10.1088/1742-6596/954/1/012030

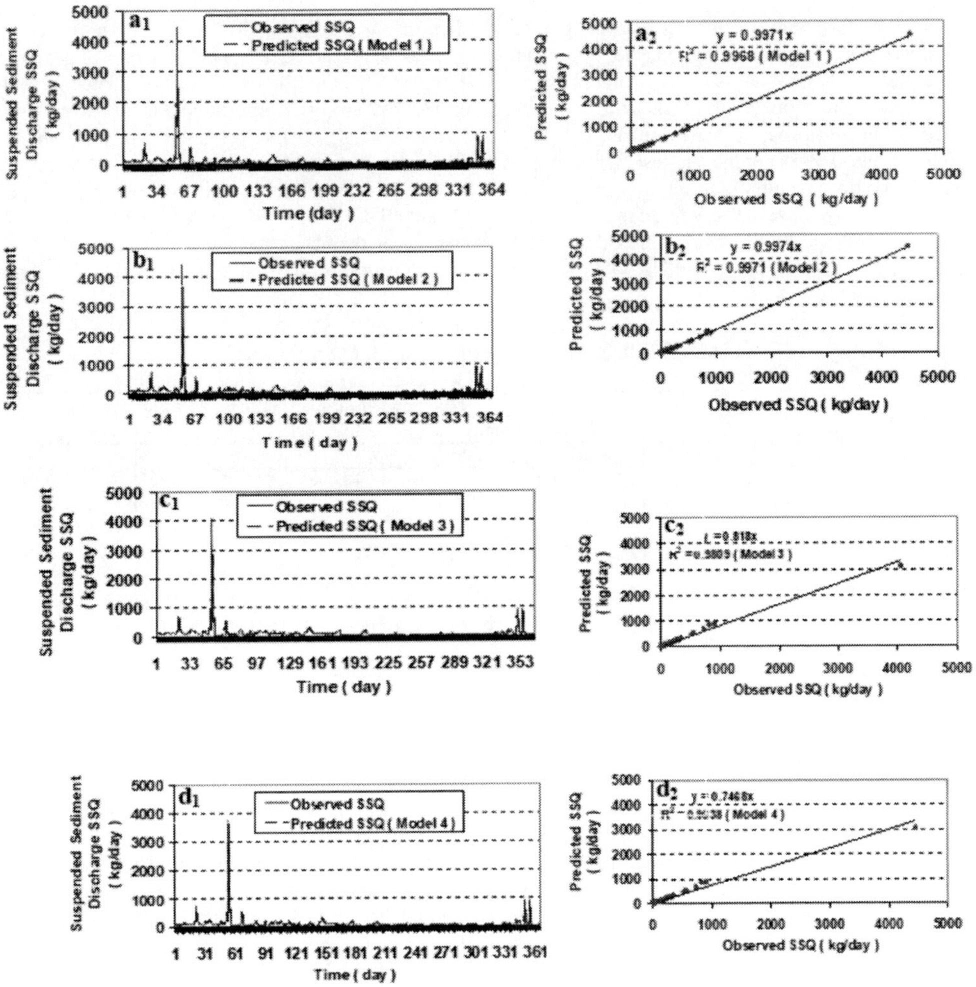

Figure 4 Comparisons between observed and predicted SSQ using MLRg

6.2. Multilayer Perceptron Feedforward

The performance of the multilayer perceptron (MLP) feed forward using the algorithms LM, BFGS-Quasi Newton and SCG using the daily data is show in Table 5,6 and 7. Based on the selected input data structure and two year of record for training and one year for testing the comparison between observed and predicted suspended sediment discharge by LM, BFGS and SCG algorithms are plotted in Figure a_1-s_2. The number nodes (neuron) in input layer are 3 (P_t, H_t, Q_t), 6 (P_t, H_t, Q_t, P_{t-1}, H_{t-1}, Q_{t-1}), 9 (P_t, H_t, Q_t, P_{t-1}, H_{t-1}, Q_{t-1}, P_{t-2}, H_{t-2}, Q_{t-2}) and 12 (P_t, H_t, Q_t, P_{t-1}, H_{t-1}, Q_{t-1}, P_{t-2}, H_{t-2}, Q_{t-2}, P_{t-3}, H_{t-3}, Q_{t-3}) respectively. Meanwhile the number nodes in hidden layer are determined by trial and error method. The number of output neurons only one because only one target output (suspended sediment discharge) from the ANN model.

The performance of the LM algorithm to prediction suspended sediment discharge based on number neuron in input, hidden and output layer is show in Table 5. The performance of statistics value (MAE, RMSE, R^2 and CE) at the processes training and testing, the best result on structure 6-10-1 (6 neurons in input layer, 10 neuron in hidden layer and 1 neuron in output layer). The performance value in testing process, MAE and RMSE (13.9107 and 18.9514) is smallest, meanwhile R^2 and CE (0.9953 and 0.9954) is the highest if it compared with the another LM model (3-7-1, 9-10-1 and 12-12-1). Testing result obtained by LM model 6-10-1 followed the observed suspended sediment discharge very closely the whole span of the testing data. The MAE LM 3-7-1 and LM 6-10-1Is a negative, this means the total value of the prediction is lower than the total value of observed. While the MAE of LM 9-10-1 and 12-12-1 is a positive, this means the total value of prediction is higher than the observed.

Table 5. The performance of statistical algorithm LM

No	ANN Model Inputs	Lavenberg-Marquard (LM)								
		Structure	Training				Testing			
			MAE	RMSE	R^2	CE	MAE	RMSE	R^2	CE
1	P_t, H_t, Q_t	3:7:1	0.5659	3.3976	0.9999	0.9998	15.9973	21.4476	0.9958	0.9928
2	P_t, H_t, Q_t, P_{t-1}, H_{t-1}, Q_{t-1}	6:10:1	0.5503	3.3953	0.9999	0.9998	13.9107	18.9514	0.9953	0.9944
3	P_t, H_t, Q_t, P_{t-1}, H_{t-1}, Q_{t-1}, P_{t-2}, H_{t-2}, Q_{t-2},	9:10:1	0.8389	3.5256	0.9999	0.9998	18.6625	25.9216	0.9907	0.9895
4	P_t, H_t, Q_t, P_{t-1}, H_{t-1}, Q_{t-1}, P_{t-2}, H_{t-2}, Q_{t-2}, P_{t-3}, H_{t-3}, Q_{t-3}	12:12:1	0.7056	3.3736	0.9999	0.9999	24.7100	25.6389	0.9915	0.9897

The results demonstrate the accuracy and reliability of the proposed LM model to prediction suspended sediment discharge in Jenderam catchments.

The performance of the BFGS Quasi-Newton algorithm to prediction suspended sediment discharge based on number neuron in input, hidden and output layer is show in Table 6. The performance of statistics value (MAE, RMSE, R^2 and CE) at the processes training and testing, the best result on structure 3-7-1 (3 neurons in input layer, 7 neuron in hidden layer and 1 neuron in output layer). The performance value in testing process, MAE and RMSE (13.5769 and 17.9011) is smallest, meanwhile R^2 and CE (0.9999 and 0.9998) is the highest if it compared with the another BFGS Quasi-Newton model (6-3-1, 9-10-1 and 12-12-1). Testing result obtained by BFGS Quasi-Newton model 3-7-1 followed the observed suspended sediment discharge very closely the whole span of the testing data.

IOP Conf. Series: Journal of Physics: Conf. Series **954** (2018) 012030 doi:10.1088/1742-6596/954/1/012030

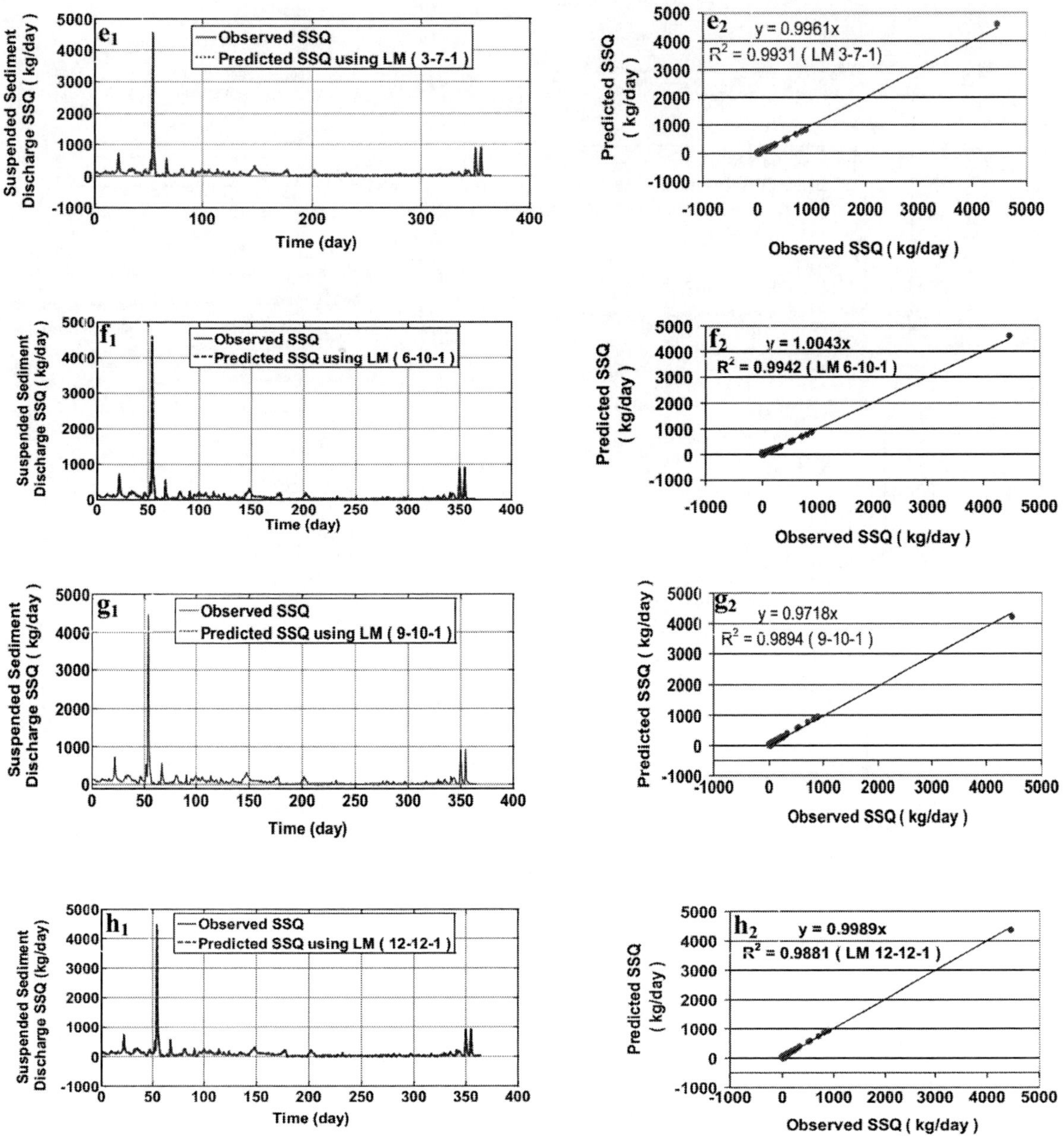

Figure 5 Comparisons between observed and predicted SSQ using LM algorithm based on the testing data.

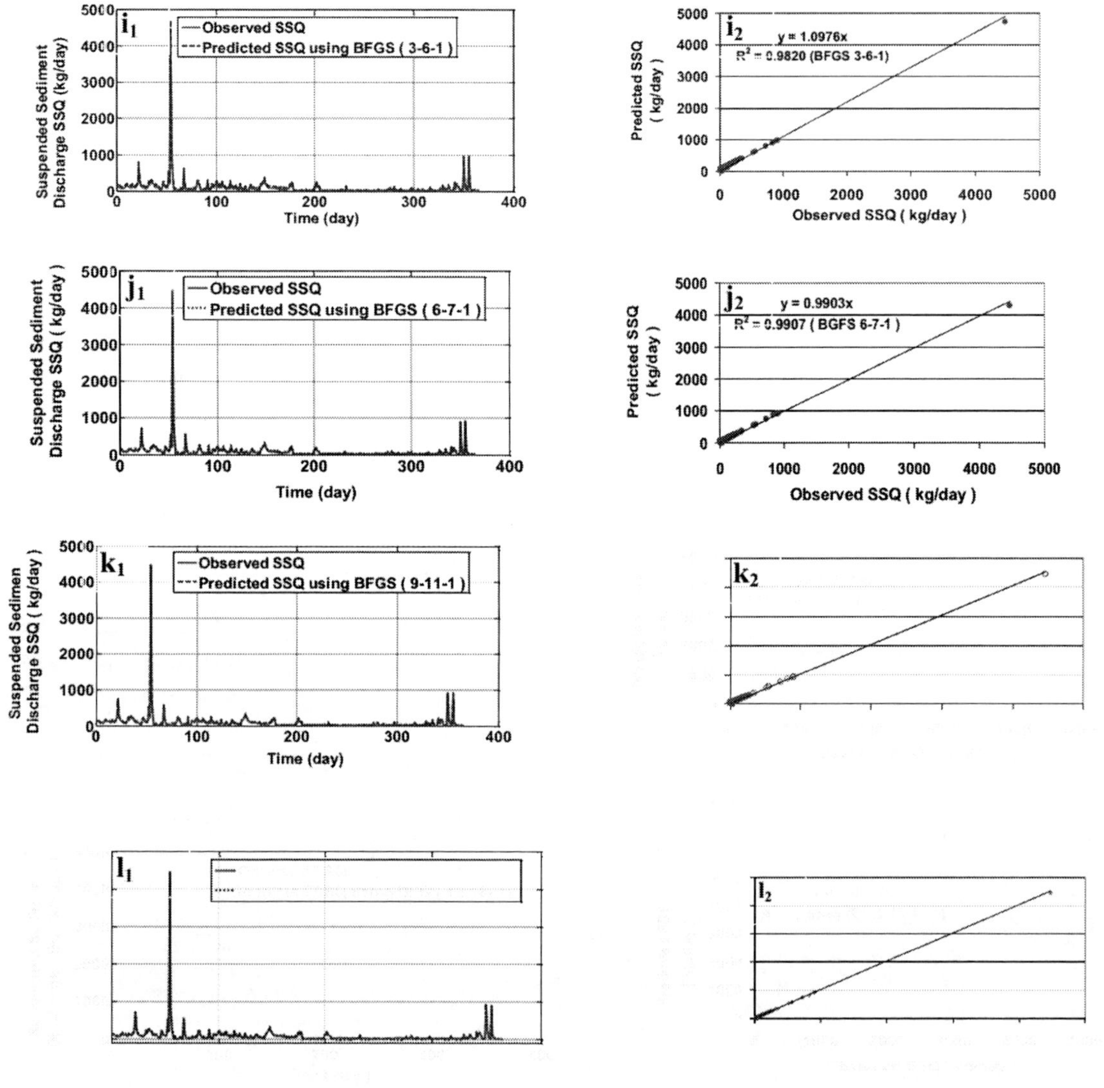

Figure 6 Comparisons between observed and predicted SSQ using BFGS Quasi Neuton algorithm based on the testing data.

Table 6. The performance of statistical algorithm BFGS Quasi-Newton

No	ANN model inputs	BFGS Quasi-Newton								
		structure	Training				Testing			
			MAE	RMSE	R^2	CE	MAE	RMSE	R^2	CE
1	P_t, H_t, Q_t	3:7:1					13.5769	17.9011	0.9999	0.9998
2	$P_t, H_t, Q_t, P_{t-1}, H_{t-1}, Q_{t-1}$	6:7:1	1.1414	3.8007	0.9999	0.9998	19.8777	23.1250	0.9950	0.9916
3	$P_t, H_t, Q_t, P_{t-1}, H_{t-1}, Q_{t-1}, P_{t-2}, H_{t-2}, Q_{t-2}$	9:11:1	0.6732	3.4310	0.9999	0.9998	25.6947	27.2228	0.9899	0.9884

No	ANN model inputs	BFGS Quasi-Newton								
		structure	Training				Testing			
			MAE	RMSE	R^2	CE	MAE	RMSE	R^2	CE
4	P_t, H_t, Q_t, P_{t-1}, H_{t-1}, Q_{t-1}, P_{t-2}, H_{t-2}, Q_{t-2}, P_{t-3}, H_{t-3}, Q_{t-3}	12:16:1	1.2361	3.9058	0.9999	0.9998	24.9865	27.4468	0.9888	0.9882

The performance of the SCG algorithm to prediction suspended sediment discharge based on number neuron in input, hidden and output layer is show in Table 7. The performance of statistics value (MAE, RMSE, R^2 and CE) at the processes training and testing, the best result on structure 12-14-1 (12 neurons in input layer, 14 neuron in hidden layer and 1 neuron in output layer). The performance value in testing process, MAE and RMSE (13.5769 and 17.9011) is smallest, meanwhile R^2 and CE (0.9999 and 0.9998) is the highest if it compared with the another SCG model (3-4-1, 6-8-1 and 9-13-1). The MAE SCG 3-4-1 and SCG 12-14-1 Is a negative, this means the total value of the prediction is lower than the total value of observed. While the MAE of BFGS is all positive, this means the total value of prediction is higher than the observed. Testing result obtained by SCG model 12-14-1 followed the observed suspended sediment discharge very closely the whole span of the testing data.

Table 7. The performance of statistical scaled conjugate gradient

No	ANN model inputs	SCG								
		structure	Training				Testing			
			MAE	RMSE	R^2	CE	MAE	RMSE	R^2	CE
1	P_t, H_t, Q_t	3:4:1	1.7751	4.4892	0.9999	0.9997	19.9337	38.8298	0.9927	0.9925
2	P_t, H_t, Q_t, P_{t-1}, H_{t-1}, Q_{t-1}	6:8:1	1.8234	4.3843	0.9999	0.9997	24.5762	26.0567	0.9908	0.9894
3	P_t, H_t, Q_t, P_{t-1}, H_{t-1}, Q_{t-1}, P_{t-2}, H_{t-2}, Q_{t-2}	9:13:1	1.6950	4.3485	0.9999	0.9998	25.4747	27.8591	0.9888	0.9879
4	P_t, H_t, Q_t, P_{t-1}, H_{t-1}, Q_{t-1}, P_{t-2}, H_{t-2}, Q_{t-2}, P_{t-3}, H_{t-3}, Q_{t-3}	12:14:1	1.8641	4.3932	0.9999	0.9997	18.2077	26.5019	0.9917	0.9890

6.3. Radial Basis Function

In this study, RBF models only apply four structures in the input, hidden and output layers, are namely: 3-3-1, 6-6-1, 9-9-1 and 12-12-1. To obtain the best performance of statistical value every RBF models structure, done method trial and error by the change spread value between 1 to 5. The best performance for structure 3-3-1 is a spread value 1, and to structure 6-6-1, 9-9-1 and 12-12-1 spread value is 2. The performance, four structure RBF models show and are summarized in Table 8. The comparison between observed and predicted is plotted in Figure (5-9). In the training and testing processes show the RBF structure 3-3-1 (3 neurons in input layer, 3 neurons in hidden layer and 1 neuron in output Layer) has a better and high prediction performance. The MAE, RMSE (0.6684 and 3.4646) training process has a smaller, meanwhile R^2 and CE (0.9998 and 0.9998) has a highest. Whereas in testing processes show the RBF model structure 3-3-1, MAE and RMSE (15.0265 and 19.6195) has a smaller, on the other hand R^2 and CE (0.9944 and 0.9940) has a highest. Based on performance the all RBF models structure input, hidden and output layer the different not significance, because difference only a slight and or performance relative similarly.

Table 8. The performance of statistical value RBF

No	ANN model inputs	structure	Spread	RBF							
				Training				Testing			
				MAE	RMSE	R^2	CE	MAE	RMSE	R^2	CE
1	P_t, H_t, Q_t	3:3:1	1	0.6684	3.4646	0.9998	0.9998	15.0265	19.6195	0.9944	0.9940
2	P_t, H_t, Q_t, P_{t-1}, H_{t-1}, Q_{t-1}	6:6:1	2	1.0877	3.4817	0.9998	0.9998	29.4478	34.2146	0.9834	0.9817
3	P_t, H_t, Q_t, P_{t-1}, H_{t-1}, Q_{t-1}, P_{t-2}, H_{t-2}, Q_{t-2}	9:9:1	2	1.2041	3.5195	0.9998	0.9998	28.5460	32.2658	0.9847	0.9837
4	P_t, H_t, Q_t, P_{t-1}, H_{t-1}, Q_{t-1}, P_{t-2}, H_{t-2}, Q_{t-2}, P_{t-3}, H_{t-3}, Q_{t-3}	12:12:1	2	1.2886	3.5433	0.9998	0.9998	26.8016	31.0912	0.9857	0.9849

In this study the all performance of the LM, BFGS, SCG and RBF the training result in training process is the better than the result in testing process. The LM learning algorithm attained the required accuracy with less number of the iteration. The LM achieved the goal before 5000 epoch and the convergent faster and shortest time than SCG and BFGS. When compared between MLRg, LM, SCG, BFGS and RBF , the MLRg model 2 with the 6 input variable is the better and more accurate to prediction suspended sediment discharge in Jenderam catchments.

The special for RBF, the MAE negative only on structure data input 3-3-1, while for structure data input 6-6-1, 9-9-1 and 12-12-1 the all MAE is a positive. Based on the performance statistics value, MLRg, LM, SCG, BFGS and RBF suitable for modeling the non-linear complex behavior of suspended sediment discharge.

7. Conclusion

In this paper, the MLRg, MLP (LM, SCG and BFGS) and RBF are used prediction daily suspended sediment discharge in Jenderam catchments Selangor Malaysia. The perfect line between observed and predicted, the MLRg model 2 with a 6 variable independent produced R^2 value 0.9974, with highest than the model 1 (R^2 0.9968), model 3 (R^2 0.9809) and model 4 (R^2 0.9638) only a slightly different. Even between MLRg model 1 and model 2 (3 and 6 input variable independent) there was no significance difference in prediction, because the value MAE, RMSE, R^2 and CE difference only slightly (Table 4). The all predicted value positive for four MLRg model. Increasing the number of independent variables in the model MLRg, suspended sediment discharge prediction have a trend less accurate, but the difference only slightly. In this study the all performance of the LM, BFGS, SCG and RBF the training result in training process is the better than the result in testing process. When compared between MLRg, LM, SCG, RBF, has a obtained the BFGS model structure 3-7-1 is the better and more accurate to prediction suspended sediment discharge in Jenderam catchments. Based on performance of MLRg, LM, SCG, BFGS and RBF accurately for modeling the non-linear complex behavior of suspended sediment responses to rainfall, water depth and discharge in small catchments area .

References

[1] Kninghton, D. 1998. Fluvial Forms and Processes: *A New Perspective*. Hodder Arnold.

[2] Mustafa, M.R., Rezaur, R.B., Saiedi, S., & Isa, M.H. 2012. River suspended sediment prediction using various multilayer perceptron neural network training algorithms-A case studi in Malaysia. *Water Resour Manage*, **26**:1879-1897.

[3] Mustafa, M.R., Rezaur, Raharhjo, H., & Isa, M.H. 2012. Prediction of pore-water pressure using radial basis function neural network. *Engineering Geology 135-136*, 40-47.

[4] Welling, D.E., & Webb,B.W. 1986. *Solute in River Systems*. Chichester Wiley.

[5] Sanusi, W., Jemain, A. A., Zin, W. Z. W., & Zahari, M. 2015. The drought characteristics using the first-order homogeneous Markov chain of monthly rainfall data in peninsular Malaysia. *Water Resources Management*, **29**, 5, 1523-1539.

[6] Maru, R. and Ahmad, S. 2015. The relationship between land use changes and the urban heat island phenomenon in Jakarta, Indonesia. *J. of Adv. Sci. Letters*. **21**, 2, pp. 150–152(3).

[7] Maru, R., Baharuddin, I.I., Umar, R., Rasyid., R., Uca, Sanusi, W., and Bayudin. 2015. Analysis of The Heat Island Phenomenon in Makassar, South Sulawesi, Indonesia. *American J. of Appl. Sci.*. **12**, 9, pp 616-626. DOI: 10.3844/ajassp.2015.

[8] Maru, R. and Ahmad, S. 2015. The Relationship between Temperature Patterns and Urban Morfometri in the Jakarta City, Indonesia. *Asian J. of Atmospheric Env.* **9**, 2, pp. 128-136.

[9] Kisi, O., & Guven, A. 2010. A machine code-based genetic programming for suspended sediment concentration estimation. *Adv. in Eng. Soft.* **41**.

[10] Sadeghi, A.H.R., Mizuyam, T., Miyata, S.,Gomi ,T., Kosugi, K., Fukushima, T., Mizugaki, S., & Onda, Y. 2008. Development, evaluation and interpretation of sediment rating curves for a Japanese small mountainous reforested watershed. *Geoderma*, **144**, pp. 198-211.

[11] Cigizoglu, H.K. 2004. Estimation and forecasting of daily suspended sediment data by multi-layer perceptrons. *Adv. in Water Resources*, **27**, 185-195.

[12] Melesse, A.M., Ahmad, S., McClain, M.E., Wang, X. & Lim, Y.H. 2011. Suspended sediment load prediction of river systems: An artificial neural network approach. *Agricultural Water Management*. **98**, 5, pp. 855-866.

[13] Kutner,M.H., Nachtsheim, C.J., Neter, J., & Li, W. 2005. *Applied linear statistics models*. McGraw-Hill Irwin.

[14] Montgomery, D.C., Peck, E.A., & Vining. G.G. 2006. *Introduction to linear regression analysis*. Fourth Edition Wiley-Interscience A john Wiley & Sons, Inc Publication.

[15] Maniquis, MC., Lee S., & Kim, L.H. 2010. Multiple kinear regression models of urban runoff pollutant load and event mean concentration considering rainfall variables. *J. of Env. Sci.* **22**, 6, pp. 946-952.

[16] Yuill, B.T., & Gasparini, N.M. 2011. Hydrologic controls on wash load sediment concentrations within a low-ordered, ephemeral watershed. *J. of Hydrology*, **410**, pp. 73-83.

[17] Roman, D.C., Vogel, R.M., & Schwarz, G.E. 2012. Regional regression models of watershed suspended-sediment discharge for the eastern United States. *J. of Hydrology*, **472-473**, pp. 53-62.

[18] Yilmaz, I., & Kaynar, O. 2011. Multiple regression, ANN (RBF, MLP) and ANFIS models for prediction of swell potential of clayey soils. *Expert Systems with App.* **38**, pp. 5958-5966.

[19] Kleinbaum, D.G., Kupper, L.L., Muller, K.E., & Nizam, A. 1998. *Applied regression analysis and other multivariable methods*. Duxbury Press

[20] Haykin, S. 1999. *Neural Networks and Learning Machines*. Prentice Hall International, Inc

[21] Fauset, L., 1994. Foundation of Neural Network. New Jersey: Prentice-Hall

[22] Beale, M.H., Hagan, M.T., & Demuth, H.B. 2012. *Neural Netwok Toolbox™ User's Guide Matlab*. The MathWorks, Inc.

[23] Liu, Q.J., Shi, Z.H., Fang, N.F., Zhu, H.D., & Ai, L. 2013. Modeling the daily suspended sediment concentration in a hyperconcentrated river on the Loess Plateau, China, using the Wavelet-ANN approach. *Geomorphology*,**186**, pp. 181-190.

[24] Moller, M.F. 1993. A scaled conjugate gradient algorithms. Neural Netw. **12**, pp. 145-151.

[25] Kegl, B., Krzyzak, A., & Niemann, H. 2000. Radial basis function networks and complexity regularization in function learning and classification, *International Conference in Function Learning and Classification*.

[26] Moradkhani, H., Hsu, K., Gupta H.V., & Sorooshian, S. 2004. Improved streamflow forecasting using self-organizing radial basis function artificial neural networks. *J. of Hydro.*, **295**, pp. 246-262.

[27] Lin, G. F., & Chen, L. H. 2004. A non-linear rainfall-runoff model using radial basis function

network. *J. of Hydro.,* **289**, pp. 1-8.

[28] Alp, M., & Cigizoglu, H. K. 2007. Suspended sediment load simulation by two artificial neural network methods using hydrometeorological data. *Env. Model Sof.* **22**, pp. 2-13.

[29] Kagoda, P. A., Ndiritu, J., Ntuli, C., & Mwaka, B. 2010. Application of radial basis function neural networks to short-term streamflow forescasting. *Phy. Chem. of Earth.*

[30] Bishop, C.M., 2008. Training with noise is equivalent to Tikhonov regularization. *Training*, **7**, 1.

[31] Rojas, R. 1996. *Neural networks: A systematic introduction*. Springer Verlag, Berlin pp 151-184.

[32] Cigizoglu, H.K., & Kisi, O. 2006. Methods to improve the neural network performance is suspended sediment estimation. *J. of Hydro.* **3**, 17, pp. 221-238.

[33] Maier, H.R. and Dandy, G.C., 2000. Neural networks for the prediction and forecasting of water resources variables: a review of modelling issues and applications. *Env. Model. Soft.*, **15**, 1, pp.101-124.

[34] Rahman, A. and Ahmar, A.S., 2017, September. Forecasting of primary energy consumption data in the United States: A comparison between ARIMA and Holter-Winters models. In *AIP Conference Proceedings* (**1885**, 1, p. 020163). AIP Publishing.

[35] Ahmar, A.S., Rahman, A., Arifin, A.N.M. and Ahmar, A.A., 2017. Predicting movement of stock of "Y" using Sutte Indicator. *Cogent Eco. & Fin.*, **5**, 1, p.1347123.